MECHANISMS OF IONIC POLYMERIZATION

Current Problems

MACROMOLECULAR COMPOUNDS

Series Editor: M. M. Koton, *Institute of Macromolecular Compounds Leningrad, USSR*

ION-EXCHANGE SORPTION AND PREPARATIVE CHROMATOGRAPHY OF BIOLOGICALLY ACTIVE MOLECULES
G. V. Samsonov

MECHANISMS OF IONIC POLYMERIZATION
Current Problems
B. L. Erusalimskii

MECHANISMS OF IONIC POLYMERIZATION

Current Problems

B. L. Erusalimskii

Institute of Macromolecular Compounds
Academy of Sciences of the USSR
Leningrad, USSR

Translated and edited by
Trans-Inter-Scientia
Tonbridge, Kent, England

CONSULTANTS BUREAU • NEW YORK AND LONDON

Library of Congress Cataloging in Publication Data

Erusalimskii, B. L.
Mechanisms of ionic polymerization.

(Macromolecular compounds)
Translated from Russian.
Includes bibliographical references and index.
1. Addition polymerization. I. Trans-inter-scientia (Firm) II. Title. III. Series.
QD281.P6E654 1986 547′.28 86-24368
ISBN 0-306-10991-3

A Division of Plenum Publishing Corporation
233 Spring Street, New York, N.Y. 10013

Printed in the United States of America

Foreword

In the last twenty years the literature on the processes of ionic polymerization has reached such a level that there is not a single question which is not covered by the information contained in the many monographs, reference books, and textbooks in this field. It is easy for the interested reader to find sources for in-depth study, for a superficial acquaintance with the fundamentals of the subject or with the general features of these processes. At the same time the field is being continually enriched by new facts which have not only broadened the data base but which influence existing concepts on the mechanisms of these reactions. Such influences often touch the very foundations of these concepts, i.e., they go beyond simple descriptions of the structure of the pre-reaction states or earlier schemes. It is therefore appropriate to attempt a critical appraisal of the modern views on the mechanisms of formation of macromolecules in ionic systems which envisages, so far as is possible, the differentiating of fundamental and hypothetical conclusions or concepts.

With this in mind we have preferred to address ourselves to the reader who is already quite well acquainted with the general literature. This has allowed us to dispense with detailed introductions to the questions discussed and to limit ourselves to brief comments on the fundamentals of the subject. Such an approach has made detailed explanations of general facts unnecesssary and the citing of the older works superfluous. It is felt that these are adequately covered in the general literature.

On the basis of such an approach attention has been concentrated mainly on the problems of the reactivity and stereospecificity of ionic active sites, the discussion of which constitutes the bulk of the present monograph (Chapters 4 and 5).

These chapters are prefaced by a short description of the processes of ionic polymerization (Chapter 1), a brief examination of the consequences which flow from some physicochemical studies on ionic initiators and growing chains (Chapter 2), and a general analysis of multicenteredness, i.e., of the origin and coexistence of different active sites (Chapter 3).

The author would like to take this opportunity to express his gratitude to his colleages E. Yu. Melenevskaya, G. B. Erusalimskii, V. N. Zgonnik, V. V. Mazurek, and A. I. Kol'tsov for their valuable discussions of some of the general questions and a number of particular problems.

Contents

Chapter 1

Characteristics of Ionic Polymerization Processes

The problem of preparing a work of a general nature may be approached in several essentially different ways. Usually the aim is to provide a review of the fundamental work in the field in its entirety, together with a consideration of the results obtained over a considerable period of time. We have chosen a different approach to the problems which comprise the subject of the present monograph. Its main purpose centers upon an attempt to examine the state of the art and possible ways of resolving the thorny or as yet unexplained aspects of the problem. Naturally, such an approach demands the bringing together of the most recent and fundamentally significant results of both theoretical and experimental research. At the same time, even though the work is aimed at the well-informed reader, it is impossible to avoid a short, however brief, account of the fundamental concepts. Naturally, it is difficult to define what is fundamental in the field of mechanisms of ionic polymerization. There is no doubt that the finer points, as well as many of the 'conceptual" elements of the different schemes which are at present accepted, will be re-examined many times by others. There is nothing surprising about such a situation because the problem relates to a branch of polymer chemistry which has been developed comparatively recently. Therefore, in such circumstances, it is better to have as a basis a thorough understanding of the current points of view, the validity of which seldom exceeds the limits of probable hypothesis.

The present chapter consists of a condensed outline of existing ideas on the mechanisms of formation of macromolecules in ionic systems. In contrast with the basic material, which is of a general nature, here we prefer to deal with certain types of ionic processes. We have in mind a short account of the main features of the systems which include anionic and cationic systems and complex initiating systems based on the transition metals. We have avoided superfluous detailed discussions and cited previous research by reference to monographs.

Instead of a rather broad comparison of the processes of ionic and radical polymerization, we have confined ourselves solely to those features which are peculiar to polymerization in anionic systems. This is determined by the fact that of all the reactions of ionic polymerization, the greatest amount of work has been devoted to the mechanisms of anionic processes.

1.1. ANIONIC SYSTEMS

A distinguishing feature of the majority of anionic polymerizations (which distinguish them from polymerization in cationic systems) is the possibility of creating conditions necessary for continuous growth and consequently the synthesis of high-molar-mass monodisperse polymers and block copolymers. A less common feature, but one common to a large range of anionic processes, is the stereospecificity of the reactions which form macromolecules. The formulation of mechanisms for these processes amounts, to a considerable extent, to establishing the causes which give rise to deviations from the optimum development of the above characteristics.

When applied to the most common growing chains of the anionic type, which are usually denoted by the symbol M_nMt (in which M is the monomer and Mt is the metal atom which functions as a counterion), these causes are due to the combined effect of a large number of different factors. The most important of these is the nature of the fragments M_n and Mt, the nature and number of other ligands in the coordination sphere of the metal atom (these are omitted in the abbreviated symbol), and the nature of the reaction medium. The effect of the conditions in which the reaction takes place (the concentration of the reactants and the temperature) is less fundamental in character. Changes in these conditions, which can often strongly influence the final result, can be regulated in each of these systems.

Given the fixed nature of the above fragments, the reactivity of the growing chains is determined by the stability of the active bond, its lack of steric hindrance, and upon the ability of the counterion to participate in donor–acceptor reactions with the monomer prior to its insertion into the chain, i.e., upon the formation of an intermediate complex of the type ASM in which AS denotes the active site. To the changes in the above features there corresponds a wide range of reactivities of the different types of active site, whose actual forms and equilibrium concentrations depend upon the nature of the reaction medium and the range of concentrations chosen. The significant role played by the concentration is due to extremely small values of the corresponding equilibrium constants (deaggregation, ionic dissociation, etc.).

The transition from one type of active site to another also has an effect upon the stereochemistry of the growth reactions, and this may lead to the dependence of the microstructure of the polymer upon the total concentration of active sites even in the same reaction medium. Nevertheless, the kinetic and structural effects are much more sensitive to the nature of the solvent and to the presence, in nonpolar media, of small numbers of electron donors which are capable of forming quite stable stoichiometric complexes with the active sites.

To a lesser degree, a relation between the structure of the macromolecules and the concentration of the monomer is found. The reasons for this, which are very varied, may be attributed to competition of the unimolecular (isomerization of the active sites and the formation of intramolecular complexes) and the bimolecular events (growth reactions). The actual limits within which the concentrations of the active sites and of the monomer may vary, in commonly employed experimental conditions, differ very considerably and this fact must be taken into consideration in any quantitative evaluation of these processes.

In the anionic polymerization of nonpolar monomers deviations from the ideal course of the growth reaction, in the sense of the formation of monodisperse polymers, are usually the result of chains of different reactivity taking part simultaneously in the growth reaction. For polar monomers, termination reactions involving irreversible reactions between the functional groups of the monomer units and counterions of the active sites are more typical. The bulk of such transformations occurs in the very first stages of polymerization. This leads to a much lower efficiency of initiation (usually of the order of 0.01) than in the polymerization of nonpolar molecules where the initiators may be completely expended in the formation of the active sites. In the main stage of the formation of macromolecules, the role of such processes becomes less significant and, if the temperature is low enough, they may be eliminated, or at least considerably reduced. This often makes it possible to synthesize polymers with a narrow molar mass distribution (MMD), even polymerizing polar monomers.

The usual functional relation between the rate of any polymerization process (R), the propagation rate constant (k_p), and the termination rate constant (k_t) is $R = f(k_p, k_t)$ (we deliberately do not include the initiation reaction in our discussion). However, the situation is more complex when there is a transition from one type of active site with fixed M_n and Mt fragments to another brought about by the introduction of additives or changes in the nature of the solvent. In such circumstances changes of the active site may affect k_p and k_t to different degrees and this has not generally been quantified. The true significance of the changes in the reactivity of the active site which is expressed by the ratio k_p/k_t for each type of site compared remains unclear. The few facts at our disposal indicate the possibility that changes in k_p and k_t may play a significant role and do not indicate any conclusions of a contradictory nature. The amount of detailed research into the kinetics of the anionic polymerization of polar monomers is as yet rather limited.

In approaching the mechanisms of stereoregulation in the polymerization of polar vinyl and vinylidene monomers, the main concept is that of the dominating role of the nature of the counterion and the penultimate unit of the growing chain. There are processes in which the second of these factors is the more important. Practically all the known schemes which are used to interpret the stereospecific effects in such systems include as a necessary element the formation of an intramolecular complex of the counterion with the penultimate unit or with more remote links of the growing chain. This is also shown by the inability of nonpolar vinyl monomers to form stereoregular macromolecules in anionic systems (styrene and its derivatives); some exceptions do not have a fundamental character in this sense (see Chapter 5). A similar analogous interpretation can be used as the cause of atacticity in the anionic polymerization to 1,2- and 3,4-dienes. However, the presence of active electron donor groups in the monomer does not always ensure the stereospecificity of the growth reaction even in favorable conditions (acrylonitrile, 4-vinylpyridine). There is an obvious relationship between the orientation of links in the macromolecule and the dominant contribution of one of the possible structural forms of the active site in the reaction.

In the anionic polymerization of dienes, only those processes which result in the formation of macromolecules which contain mainly cis-1,4 or trans-1,4 units can be considered to be stereospecific in the strict sense of the word. In the different variants of the mechanisms of selection of one of these forms, competition between isomerization of the final group and addition of the next monomer molecule to it predominates. According to this concept, the structure of the chain is determined by the end of the active site. Selection in the sense of the formation of 1,4 links or of 1,2 or 3,4 links is determined by factors which favor mainly the interaction of the monomer with the C_α or C_γ atoms of the final group. In the final analysis these factors amount to the electronic and spatial features of the active group; there is as yet no common view as to the cause of the relative role of these. Nevertheless, the lack of steric hindrance of the atoms in the final link of the active sites which are attacked by the monomer seems to be especially essential for the final result.

1.2. CATIONIC SYSTEMS

The drawing of parallels between cationic and anionic processes amounts to accentuating the differences in the selectivity of the corresponding active sites with respect to monomers. If these well-known features are put to one side, the main distinguishing feature of growing cationic chains is their increased tendency to deactivation. The difference between growing carbenium ions and carbanions is particularly great. Chain transfer is common in cationic polymerization; termination plays a minor role. This feature is seen in the wide distributions of molar masses which are typical of cationic polymerization, in some cases in extremely low molar-mass values and in others the total cessation of polymerization long before the monomer is exhausted. The tendency to deactivation is reduced considerably at temperatures in the region of −100°C. The high reactivity of certain types of cationic sites thus makes it possible to synthesize high-molar-mass polymers from certain monomers. The possibility of suppressing such reactions by changes in the nature of the initiator and the reaction medium is not excluded but is less typical.

The amount of information on the physicochemical features of cationic agents, particularly of growing chains, is very limited. This makes the interpretation of the mechanism of cationic processes more difficult than those in anionic systems. Models similar to those which are of wide application in the investigation of anionic active sites are practically excluded here; the problem lies in the extremely low stability of the carbenium derivatives. Information on active oxonium centers which participate in the polymerization in oxygen-containing heterocyclic compounds and whose stability is close to that of anionic "living" chains is much more readily accessible. This information is gradually approaching a level necessary for quite broad generalizations. Nevertheless, the problem of the geometry and electronic structure of active sites at the stages of addition and ring opening, which have long been under discussion, continues to be controversial.

The lack of information on cationic agents also extends to initiators. The simplest of these, the protic acids, are sources of quite complex processes and have specific characteristics which are, in particular, associated with the ability of the corresponding ion pairs to convert to the ester form. Furthermore, of the great variety of initiators of cationic polymerization, protic acids are of relatively limited significance. Systems of the general type EX_nB where E is boron, aluminum, titanium, etc., X is a halogen, and B is a Lewis base (H_2O, ROH, R_2O, RX, etc.) play a much greater role. There is no direct experimental basis for the ionic structures which are usually written $H^+[EX_nOH]^-$, or $R^+[EX_nOR]^-$, etc. The basis of such concepts is provided only by indirect evidence; this is particularly true for the end groups in macromolecules which are produced by these initiators.

Heterocyclic monomers are generally more suitable for cationic rather than anionic polymerization. Thus anionic initiators are quite active with oxiranes but not with larger cyclic compounds.

There are no such restrictions on cationic agents and this is obviously due to the tendency of oxygen cyclic compounds to form oxonium derivatives. The same selectivity is also a feature of polyoxygen monomers such as the cyclic acetals. A similar absence of preference in other heterocyclic compounds, such as the lactones which are capable of polymerization by both cationic and anionic mechanisms, may be ascribed to the prior activation of the monomer in these systems (here Ct^+ and An^- are the attacking cationic and anionic agents respectively) in the following manner:

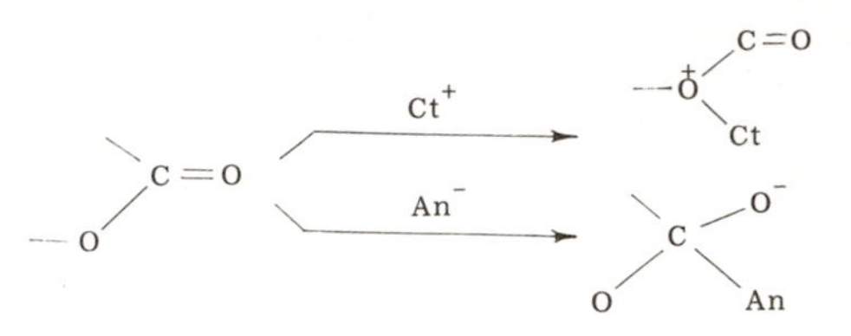

The stereospecificity of cationic active sites, which is clearly evident for monomers which are especially inclined to (α-methylstyrene and its homologs), or capable only of, cationic polymerization (alkenyl ethers) but not of anionic polymerization, does not extend to the polymerization of dienes. Variations in the nature of the counterions only slightly affect the microstructure of polydienes whose chief feature in cationic initiation is the absence or low level of cis-1,4 units. Of the remaining structural units the trans-1,4 links are rather more dominant, however, not so dominant that these polymers may be regarded as stereoregular.

A possible explanation of such a limited stereochemical selectivity in the polymerization of dienes is the absence of accepting properties in the counterions of cationic active sites. This requires further study into the particular causes which lead to the formation of stereoregular macromolecules in cationic systems, including other monomers. Participation of the counterion in the stereoregulation processes, as an electron acceptor with respect to units of the growing chain, is a necessary element in the majority of generally accepted schemes for stereoregulation in anionic polymerization. With unsaturated ethers, it is the carboxonium nature of their active sites which acts as the driving force in stereoregulation. The chelate-like structure of the end group counterion system is an orienting factor when an active site interacts with the monomer. It is more difficult to consider the mechanism in the cationic polymerization of α-methylstyrene, especially since in anionic systems this monomer has no tendency to stereospecific polymerization. The differences in actual mechanisms of stereoregulation in cationic polymerization processes are demonstrated by the tendency of vinyl ethers to form isotactic sequences while α-methylstyrene and its analogs tend to form syndiotactic sequences. The sensitivity of the chain structure to the nature of the initiator and the solvent is common to each type of monomer, thus increasing the polarity which favors the formation of syndiotactic sequences. However, this leads to syndiotactic polymers only in the case of α-methylstyrene; for vinyl ethers in polar media the growth of heterotactic triads is more typical.

One of the possible reasons for the differences observed is the probability of intramolecular interactions in the growing chains of vinyl ethers which are of the "oxonium transfer" type such as well-known reactions for the cationic polymerization of cyclic oxides and acetals. In the two latter cases such events are often terminated by irreversible reactions (the splitting off of cyclic compounds which are different from the original monomer), which make them easier to detect. For vinyl ethers whose active sites are not so clearly oxonium in nature, only relatively weak interaction between the cation and the oxygen atoms of the growing chain is feasible. Nevertheless, such an interaction may influence the structure of the macromolecule.

There is a special type of structural effect which is typical of the cationic polymerization of α-olefins; here polymerization is accompanied by isomerizations which are the result of the tendency of the chain ends to pass to the most stable form, i.e., to tertiary carbenium ions. The mechanisms of such events more often than not are hydride shifts. The concept of the regularity of macromolecules in these cases relates more to the degree of uniformity of the structure of the individual links rather than to their succession. The most regular chain occurs, first, when there is a total absence of isomerized links (this is a consequence of either the formation of tertiary carbenium ions at the very stage where monomer joins the active site or of the choice of special reaction conditions) and second, when the total exclusion of "normal" links is brought about by a particularly high rate of isomerization. Isolated examples of such extreme cases are known, but more often polymer chains of a mixed structure are formed. The nature and content of the different structural units can be usually controlled in some way, particularly by varying the temperature over a fairly large range.

Analogous isomerizations are probable in the cationic polymerization of dienes. The considerable reduction in unsaturation of polydienes formed during cationic initiation is obviously the result of the interaction of the cations with the side vinyl (or vinylidene) groups.

1.3. SYSTEMS BASED ON TRANSITION METALS

Below we concern ourselves with some of the questions relating to the polymerization of unsaturated nonpolar monomers by Ziegler–Natta catalysts and π-allyl complexes.

1.3.1. Ziegler–Natta Catalysts

The typical Ziegler–Natta catalysts are formed by mixing a derivative of a metal from groups IV to VIII with an organometallic compound of groups I to III. They contain transition and nontransition metals which are linked via halogen atoms and hydrocarbon ligands coordinated with the transition metal.

The great range of components which are suitable for the production of such complexes leads to an enormous variety of these catalysts. The nature of the final products depends, in particular, on the valence state of the transition metals, and on the ratio of the initial components. A common feature of their behavior in polymerization is in promoting initiation and growth via the formation of intermediate complexes with the monomer. The necessity of this stage is demonstrated by the inhibition of polymerization on introducing independent electron donors which block the vacant sites in the catalytic complexes. This is the fundamental feature which distinguishes such complexes from the usual anionic active sites; in the latter the absence of vacant sites in the counterion does not rule out initiation and growth. With the favorable geometry of the complex ASM, the monomer (M), which is a ligand for the transition metal, undergoes a selective preorientation which ensures the strict uniformity of structure of the growing chain and the formation of macromolecules with a high degree of stereoregularity. This feature, which is often regarded as one of the main properties of Ziegler–Natta catalysts, is far from universal; there are many more nonstereospecific catalysts of this type than stereospecific. However, the above diversity favors the possibility of selecting systems which can ensure the required structure of the macromolecule. Such a choice is, as a rule, very selective. One and the same system comparatively rarely exhibits stereospecificity with respect to different monomers. Some complexes of this type also differ in their selective initiating capability. Among known cases of a similar type are the passivity of the systems Et_2AlCl–MtX_2, where Mt is Ni or Co (which initiate polymerization of dienes), with respect to α-olefins and the inability of the system Et_2AlCl–Cp_2TiCl_2 (which polymerizes ethylene) to form polypropylene.

The dominant concept in interpretation of the mechanism of the processes which are initiated by Ziegler–Natta catalysts is the essential role of the activation of the monomer at the stage of complex formation. This causes stretching of the C=C bond and facilitates the insertion of monomer into the carbon–transition metal bond. The length of this bond in olefin complexes with transition metal derivatives reaches 1.40-1.47 Å. Nevertheless, a universal view of the energy contributions of each of these stages does not exist. The known magnitudes of the activation energy of the growth reactions in the various Ziegler–Natta processes are total values. Attempts to determine their component parts which are based either on indirect data or on theoretical considerations lead to results which, as yet, cannot be regarded as unambiguous. The absence of clarity also extends to the problem of which of the two stages of the growth reaction is rate determining.

It does not seem sound to search for a common approach to all processes of this type. The number of known catalysts is extremely large and so many possible situations may exist. The nature of the metals and ligands determine both the degree to which bimetallic complexes accept electrons from the monomer π-system, and also the stability of the carbon—transition metal bond. Apart from the usual donor-accepting bonds, the back donation from the transition metal to the olefin is also possible; here the monomer acts as a donor with respect to the d-orbitals of the transition metal. Varying the components of the complex with the aim of influencing the above features may, in principle, be accompanied by changes in the relative roles of these characteristics.

The environment of a transition metal containing an organic ligand plays a decisive role in determining the possibility of inserting the monomer into the carbon—transition metal bond. For this event to be repeated many times, it is necessary for the bond to be moderately strong and for the complex active site to be fairly stable.

The bridge—bond system between the metal atoms of the catalyst was initially thought to play a vital role in creating the necessary conditions. This idea was used to explain the necessity for the presence of a second component in the catalyst, i.e., the nontransition metal alkyl or alkyl halide. The primary function of such a component is the alkylation of the transition metal, but the activating or stabilizing influence of the fragment containing the nontransition metal on the bond between the organic ligand and the transition metal may be regarded as of no less importance.

The bimetallic nature of the active sites which take part in the processes under discussion continues to be supported by a number of researchers. However, the monometallic mechanism put forward in the middle sixties and which has been developed in recent years is, at present, the most commonly accepted. According to this, the second component of the catalyst has only an alkylating function. Various arguments based on experimental determination of the features of the individual system may be presented in support of both these ideas. Hence the choice between one or another mechanism must be regarded as being dependent upon the particular system.

One of the important features of Ziegler—Natta catalysts is their ability to form linear high-molar-mass poly(α-olefins) (with $\bar{M}$ of the order of 10^6). This is well illustrated by the polymerization of ethylene and propylene. This is a property possessed by many heterogeneous catalysts of this type and when these are used in polymerization, the role of growth-limiting events is minimal. On the other hand, growing chains formed in homogeneous systems of a similar nature are more often than not distinguished by a considerable tendency to side reactions which prevent a high-molar-mass product from being obtained. Nevertheless, recent research has led to the formulation of homogeneous Ziegler–Natta systems which are extremely effective in polymerizing α-olefins. These include the important groups of catalysts based on zirconium and aluminooxane derivatives.

In any discussion of the reactivity of such sites, the question of the reliability of the value of the propagation rate constant (k_p) appearing in the literature often arises. There is no common view on the basis of the published values since the accuracy necessary for the determination of the number of active sites is debatable. In the case of heterogeneous systems, the various methods used to determine the number of active sites involved often lead to results which differ widely. Values of k_p calculated for some homogeneous systems are more reliable. Their very large range of values may be regarded as evidence of the great variety of roles played by the above factors when the components of the catalytic systems are varied.

1.3.2. π-Allyl Complexes

Many of the features of the systems described above are also present in the π-allyl transition metal complexes, the main use of which is in the polymerization of dienes. The π-allyl derivatives of nickel, cobalt, etc., in combination with halogens of various metals (both transition and nontransition), are typical in this respect. The action of these systems (just as in the Zielger–Natta catalysts) is considered to proceed in two stages: the coordination of the monomer on the transition metal, followed by its insertion into the carbon–transition metal bond.

In systems of the above type, the homogeneity of many of the solutions of stereospecific complexes and the possibility of their physicochemical properties being known in considerable detail permit a somewhat closer approach to the mechanism of stereoregulation. In this respect, π-allyl complexes are definitely superior to the majority of those Ziegler–Natta catalysts which have activity and stereospecificity; a similar situation for some soluble Ziegler–Natta complexes gives information which may be applied only to a limited extent to heterogeneous systems of the same type.

Some π-allkyl complexes are often regarded as single-unit models of propagating diene chains. Their stuctural features, in particular the isomeric forms of the alkene substituent in the original complex and the isomeric composition of olefins which are formed in hydrolysis, are commonly used for the interpretation of the structural effects which are observed in polymerization. The discovery of some correlations which are inaccessible in the case of Ziegler–Natta catalysts here becomes possible in principle. The correctness of the resulting conclusions, especially those based on the structure of the products of deactivation of π-allkyl complexes, is not always obvious.

The formation of polydienes with a high degree of regularity when polymerized using π-allyl complexes with a definite central atom is achieved by the selection of ligands whose variation may favor a given type of structure for the elementary link. The heterogeneity of the structure of the chain, in the sense of the coexistence of both forms of 1,4 isomers, is usually regarded as the result of competition between growth events and isomerization of the final link.

The 1,2 and 3,4 polydienes formed with certain π-allyl complexes do not as a rule differ in stereoregularity. High syndiotacticity is found in only a few cases. This gives rise to the concept of monodentate forms of the intermediate monomer complexing with active sites and excluding the formation of 1,4 links in the macromolecules. The problem of the conditions necessary for the selective joining of iso- or syndiotactic sequences is not considered here. Since isotatic or syndiotactic growth may proceed in Zieger–Natta systems, it is necessary to look for those conditions which determine which particular polymer chain structure is formed. Apparently, in the synthesis of isotactic 1,2- or 3,4-polydienes, the deciding factor is the heterogeneity of the systems.

In the contemporary litrature devoted to the processes treated in the present monograph, there occurs the subdivision of polymerization reactions into "truly ionic" and those which proceed as a result of breaking bonds in active sites which are of essentially covalent nature (for example carbon–magnesium, carbon–lithium bonds, etc.). We do not wish to become involved with such an approach, but prefer to regard as ionic those additive polymerization processes which are distinctly not free-radical processes.

This brief description of ionic systems does not exhaust the diversity which is typical of all types. They are discussed in greater detail in the following chapters. Further discussion is based on the principle of making more familiar certain general problems which anticipate fundamental factual material.

With regard to the absence of detailed references in this chapter, we would point out that all the problems we have touched upon here, including monographs, reviews, and original material up to the mid-seventies, can be found in [1, 2].

REFERENCES

1. A. D. Jenkins and A. Ledwith (editors), Reactivity, Mechanism, and Structure in Polymer Chemistry, Wiley (1974).
2. R. N. Haward (editor), Developments in Polymerization, Vols. 1 and 2, Burgess International Ideas, Philadelphia (1979).

Chapter 2

The Informativeness of Research Methods into Ionic Active Sites

The methods used for the study of non-free-radical active sites which are responsible for the formation of macromolecules are mainly spectroscopic, electrochemical, and quantum-chemical. Among these, spectroscopic methods are used extensively, their application to the study of such species already becoming routine by the 1960's. The parallel electrochemical methods are important in a small range of suitable systems. Quantum-chemical studies began to advance to general systems only by the mid-seventies.

Many of the results obtained using these methods have been fully discussed in monographs and reviews and there is no need to summarize them afresh here. The aim of the present chapter is to attempt to evaluate the merits of the existing information on the physicochemical features of ionic active sites in order to understand the detailed mechanism of the elementary stages of the polymerization processes.

Although much more data have been obtained spectroscopically than by using any other technique, it is more appropriate to consider the electrochemical features of ionic systems first.

2.1. ELECTROCHEMICAL STUDIES

The general principles of the approach to the study of ionic initiators (R^*Y) and the growing chains (M^*Y) using electrochemical methods are discussed in detail in [1, 2], which also contain the most important results of work carried out over a considerable period. Therefore, we will try to keep our comments on these concepts to a minimum.

We denote both growing chains (or the initiating components) and the counterions of the active site by Ct^+ and A^-. Then the ionic equilibria can be expressed in all possible systems by the simplified equation

$$Ct^+, A^- \underset{}{\overset{K_{diss}}{\rightleftarrows}} Ct^+ + A^- \qquad (2.1)$$

Several issues may be distinguished which are appropriate for study by electrical conductivity measurements. These are, firstly, the determination of the dissociation constant K_{diss} [see Eq. (2.2)], secondly, the evaluation of the interionic distance in the ion pair [see Eq. (2.3)], and thirdly, the presence in the system of ionic triplets, e.g., $Ct^+A^-K^+$ and $A^-Ct^+A^-$ [see Eq. (2.4)]:

$$F/\Lambda = 1/\Lambda_0 + f^2c\Lambda/FK_{diss}\Lambda, \qquad (2.2)$$

$$K_{diss} = \frac{3000}{4\pi Na^3}\exp(-e^2/a\varepsilon kT), \qquad (2.3)$$

$$\Lambda c^{0.5}g(c) = \Lambda_0 K_{diss}^{0.5} + (\lambda_0 K_{diss}^{0.5}\, k_{diss}^{-1})(1 - \Lambda/\Lambda_0)c \qquad (2.4)$$

where Λ is the equivalent conductivity, Λ_0 is the equivalent conductivity at infinite dilution, F is the Fuoss function, f is the ionic activity coefficient, c is the total concentration, N is Avogadro's number, a is the distance between the ions in an ion pair, ε is the permittivity, k is Boltzmann's constant, T is the absolute temperature, λ_0 is the equivalent conductivity of the hypothetical electrolyte $(Ct_2A^+)(CtA_2^-)$ at infinite dilution, k_{diss} is the ionic dissociation constant of the hypothetical electrolyte, and g is a function given by $-\log_{10} g(c) = (1000K_{diss})^{0.5}(0.4343 - \beta\Lambda_0)\Lambda_0{}^{3/2} - 0.2171\Lambda/\Lambda_0$, where $\beta = (0.4343e^2/2\varepsilon kT)(8\pi Ne^2/1000\varepsilon kT)^{0.5}$.

These studies make it possible to separate the kinetic and structural effects inherent in ionic polymerization.

Most systems of type M_n^*Y which have been studied using electrochemical methods are anionic. This is due mainly to the high stability of carbanions in comparison with carbenium ions in growing chains. As a result there has been more research on the electrochemical features of anionic active sites and an increase in their use for the interpretation of the mechanisms of the corresponding polymerization processes.

2.1.1. Anionic Systems

Equilibria (2.1) have been studied in anionic systems; examples have included mainly initiators and growing chains and alkaline earth organometallic compounds and certain alkali metal alkoxides. Without recourse to repeated quotations of actual results obtained in the period up to the middle 1970's which are available both in monographs and in textbooks, the orders of magnitude of K_{diss} for several chain ends with alkali metal counterions in solution in tetrahydrofuran (THF) are given below:

Monomer	Styrene	Isoprene	Methyl meth-acrylate	Acrylo-nitrile*	Ethyl-ene oxide
K_{diss} (moles/liter)	10^{-7} (25°C)	10^{-10} (30°C)	10^{-10} (−78°C)	10^{-9} (25°C)	10^{-10} (20°C)

*Polystyrene chains with acrylonitrile end units were used for conductivity studies [3].

The dependence of K_{diss} on the solvent is illustrated by the values obtained at the same temperature for living chains of polystyrene in dimethoxyethane (10^{-6}) and polyethyleneoxide in dimethyl sulfoxide (10^{-5}).

Examples of systems which were investigated later, in which, together with K_{diss}, the interionic distances [Eq. (2.3)] and the magnitudes of k_{diss} for the ionic triplets [Eq. (2.4)] were evaluated, are included in Table 2.1.

The value of K_{diss} enables the concentration of free ions to be evaluated in any given experimental conditions, so that independently of the kinetic data one can evaluate the probability that such active sites take part in polymerization. Under the usual conditions of polymerization, this is significant when $K_{diss} > 10^{-8}$. This applies to systems in which free ions are distinguished by an especially high reactivity (see Chapter 4). Quantitative conclusions on this require values of rate constants for ion pairs and free ions. For propagation in anionic systems, the simplest equation which connects these and other values is

$$k_c = k_{\pm} + k_- (K_{diss}/C)^{0.5}. \qquad (2.5)$$

Here k_c, $k_{\pm}$, and k_- are the rate constants for the growth reaction as a whole, on ion pairs and on free anions, respectively, and C is the concentration. In this sense, the magnitude of K_{diss} is the most important of the electrochemical characteristics used for the study of the mechanism of ionic polymerization. It is unnecessary only in processes which are initiated by ionizing radiation, i.e., free from ion pairs. The problem, which often arises in these cases, of the differentiation of ionic and free-radical reactions is not directly connected with electrochemical characteristics.

There are considerable divergences between the magnitudes of K_{diss} and between the rate constants in Eq. (2.5) reported by different laboratories. Ignoring differences in the accuracy of electrochemical measurements, the causes of such effects may be attributed either to lack of sufficient stability or to insufficient pre-

TABLE 2.1. Electrochemical Characteristics of M_nMt Chains

M	Mt	Solvent	t, °C	K_{diss}, moles/liter	k_{diss}, moles/liter	a, Å	Ref.
Butadiene	Lithium	THF	−70	9×10^{-10}	8×10^{-3}	3.02	4
		DME[a]	−60	1×10^{-8}	3×10^{-3}	4.00	
Isoprene	Lithium	THF	−70	3×10^{-8}	5×10^{-3}	2.88	4
		DME	−60	8×10^{-8}	2×10^{-3}	4.00	
Polystyrene chain with oxide end	Lithium	THF	25	1×10^{-10}	9×10^{-5}	2.9	5
	Sodium	THF	25	8×10^{-10}	$8 \times 7\ 10^{-5}$	3.2	
	Potassium	THF	25	5×10^{-9}	—	3.6	

[a]DME = dimethoxyethane.

cision in evaluating concentrations. Both of these are surmountable problems.

Additional but less significant inaccuracies are due to the formation of ion triplets which contribute to the conductivity of the systems, but are kinetically approximate to ion pairs. For example, with polybutadienyllithium in dimethoxyethane (DME) (see Table 2.1), at the common total working concentration of 10^{-3} mole/liter the concentration of ion triplets is about 0.01% [4].

In obtaining data on the electrical conductivity of chains M_nMt in ether media and their use in studying polymerization mechanisms, the recently discovered side reactions which take place in these systems must be taken into consideration. The reaction of active species with the solvent leads to the formation of alkoxide derivatives M_nOMt or ROMt (where R is the fragment of a solvent molecule), leading one to expect a reduction in the electrical conductivity (κ) when the system is kept for some time. Such a conclusion follows from the comparison of the values of K_{diss} for the corresponding living chains (see Table 2.1) due to the substitution of $\rangle$CMt by —OMt. However, in the intermediate stages, changes of a much more complex character are possible. This is shown by the results obtained for the M_nLi—THF and M_nLi—DME systems [6], which are shown in Fig. 2.1. Naturally, the data relating to the rising part of the curves (the shape of which is obviously the result of the formation of mixed aggregates $pM_nLi{\cdot}mROLi$) may lead to values of K_{diss} which cannot be usefully used in equations such as Eq. (2.5). It cannot be excluded that some of the divergences between the data in the literature are the result of this situation.

The state of the ion pairs in the sense of whether they are contact ion pairs (IP_c) or separated ion pairs (IP_s) may be found without recourse to an electrochemical method [see Eq. (2.2)]. Nevertheless, if values of K_{diss} are available the interionic distance (a) can be calculated from Eq. (2.3). The values found in this way sometimes prove to be intermediate between those of contact and separated ion pairs. For example, in the case of polydienyllithium chains in DME, the value of a is found to be 4 Å (see Table 2.1). Such intermediate values which are qualitatively characterized by the state of the equilibrium

$$IP_c \rightleftharpoons IP_s \qquad (2.6)$$

can be used for the approximate evaluation of the equilibrium constant $Q = [IP_s]/[IP_c]$ provided the corresponding limiting values are known.

From Q the rate constants may be evaluated on each type of ion pair if the corresponding limiting values are known. However, Q

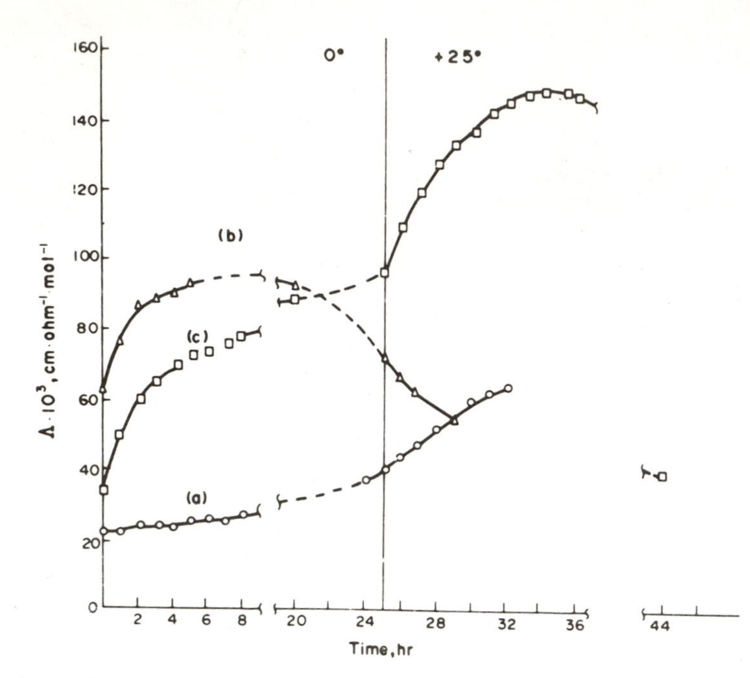

Fig. 2.1. Changes in the equivalent conductivity of polybutadiene in electron-donor solvents at 0° and 25°C. Initial concentration of PBLi, moles/liter: a) $3.4 \cdot 10^{-3}$; b) 5.06×10^{-3}; c) $4.80 \cdot 10^{-4}$.

cannot be found in cases as those considered above by spectroscopy which is suitable for this purpose in many other systems (see Section 2.2). The differentiation of the individual contributions of IP_c and IP_s to the total growth rate on the basis of purely kinetic data is then not possible [1]. At the same time, the combination of electrochemical and kinetic measurements can give the approximate answer to this question (see Chapter 4). Such a method may be regarded as useful in situations where there is an experimental intermediate value of a and there are reliable data on the interionic distances for both IP_c and IP_s.

In recent times the mechanism of dissociation of ion pairs to free ions has been made more precise. This has been achieved due to the combination of electrochemical and spectroscopic data. The individual constants of the equilibrium of the IP [Eq. (2.1)], i.e., the rate constant for the dissociation of IP_c (k_d) and the rate constant for the recombination of free ions (k_{rec}), have been used. In order to obtain these a special method was devised by Persoons [7] for media of low polarity. This involves measuring the relaxation following the application of a strong electric field to the electrolyte solution, thus shifting Eq. (2.1) towards dissociation.

TABLE 2.2. Kinetic Features of the Equilibrium between Ion Pairs and Free Ions in Fluorenyllithium–Diethyl Ether–Tetrahydrofuran Systems [8]

Temperature, °C	−20°		−40°		−55°	
THF content, moles/liter	0.43	1.38	0.43	1.38	0.43	1.38
k_d, sec^{-1}	10.96	54.9	15.28	83.9	18.84	89.2
k_{rec}, liter/mole/sec^{-11}	2.58	1.51	1.72	0.98	1.26	0.73
$Q \times 10^2$ [a]	(3.8) [b]	13.2	6.6	22.9	10.7	37.2
k_d'	(297) [b]	471	247	450	195	329

[a] From the data of [9].
[b] Considerable error found in the value of Q lying between 0.03 and 0.04.

The variation in the electrical conductivity of the system caused by the application of the field is determined experimentally as a function of the frequency, and from this is found the relaxation time (τ), which is a function of the individual constants k_p and k_{rec} and the equilibrium constant K_{diss} [Eqs. (2.7) and (2.8)]. Here C is the initial concentration of the dissociating agent:

$$1/\tau = 2(k_{rec}k_d)^{0.5}C^{0.5} \quad (2.7)$$

$$1/\tau = 2(k_{rec}K_{diss}^{0.5})C^{0.5} \quad (2.8)$$

These equations hold when the concentration of free ions, which is extremely small in these systems, is neglected. The constant k_{rec} [Eq. (2.8)] and thence k_d were calculated from the inverse dependence of relaxation time on $C^{0.5}$ (which has, in certain circumstances, a linear character) and the independently established value of K_{diss}.

This method was applied to fluorenyllithium by Persoons and van Beylen [2], using diethylether (DEE), THF, and mixed DEE–THF solvents of different compositions at temperatures from −55°C to 20°C, who found that the value of k_d increased in each case. This was attributed to the two-stage mechanism of dissociation of IP_c, which went through an intermediate stage of forming IP_s:

$$Ip_c \rightleftharpoons IP_s \underset{k_{rec}}{\overset{k'_d}{\rightleftharpoons}} \text{free ions} \quad (2.9)$$

Such a mechanism leads to the following relation between k_d and k'_d:

$$k_d = k'_d \frac{Q}{1 + Q}. \quad (2.10)$$

In order to evaluate the magnitudes of the individual constants k'_d, the authors made use of the spectroscopic characteristics of the corresponding ion pairs (see Section 2.2), which are used to determine the ratio Q. Some of the results obtained are shown in Table 2.2.

According to scheme (2.9), the constant k_d is the overall rate constant and, consequently, the ratio k_d/k'_d is the individual rate constant for the transformation stage IP_c to IP_s. The data given in Table 2.2 enable one to see that this constant in simple equilibrium conditions increases noticeably with decreasing temperature. In particular, in the system containing 1.38 moles/liter THF, it has values of 0.15 and 0.27 sec^{-1} at −20°C and −55°C, respectively. Thus, the negative activation energy for k_d [8] is due to the first stage of the total transformation [Eq. (2.9)].

As far as the differences in the magnitudes of k_d and k'_d themselves are concerned, they are determined by the amount of IP_s present in any given experimental conditions. Increase in this quantity leads to the convergence of values.

Until recently the generally accepted view of the complex equilibrium which characterizes the dissociation of IP_c was as shown by the scheme

$$IP_c \underset{a}{\rightleftharpoons} IP_s, \quad IP_c \overset{b}{\rightleftharpoons} \text{free ions}, \quad IP_s \overset{c}{\rightleftharpoons} \text{free ions} \tag{2.11}$$

Essentially, scheme (2.9) casts doubt upon the possibility of the direct dissociation of IP_c [direction (2.11b)]. Several years ago, Fuoss [10], confining himself to the applicability of his previous concept [as expressed by Eqs. (2.2) and (2.3)], in which the transformations of Eq. (2.6) were not considered, paid special attention to this idea. A mechanism corresponding to scheme (2.9), according to which the actual interionic distance (r) lies within the limits $a \leq r < R$, where a corresponds to IP_c and R is the radius of the solvated ion, i.e., the solvation sheath surrounding the free ions, was presented in [10]. It is accepted that Eq. (2.3) is applicable only for the systems in which the dimensions of the ions are considerably greater than those of the solvent molecules. The ion pair $R_4N^+{-}BF_4^-$ in the mixed solvent $CH_3CN{-}CCl_4$ is given as an example. Equation (2.3) does not apply when the ratio of ion to solvent molecule is small or inverse. The breakdown of the relationship between K_{diss} which corresponds to this equation and the magnitude of ε has been established in studies of the electrical conductivity of solutions of potassium iodide in mixtures of ethylene carbonate and dimethyl sulfoxide with water. As shown, various isodielectric mixtures give different values of K_{diss}.

On the basis of the mechanisms given by Eq. (2.9), Fuoss came to the following final view of the equation relating K_{diss} with the particular equilibrium constants K_R and K_a which are valid for the recombination stages for free ions and the formation of IP_c, respectively:

$$K_{diss} = 1/K_R(1 + K_a). \tag{2.12}$$

Although the basis for concluding that Eq. (2.12) holds is the experimental data for systems which are far removed from our present discussions, the scheme represented by Eq. (2.9) may be regarded as being of quite general significance. Essentially, its application to agents of the type M_n^*Y leaves open the question of the strictly individual nature of the growth rate constants calculated for IP_c from Eq. (2.5). According to scheme (2.9), the separated ion pairs must also exist in those systems which contain free ions. Attempts to find them experimentally in such systems have been unsuccessful. This may be a result of the methods used being insufficiently sensitive. Obviously, the limit of detection lies in the range $Q > 0.01$. When IP_s is much more reactive than IP_c (e.g., polystyryl chains), the kinetic contribution of IP_s can be considerable even though the value of Q is low. Thus the difference between the schemes (2.9) and (2.11) affects not only the mechanism of dissociation, but also the physical sense of the rate constants which are evaluated in the processes of ionic polymerization by the usual methods.

2.1.2. Cationic Systems

Of the first attempts to apply electrochemical methods to the study of the cationic polymerization processes, those of the Czech school deserve special mention. These were concerned with the initiating systems $ROH{-}AlX_3$ (where X is a halogen, R is H, an alkyl, acyl, etc.) and date back almost 25 years. The results obtained by Zlamal, Ambrozh, et al., enabled the formulation of an indisputable connection between the electrical conductivity of such systems in halogenated solvents and the molar mass of the polyisobutene formed. A quite detailed analysis of these data is contained in the recently published review by Gandini and Cheradame [11]. We will therefore only note that the agents chosen are convenient for electrochemical study, i.e., are comparatively easily ionized and practically insensitive to microscopic traces of Lewis bases of the general type ROH. In the above experiments the ratio of the original components themselves lay within the limits $0.5 \leq ROH{-}AlX_3 \leq 4.0$. In such solvents as CH_2Cl_2, $CH_2Cl{-}CH_2Cl$, etc., these systems are distinguished by a larger and more easily reproducible electrical conductivity than in the corresponding Lewis acids. Obtaining well defined electrochemical characteristics for the latter is more difficult and is connected with the especially high demands on the cleanliness of the apparatus and the purity of the reagents.

Similar difficulties were overcome later by Grattan and Plesch [12, 13], who chose relatively simple cationic systems. These authors studied the electrical conductivity of solutions of $AlCl_3$ in CH_3Cl and $AlBr_3$ in CH_3Br. This is of interest in connection with the kinetic data of another group of Czech workers, Chmelir et al. [14-16], which led to the conclusion of autocatalysis typical of such agents, i.e., the conclusion that their ability to generate cationic active sites which initiate polymerization independently of additional catalysis is due to the following equilibrium:

$$2AlX_3 \rightleftharpoons AlX_2^+ + AlX_4^- \qquad (2.13)$$

$$AlX_2^+ + M \rightleftharpoons X_2Al{-}M^+ \qquad (2.14)$$

Analysis of earlier electrochemical research (see [13] for bibliography) led Grattan and Plesch to modify former experimental methods, and, in particular, to account for the presence of microadditives in the solvents according to the dependence of the electrical conductivity on the aluminum halide concentration; the linear nature of this dependence is evidence of the absence of microadditives. As a result, they succeeded in obtaining new and more precise data; this is shown partially in Table 2.3.

From the point of view of the informative use of these results for investigating the mechanism of cationic polymerization which proceeds in AlX_3–RX systems, the actual ability of the original Lewis acids to ionize spontaneously must be emphasized. The true nature of the ions may differ somewhat from that shown in Eq. (2.13); this is shown in Eqs. (2.15) and (2.16):

$$2(AlX_3)_2 \rightleftharpoons Al_2X_5^+ + Al_2X_7^- \qquad (2.15)$$

$$2AlX_3 + RX \rightleftharpoons AlX_2^+\cdot RX + AlX_4^- \qquad (2.16)$$

$$AlX_2^+\cdot RX \underset{}{\overset{AlX_3}{\rightleftharpoons}} Al_2X_5^+\cdot RX \qquad AlX_4^- \overset{AlX_3}{\rightleftharpoons} AlX_7^-$$

The first of these has already been considered in previous works (see [14-16]). The idea expressed in scheme (2.14) was presented by Grattan and Plesch in connection with the analysis of thermodynamic data [12]. Without making a precise choice between the structural variations of the above ions, the authors expressed the opinion that the probability of the existence in an RX medium of a "free" AlX_2^+ ion is extremely small and gave preference to the complex $AlX_2^+\cdot RX$. In discussing the nature of the ions which exist in these systems, the authors took into consideration the data available in the literature. According to these the positive and negative ions are similar in size and have practically the same mobility.

The sharp differentiation of the electrochemical effects which are the result of equilibria of the types (2.13) and (2.14) are related to the important elements discussed in [12] and also those which are caused by microadditives. Their absolute elimination does not seem to be achieved even in experiments prepared with extreme care.

The data of these authors, obtained from electrical conductivity of these systems in the polymerization of isobutene, are of special interest [13]. Before addressing ourselves to these data,

TABLE 2.3. Dissociation Constants for Aluminum Halides in Alkyl Halides [12]

Halide	Solvent	t, °C	K_{diss}, moles/liter
$AlBr_3$	CH_3Br	−23	2.2×10^{-8}
		−78	2.4×10^{-7}
$AlCl_3$	C_2H_5Cl	0	2.4×10^{-6}
		−78	3.0×10^{-4}
$AlCl_3$	CH_2Cl_2	0	2.3×10^{-4}
		−78	3.3×10^{-3}

it is necessary to note that effective polymerization is achieved only when the monomer is introduced quickly into the solution of initiator. An example is quoted in which the monomer, introduced over a minute in a quantity corresponding to the final concentration of isobutene, 2 moles/liter, produces 100% conversion with $AlBr_3$ in CH_3Br at −78°C. In closely similar reaction conditions, but with the monomer introduced over a period of 1.5 h, polymer is not formed at all.

These features are ascribed to competition between the two reactions

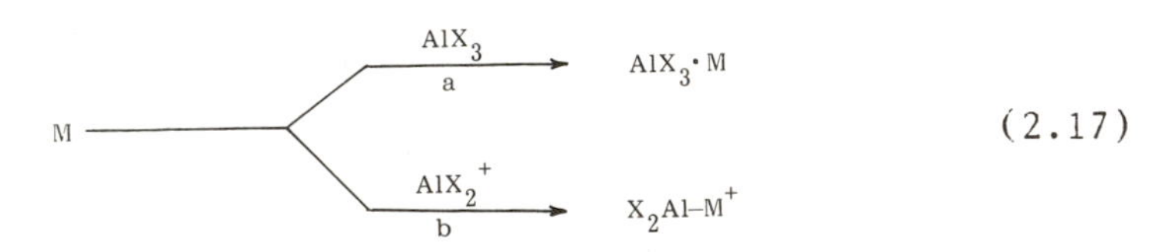

(2.17)

In order to realize reaction (2.17b) a sufficiently high concentration of monomer is required. On the other hand, the gradual addition of small portions of monomer create (due to the large excess of AlX_3 over AlX_2^+) a significant advantage for reaction (2.17a), which passivates the monomer as a polymerizing compound. Judging by the data of [12], the fraction of cations of the type AlX_2^+ in conditions similar to those described constitute about 0.001% of the original concentration of AlX_3. These facts allow one to exclude the previous suggestion of the possibility of initiating the process of polymerization in such systems by means of the direct interaction of the monomer with the nonionized aluminum halides.

We note the following electrochemical characteristics of the isobutene–$AlBr_3$–CH_3Br system [13]. At the moment of introduction of the methyl bromide solution containing the monomer into the aluminum bromide solution the electrical conductivity of the system

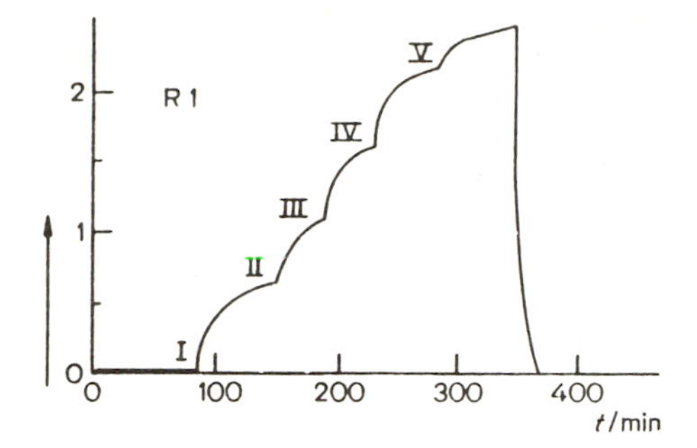

Fig. 2.2. Changes in the conductivity (κ) during polymerization in an isobutylene—methyl bromide—aluminum bromide system at 0°C. Initiator concentration $3.6 \cdot 10^{-2}$ moles/liter. Quantities of monomer introduced over a measured time interval (in moles):

I	II	III	IV	V
2.39	2.05	2.21	3.1	2.65

increases sharply; the duration of the "κ-jump" may be taken as the duration of polymerization (several seconds). Later, the magnitude of κ decreases smoothly for some time, reaches some constant value, and then remains stable for a long period, up to several days. A similar picture is produced with each introduction of a fresh quantity of isobutene into CH_3Br, as is shown for one of the series of experiments in Fig. 2.2. The phenomena described, which may be distinguished by the processes of polymerization and by particular "post-effects," are qualitatively contrary to those observed when the monomer is added slowly. In these cases the initial value of κ falls especially rapidly when the isobutene is first added. In these experiments the monomer was added by slow recondensing from the measuring vessel to the reaction vessel. A precise evaluation of the quantity of isobutene which produces the maximum effect is not possible from the data given in [13]. From these graphs it can be seen that this corresponds to the condition $[M] \simeq 0.005$ mole/liter.

The interpretation proposed by the authors for these data can be reduced to the following.

The initial jump in electrical conductivity during polymerization is caused by the formation of the growing chains with tert-butylium end groups. This is confirmed by the results of the study of the electrical conductivity of the model system $(CH_2)_3CBr$—$AlBr_3$—CH_3Br in which κ increases linearly with the concentration of $(CH_3)_3CBr$. The cause of this further increase in κ after polymerization is complete (see Fig. 2.2) is the interaction of the AlX_2^+ ions with the unsaturated final links of the macromolecules which are formed as a result of transfer reactions which are typical for the polymerization of isobutene, for example:

$$\sim CH_2\overset{+}{C}(CH_3)_2 + M \longrightarrow \sim CH_2-\overset{\overset{CH_3}{|}}{C}=CH_2 + HM^+ \quad (2.18)$$

further transformation gives

$$\sim CH_2-\overset{\overset{CH_3}{|}}{C}=CH_2 + AlX_2^+ \longrightarrow \sim CH_2\overset{+}{C}(CH_3)CH_2AlX_2 \quad (2.19)$$

which takes place at a slower rate compared with polymerization, and slowly raises the total concentration of ions as a consequence of the corresponding displacement of the initial equilibrium (2.13). This conclusion is based on the data obtained by the deactivation of the reacting mixture using tritiated water. In spite of the well-known low efficiency of initiation of cationic polymerization of isobutene, which is the result of reactions of the type (2.18), etc., tritium has been found in each of these polymer chains.

Data obtained in the same experimental conditions in a model $CH_3Br-AlBr_3-$polyisobutene system show that Eqs. (2.18) and (2.19) are responsible for this result. Thus from these it follows [13] that unsaturated end groups of the macromolecules can add to aluminum halide fragments.

As for the reduction in the value of κ when the monomer is added in microscopically small amounts, this can be explained by the displacement of the equilibrium (2.13) towards the nonionized form of the aluminum halide as a result of the formation of new $AlX_3 \cdot M$ complexes.

Thus reaction (2.17a) exerts an inhibiting influence on the process of polymerization due to: 1) the passivating of the monomer, which is caused by the reduction (apparently very considerable) of its nucleophilicity, and 2) the reduction in the concentration of the initiating ionic agents.

We have paid special attention to the work of Grattan and Plesch [12, 13], since in the first place it enables the situation, which for a long time provided grounds for contrary opinions, to be explained. In the second place, for the examples considered it is possible to be convinced of the actual efficient application of the methods of electrochemistry to the establishment of a detailed mechanism of polymerization in processes which are especially difficult to study. We emphasize that the number of questions which require explanation has gone far beyond the limits of obtaining initial data for the calculation of the ion-pair dissociation constants, which is a routine experimental task in most other cases. This remark is in no way directed at denigrating the role of these studies. However, they are directed mainly to the evaluation of the relative contributions of simultaneously reacting free ions and ion pairs. In the case of the AlX_3—RX systems the questions are of a different nature, since the very existence of free ions in these systems which are responsible for initiation but are not formed due to the presence of microadditives or by secondary reactions between the components of the system is open to doubt. We also note that earlier research into cationic systems using electrochemical methods during the course of polymerization (but not in model systems with monomer absent) has been essentially on pseudocationic processes, the specificity of which is not connected with the parallel electrochemical effects; see Section 4.12.

The amount of information defining the types of cationic active sites on the basis of electrochemical features is concentrated mainly on initiators of polymerization. The existing information on growing chains, which as yet is very limited, concerns only the cationic active sites of heterocyclic monomers. This unsatisfactory situation is somewhat mitigated by the following. It does not seem that the value of K_{diss} for anionic growing chains is determined by whether the agents being compared are low- or high-molecular-weight compounds (see Section 2.1). Apparently, it is the nature of the carbanionic fragments of the ion pair which is decisive in this respect (i.e., primarily the end group of the growing chain). The evidence for the dependence of K_{diss} on the nature of the fragment R^* of the initiator R^*Y and of the end group of M_n^*Y is quite weighty and is underlined by the data for certain cationic systems.

This question was discussed recently by Ledwith and Sherrington in their review [17], which concentrated on stable salts of organic cations and which covered the literature up to 1975. Some of the data cited in [17], augmented by more recent results, are given in Table 2.4.

In the same review opinions are expressed on the probability that the active sites existing in reaction media which are common in cationic polymerization are mainly in the form of separated ion pairs. More definite conclusions on this have been drawn by Penczek et al. [18-21] who evaluated the interionic distance in growing THF and oxepane chains with SbF_6^- counterion. When nitromethane was used as a solvent for THF, nitrobenzene for oxepane, and CH_2Cl_2 for both types of chain, the authors obtained values of a for the first and second ion pairs of 9.0 and 11.9 Å at 25°C. These values were calculated from the dependence of K_{diss} on $1/\varepsilon$ (see Eq. 2.3), which has been studied by varying the composition of mixed CH_2Cl_2—nitrocompound solvents with ε in the range from 7.4 to 19.1 for oxepane and from 8.2 to 22.8 for THF; the differences in the values of ε quoted for solutions of THF and oxepane in CH_2Cl_2 are the result of different monomer contents in these systems.

TABLE 2.4. Dissociation Constants for Some Ion Pairs

Cation	Counter-ion	Solvent	t, °C	K_{diss}, moles/liter	Ref.
$Et_3\overset{+}{O}$	BF_4^-	CH_2Cl_2	−0.5	4.4×10^{-6}	17
$\sim\overset{+}{O}$ (ring, 5)				5.4×10^{-6}	
$Et_3\overset{+}{S}$	BF_4^-	$PhNO_2$	20	1.3×10^{-2}	17
$\sim\overset{+}{S}$ (ring, 6)				1.6×10^{-2}	
Ph_3C^+	SbF_6^-	CH_2Cl_2	25	1.7×10^{-4}	17
	ClO_4^-	$C_2H_4Cl_2$	21.5	1.3×10^{-3}	
$\sim\overset{+}{O}$ (ring, 5)	SbF_6^-	CH_2Cl_2	25	1.5×10^{-5}	18
		CH_3NO_2		2.0×10^{-3}	19
$\sim\overset{+}{O}$ (ring, 7)	SbF_6^-	CH_2Cl_2	25	2.8×10^{-5}	20
		$PhNO_2$		1.6×10^{-3}	

The linear character of this dependence is evidence of the constancy of the ion pair type over the whole range of ε.

The value of a obtained greatly exceeds the sum of the radii of the components of the ion pairs studied; according to the data in review [21], the interionic distance in the oxepane cation—SbF_6 contact ion pair in CH_2Cl_2 is 5.9 Å. This enables one to state that the above active site is of type IP_s, or at least that this type makes the greatest contribution.

The features put forward for growing THF and oxepane chains can bring into question the causes of the great sensitivity of the magnitude of K_{diss} on the nature of the solvent while the type of ion pair remains constant. In connection with this, it is necessary to note that in the work of Fuoss [10], which we have already cited, it is concluded that K_{diss} depends not only upon the dielectric constant of the medium but also upon other parameters. The latter include the value of R (which depends, in its turn, upon the dipole moment, the polarizability, the shape and dimensions of the solvent molecules, and also the charges and dimensions of the ions) and the difference between the energies of IP_c and IP_s which is a function of the interaction of the ions with the solvent and with each other (when r = a; see Section 2.1.1 [17]). Consequently, the dependence of K_{diss} on the nature of the solvent may be ascribed to the sensitivity of the rate constants [Eq. (2.9)] to solvation which is the result of the action of all these factors. The results which describe the behavior of fluorenyllithium in various media agree qualitatively with this conclusion (Table 2.2).

There are no values of these constants for any growing chains of this type. We will draw a more distant example, relating to the ionization of aluminum chloride in alkyl halides [see Eqs. (2.13), (2.15), and Table 2.3]. In [12] values of the constants k_d and k_{rec} at 0°C show that there is considerable dependence of the latter on the nature of the solvent while k_d is not at all sensitive to solvation effects.

Solvent	k_d (liters/mole/sec)	k_{rec}
Ethyl chloride	5.5×10^{-4}	2.3×10^{2}
Dichloromethane	9.2×10^{-4}	4

We note that equilibria (2.13) and (2.15) cannot be completely identified with equilibria (2.1) and (2.9), since in the former the stage of ion pair formation is absent. Its inclusion could lead to the following modification of equilibrium (2.13):

$$2AlX_3 \underset{a}{\rightleftharpoons} \begin{matrix}\text{contact}\\ \text{ion pair}\end{matrix} \underset{b}{\rightleftharpoons} \begin{matrix}\text{separated}\\ \text{ion pair}\end{matrix} \underset{c}{\rightleftharpoons} AlX_2^+ + AlX_4^- \qquad (2.20)$$

Such a modification affects the physical sense of the above values of k_d, which in [12] is regarded as a reaction constant of the second order, i.e., applicable to the latter scheme as the rate constant for the forward reaction of Eq. (2.20a). It is possible that the above experimental value is connected with the unimolecular stage of the dissociation of ion pairs [Eq. (2.20b)]. For the present discussion this is not so important and so we will not discuss it in detail. Nevertheless, it seems necessary to note that the ionization schemes for aluminum halides which do not take the intermediate stages into account [Eqs. (2.20a, b)] are simplified. This is essential when considering the possbility of the products formed at these stages taking part in the initiation simultaneously with reaction (2.14), induced by the free ions.

We note that Grattan and Plesch in a special report [22] on the ionic equilibria of the general type

$$2MtX_x \rightleftharpoons MtX_{x-1}^+ + MtX_{x-1}^- \qquad (2.21)$$

consider both the case of direct ionization of the original compounds and that which proceeds via free ions. However, the systems which are discussed in [12, 13] are considered by the authors to belong to the first type. From our point of view, this is difficult to accept as a rigorously based conclusion.

Concluding this section, we emphasize that the application of schemes (2.9) to polymerization processes demands a certain caution. The occurrence of IP_s is not necessarily accompanied by their participation as active sites. Their participation may be hindered by the extremely small time for which they exist as independent entities which is insufficient for the occurrence of propagation events. The existence of a system in which the frequency of exchanges between IP_s and IP_c exceeds the time necessary for the propagation events is perfectly probable.

At first sight such a situation is rather difficult to achieve, especially from data for free ions whose lifetimes may be controlled by diffusion (see the value of k_{rec} in Table 2.2). However, their decay is the result of a bimolecular mechanism and so it is possible to control the rate of such events, whereas the transition of IP_s to IP_c may, in certain circumstances, take place spontaneously and consequently account for the reduction in the lifetime of IP_s as compared with free ions. This, together with the smaller kinetic activity of IP_s, may mean that IP_s occurs but does not take part in polymerization.

2.2. SPECTROSCOPIC STUDIES

The application of various spectroscopic methods is distinguished by its universality in comparison with other approaches used for establishing the structural parameters of ionic agents. The information available in this field is extremely circumstantial and has been summarized many times in various papers (see for example [1, 23]) and in comparatively recent reviews [21, 24, 25]. Basic material may be more appropriately discussed with reference to spectroscopic data obtained more recently. Here we note only those data which for one reason or another (basically new facts and the solution of difficult problems) can be conveniently examined independently.

We note that the results of actual methods of spectroscopic studies discussed below can be applied directly to active sites only to a limited extent. Thus, when using IR or NMR spectroscopy there may be considerable differences (often 2-3 orders of magnitude) between the concentrations of agents in the solutions studied by these methods and in the system usually employed in polymerization. This may be of importance for compounds which are inclined to aggregation and solvents which favor aggregation.

Since we are forced to study low-molar-mass analogs, it is difficult to draw conclusions as to the structure of real active sites. In such studies adducts containing one monomer unit RM*Y are used; here penultimate units which might complicate the spectrum are absent. However, even in such cases the substituent R may play a significant role since any variation in this quantity (including transition to the RRMM*Y models which contain two monomer units) may be reflected in the spectrum of M*.

2.2.1. UV Spectroscopy

The great sensitivity of this method, which is used in the study of materials in the range of concentrations common to polymerization processes, and the independence of the electronic features of the active link of the length of the growing chain frees the results obtained from the limitations discussed above. On the other hand, far from all the compounds which fall into the category of the active agents under dicussion have UV spectra which are sufficiently clearly differentiated. Without resorting to a large number of actual examples (these may be found in [6, 24, 25]), we note the differences in the sharpness of the UV spectra, which are already known for growing anionic chains and which were obtained for fluorenyl derivatives (FlMt) by Hogen-Esch and Smid. The changes observed with FlMt compounds arising from changes in the nature of the metal, temperature, and reaction medium, including the introduction of small amounts of strong ED, were the basis for the first experimental indications of the existence of two types of ion pairs. The displacement of the absorption bands in the long-wave region which accompanies the transition IP_c to IP_s (λ_{max} values 349 and 373 nm, respectively) was used later as a unique "reference shift" when studying other anionic agents. However, in the case of M_nMt chains, where M is a diene or styrene, instead of a "jump" from one absorption band to another which is typical of various systems including fluourenyl derivatives [9, 23], a simple displacement of the λ_{max} of the comparatively broad absorption band takes place.

The lack of clear differentiation of the UV spectra of living chains obviously may be ascribed to the structural nonhomogeneity of the end links. The results of the analysis of the spectra of polybutadienyllithium—THF systems also point to such a conclusion; for more detail see Section 5.2. As for solving the problem of the type of ion pair, in this respect NMR spectroscopy is more informative, especially in combination with electrochemical methods. Nevertheless, UV spectroscopy is an important method of detecting and investigating ionic AS under actual conditions. This, when combined with kinetic data, enables one to reveal the simultaneous action of ionic and covalent growing chains (see Chapter 4). More generally, UV spectroscopy makes it possible to estimate changes in the concentration of AS (due to generation, consumption, and isomerization).

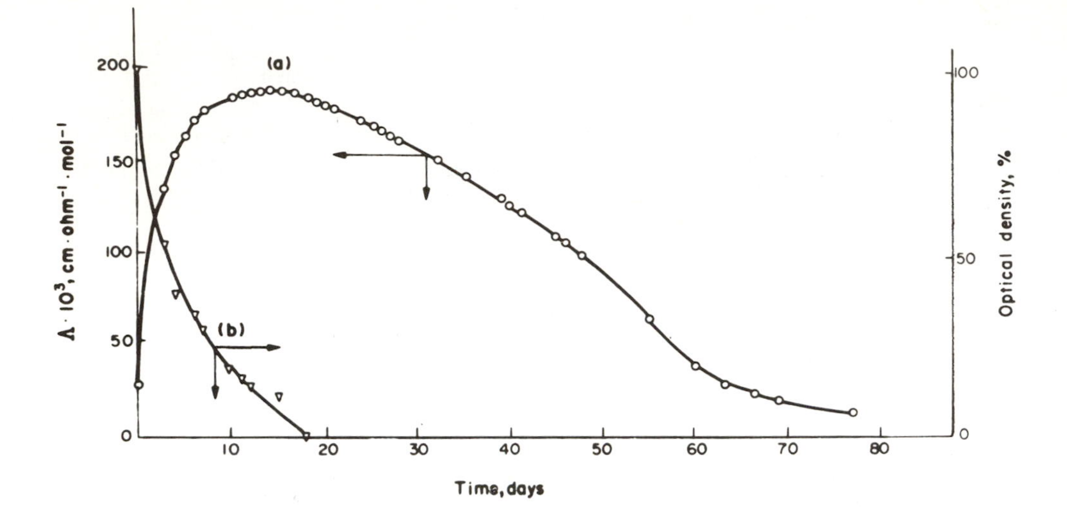

Fig. 2.3. Changes in the equivalent conductivity (a) and optical density of the absorption band in the UV spectrum at λ_{max} = 290 nm (b) in a polybutadienyllithium–tetrahydrofuran system at 20°C [6]. PBLi concentration, $3.3 \cdot 10^{-3}$ mole/liter.

Many examples could be cited to illustrate the importance of the UV method, but we will consider only the results of UV spectroscopy and electrochemical studies of the polybutadienyllithium–THF system (see Fig. 2.3). Only a combination of UV and electrochemical data allows one to assign the increase in conductivity to the formation of mixed $M_nLi \cdot ROLi$ associates.

Information that differs from this essentially analytical data is comparatively rare in studies of the UV characteristics of ionic AS. Therefore, it is especially interesting to mention the recent work of Andresen [26] devoted to the study of an effective Ziegler–Natta type catalytic system. The system was based on dicyclopentadienyldimethyltitanium and aluminooxane obtained from trimethylaluminum (the polymerization induced by some aluminooxane systems is discussed in Chapter 4).

Figure 2.4 shows the effect of ethylene introduced into the hydrocarbon solution of the above-mentioned system. The author assigned the absorption band appearing in the 395 nm region to the formation of the AS·M σ_1-complex. The following facts should be taken into consideration in confirming this conclusion. First of all, the system being studied is very active with respect to ethylene and does not polymerize propylene. In the second place, the presence of propylene does not affect the UV spectrum of the origi-

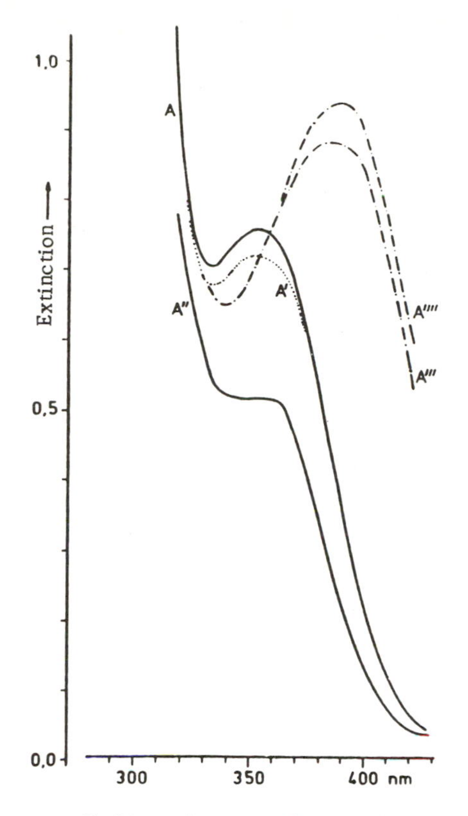

Fig. 2.4. UV spectra of dicyclopentadienyldimethyllithium-based systems (DCMT [25]). A) DCMT in toluene; A') DCMT–trimethylaluminum (TMA) (Ti/Al = 0.5); A") DCMT–TMA–H_2O (Al/H_2O = 2) 25 min after introduction of H_2O; A"') and A"") system A" 5 and 25 min, respectively, after the introduction of ethylene.

nal system. Finally, in the third place, the presence of propylene in the reaction mixture does not prevent the appearance of the aforementioned absorption bands when ethylene is introduced into the system.

It may be assumed that the superiority of ethylene as a participant with the catalytic complex compared to propylene is the result of either a spatial factor or the decisive role of the donor bond, for the formation of which it is essential that the monomer π-system should have acceptor properties. However, the reason why the feature attributed to the AS·M complex is found only in the case of titanium aluminooxane complexes but not in other related catalytic systems remains an open question. This circumstance forces one to refrain from drawing any final conclusion as to the accuracy of assigning the bands at 395 nm to the π-complex of ethylene with titanium aluminooxane.

2.2.2. IR Spectroscopy

The IR spectra of the agents under discussion are distinguished by the large amount of information they contain in comparison with other weak donor—acceptor interactions, which are accompanied by rapid mutual transition. This feature has been widely used for the identification of complexes of organometallic initiators and living chains. The results of the research published in this field up to the mid-seventies can be found in review [24].

This contains facts which indicate the possibility, using this method, of finding important structural details which remain hidden when the same systems are studied using NMR. In particular it deals with the equilibrium of the general form

$$(RMt)_n \rightleftarrows (RMt)_{n-x} + (RMt)_x \rightleftarrows nRMt$$

and the discovery of the step mechanism of the formation of organolithium complexes with electron donors, which leads to the evaluation of the constants of particular equilibria of the type

$$(RLi)_n + D \rightleftarrows (RLi)_n \cdot D$$

$$(RLi)_n \cdot xD + D \rightleftarrows (RLi)_n \cdot (x + 1)D$$

etc. It is appropriate to note here that the IR spectra of the oligobutadienyllithium (OBuLi)—sec-butyllithium system were the first experimental data on the formation of mixed $(M_nLi)_m \cdot (RLi)_p$ associates. This follows from the low-frequency shift of the absorption bands of the end C=C bond of OBuLi in the presence of metal-alkyl. Similar results were later obtained for oligopentadienyllithium—sec-butyllithium systems [27] (see Fig. 2.5).

The structural feature of the M_nMt active agents which is the source of the IR spectra is usually the end unit in living RM_mMt oligomers (degree of polymerization $m \approx 6$-8). Zgonnik, Kalnin'sh, and others have made a particularly detailed study of compounds of this type using dienyl and styryl lithium derivatives; for bibliography see [24]. In comparatively rare instances, such information is amplified by data on the effects of the interaction of the counterion with more distant links of the growing chain. The appearance of such effects was established by Tsvetanov et al. from a study of the IR spectra of $M_nM'_xMt$ chains where M_n is a polystyrene chain, M' is acrylonitrile [28] or methacrylonitrile [29], and x varies from 1 to 6; the counterion is lithium, sodium, or potassium.

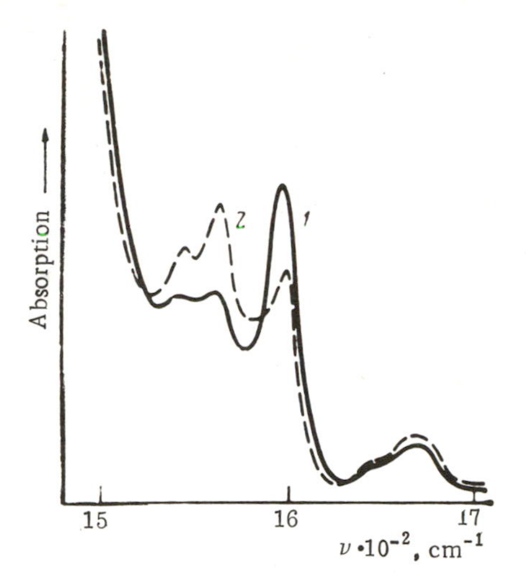

Fig. 2.5. IR spectra of a oligopentadienyllithium—sec-butyllithium—isooctane system [27]. Temperature, °C: 1) −12; 2) −74. Mixed associates are characterized by absorption bands at 1563 and 1542 cm^{-1}, which are absent in the IR spectra of the original reagents.

These results are interesting both in themselves and from the point of view of method. It must be borne in mind that in the study of such model active sites as $M_n(CH_2CHCN)_x \cdot Mt$, M is the nonpolar monomer (butadiene or styrene) required to obtain soluble organometallic chains with acrylonitrile end groups. The initial problems, which were solved by studying chains of this type, were connected with the electrical conductivity, the type of ion pair, the degree of association, etc. [3]. The possibility of ascertaining long-range effects arose in the course of studies of the IR spectra of these agents when the number of end links of acrylonitrile or methacrylonitrile was varied.

Qualitatively similar effects have been established by studies of RM_mMt compounds modelling active sites of the methacrylate series. For polar monomers the synthesis of the simplest adducts of this type is made difficult by side reactions. In these circumstances metallized products of saturated analogs of the polar monomers or their oligomers were used as low-molar-mass models. The work of Lochmann and Trekoval [30] is of great significance in this respect. It consists of a lengthy series of studies of methacrylate models of this type (for bibliography see [30]). It contains an analysis of the IR spectra of $(CH_3)_2C(COOR)Mt$ [(2.I)] and $(CH_3)_2C(COOR)CH_2C(CH_3)(COOR)Mt$ [(2.II)] and their more complex analogs; here R is CH_3 or C_2H_5 and Mt is Li, Na, or K. Compounds (2.I) are distinguished by

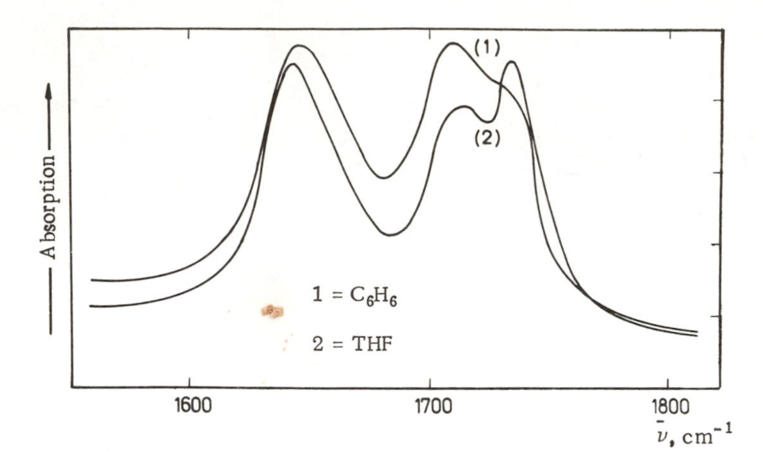

Fig. 2.6. IR spectra of two-unit lithium model (2.II) (see text): 1) in benzene; 2) in tetrahydrofuran [30]. Temperature, 25°C.

a single absorption band in the region associated with the ester group. In the (2.II) compounds an additional band is also to be found in the same region. Both this and their bands differ by having a "free" ester group peak shifted to the lower frequency. The IR spectra of the above nitrile [28, 29] and ester [30] models are compared (see below) with the characteristics which are known for the same unperturbed polar groups. The interaction of the counterions with penultimate links are shown symbolically in the simplified formula (2.III) (cm^{-1}):

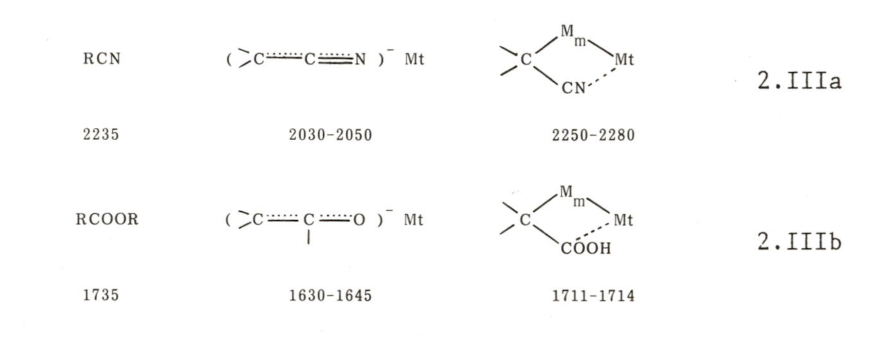

As can be seen from the IR spectrum of the two-unit lithium model, the absorption band of the intramolecular complex (2.IIIb), in benzene, has the same intensity as that for the final unit. When THF is the solvent the intensity of the band at 1712 cm^{-1} is considerably reduced, but it does not disappear completely (Fig. 2.6). This observation is all the more interesting since the NMR spectrum of the same compound in THF shows no trace of intramolecular interaction for the lithium counterion with the polar group of the penultimate link; see Section 2.2.3.

Apparently for anionic chains, in particular those of type IP_c and in media with low solvating ability, such interactions are quite common, but in order to detect them a proper selection of model compounds RM_mMt which differ in m value is necessary. The consequences of the formation of intramolecular complexes of the types (2.IIIa, b) for real polymerization processes are examined in Chapters 3-5.

The considerable amount of data on IR spectra of anionic active site—ED complexes [24] has been recently supplemented by some interesting facts established by Zgonnik, Davidyan, et al. [31] in their studies of hydrocarbon solutions of polydienyllithium (PDLi)—ED—monomer systems. Phenomenologically the effects consist of an alteration in the equilibrium concentration of "free" ED when the monomer is introduced into the reaction mixture. For example, with the ratio PDLi:THF:diene of 1:1:4, the concentration of free THF grows approximately 20% compared with the magnitude found when the monomer is absent. These data which relate to butadiene and isoprene are qualitatively the same when THF is replaced by DEE.

The observed changes in the IR spectra which show the "squeezing out" of the ED from the complex by PDLi·ED complexes enable the nature of the phenomenon to be investigated. An analysis of the results led to the conclusion that the acceptor ability of PDLi in relation to ED depends to a noticeable degree upon the structure of the final unit of the growing chain. In all the above cases polydienyllithium synthesized in hydrocarbons were used as the original agents. The primary effect in the three component systems is the polymerization of diene which, thanks to the presence of ED, leads to the formation of end units which differ from the original in their structure. Corresponding changes are detected by IR and ^1H-NMR methods [31].

The correctness of the proposed interpretation is supported by the absence of the effects of "squeezing out" when the monomer is introduced into the PDLi—ED system in conditions which exclude polymerization.

2.2.3. NMR Spectroscopy

NMR spectroscopy is the source of particularly valuable information on the structure of ionic agents in solutions; parallel use of different nuclei gives it great objectivity. For materials which we are considering at present ^{1}H- and ^{13}C-NMR data are combined in most cases. However, independent data from these and other spectra often lead to very detailed structure of the compounds studied.

The possibility of solving various problems (the geometry of the molecule, its electronic structure, the conformation of systems which contain conjugate bonds with retarded rotation, the evaluation of the magnitudes of the corresponding rotation barriers) is reflected differently in actual studies. The results of several studies, mainly relating to metal-alkenyl compounds, are given below. From the examples of analogs of active sites chosen (allyl and benzyl derivatives) or 1- and 2-unit models of growing chains it is possible to see a change in the method of approach to the study of related compounds.

2.2.3.1. Allyl Compounds

A condensed summary of the works containing NMR spectra obtained on the nuclei of ^{1}H, ^{13}C, and ^{23}Na for allyl derivatives of alkali metals supplemented by new data is given by Braunstein et al. [32]. It has been shown that freshly prepared C_3H_5Mt compounds in a solution of THF at a temperature of −20°C are found in the metastable state A_4X. When these solutions are preserved for some time a gradual transition of the original form to an unstable form with hindered rotation (type AA'BB'X) is illustrated by the data for allylpotassium in Fig. 2.7. An increase in temperature to some level which depends upon the nature of Mt causes the signals corresponding to the protons of the methyl group to merge. In particular, for allylsodium this is observed at 5°C and for allylpotassium, at 65°C; the corresponding energies of activation are found to be 11.5 and 14.3 kcal/mole. Values of spin–spin splitting constants $J_{C,H}$ are smaller than expected for sp^2 hybridization, but the reason for this is not discussed by the authors.

The question of the physical meaning of the deviation of J from the expected values has been examined in some detail by Schlosser et al. [33]; they came to the conclusion that π-allyl derivatives of alkali metal derivatives are nonplanar. Before turning to the considerations upon which this is based, we will mention previous studies by the same authors on the nature of the C–metal bond in allylmetal–THF systems. According to the ^{13}C-NMR spectra at a frequency of 15.08 MHz [34], in allylmagnesium bromide this bond has a σ-nature. For the alkali metals, spectra were obtained which indicate π-character. The nature of the metal determines its orientation relative to the allyl group. Nonsymmetry of the π-system was found only for allyllithium; the higher alkali metals were oriented symmetrically with respect to the methyl fragments of the allyl group. Further studies of the same compounds in conditions of higher resolution of the ^{13}C-NMR spectra (90.55 MHz) [35] confirmed the

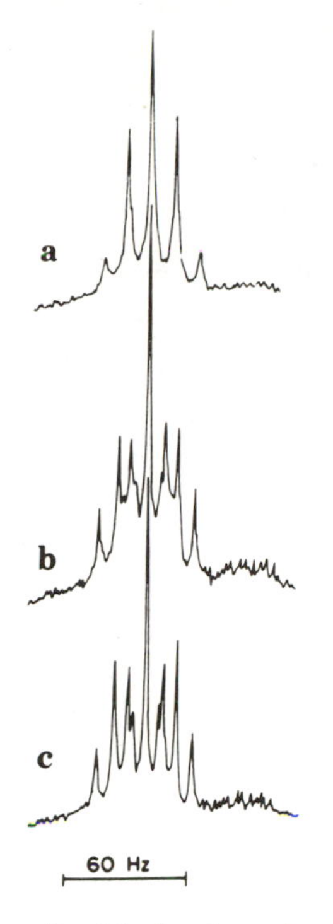

Fig. 2.7. ^{1}H-NMR spectra of allylpotassium in tetrahydrofuran at −20°C [32]. Spectra taken: a) 30; b) 60; and c) 75 min after synthesis.

previous conclusions on the structure of magnesium and lithium derivatives; in allylsodium and allylpotassium a slight deviation from strict symmetry was observed.

Turning to the spin–spin splitting constants, the authors note that in conditions which are natural for π-allyl compounds with flat structure, values of $J_{C,H}$ of the order 145-165 Hz ought to be expected. According to the experimental values found [33], only the external hydrogen atoms (H_a) may be taken as lying in the plane of the carbon backbone; the internal hydrogen atoms (H_i), and in particular, the central atoms (H_m) protrude from this plane. As an example, we present data for allylpotassium (THF, −20°C):

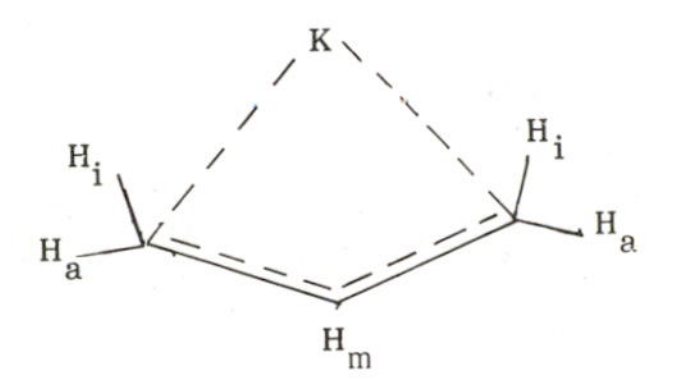

Atomic group	C,H_a	C,H_i	C,H_m
$J_{C,H}$, Hz	154.9	142.8	131.8
Fraction, %	31.0	28.6	26.4

It will be recalled that the σ-character of the bonds under consideration amounts to 1/5 the corresponding J constants.

These results are interpreted by the authors as the emergence of H_i atoms from the C-plane towards the side opposite to that of the H_m atom. Such a condition is regarded as necessary for the preservation of antiperiplanar disposition being able to ensure a high value of J_{H_i,H_m} which is equal to 14.7 Hz. As is generally known, the magnitudes of $J_{H,H}$ for H–C=C–H (cis) and H–C=C–H (trans) lie in the ranges 7-12 and 13-18, respectively.

In the authors' opinion, the discovery of a nonplanar structure causes a distribution of the electron density which makes the bonding of the metal atom with the carbon backbone stronger.

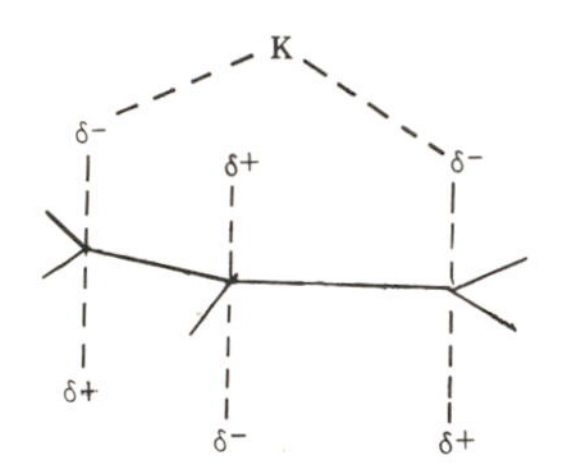

In one of the studies quoted [35], the question of the possible effect of the aggregation of the compounds studied on the NMR spectra obtained is touched upon. The high concentration of compounds which are studied by NMR (usually 0.4-0.5 moles/liter) does not allow the effects of aggregation to be neglected. For example, according to West et al. [36], the average value of the degree of association of allyllithium in THF is 1.4 at a concentration of 0.8 moles/liter (25°C). Because of the absence of corresponding data for allylpotassium, Stähle and Schlosser compared the NMR spectra of this compound and its 2-methyl derivative with that of an equimolar mixture of both. The spectrum of the mixture simply indicates the superposition of the independent features of the components and consequently shows the absence of aggregation in these conditions.

The work of Boche et al. [37], which concentrates on the lithium derivatives of the general type PhCH=CH(R)CH(Ph)Li, is an interesting example of the detailed study on the configuration effects which cause the restriction of the rotation around the "$1\frac{1}{2}$-order" C=C bonds in π-allyl compounds; for this see the references in [37] which contain earlier studies of $C_3H_4(R)Mt$ and $C_3H_3(R,R')Mt$ com-

TABLE 2.5. Conformation of PhCH=CRCH(Ph)Li Compounds in THF (%) [37][a]

R	exo, exo	exo, endo	endo, endo
H	92	8	-
CH_3	8	92	-
C_2H_5	15	68	17
i-C_3H_7	-	38	62
t-C_4H_9	-	-	100

[a]Measurements carried out in the range −46° to −20°C. Values are not given for these differences since in this range these values vary very little (±4%).

pounds and distinguish the natures of the substituent and the metal. The authors directed their attention mainlv to the dependence of the fraction on each of the three possible conformations of the allyl ion on the substituent R:

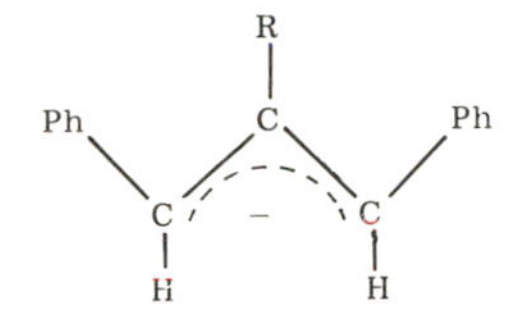

(exo,exo)

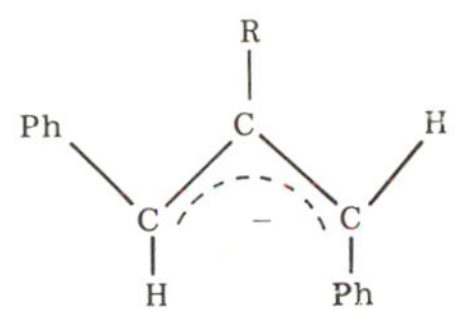

(exo,endo)

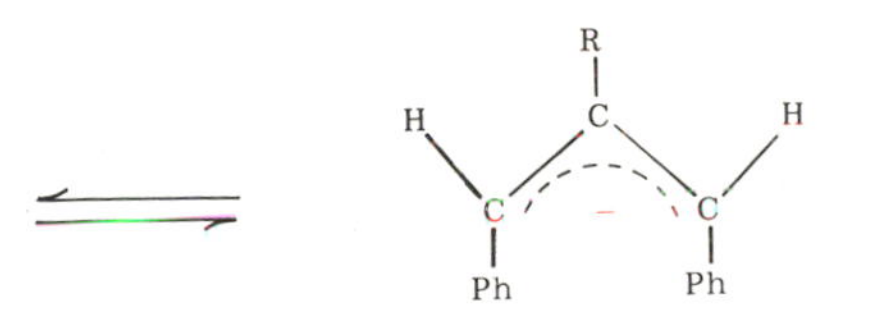

(endo,endo)

The use of ^{1}H-NMR methods and the various deutero-analogs of the compounds studied has led to quantitative estimates, some of which are shown in Table 2.5. The only thermodynamic feature of the rotation of the barriers during the mutual transitions is the free energy of activation. This has been evaluated for the majority of diphenylallyl derivatives. In particular for the reversible rotation,

$$\text{exo,exo} \rightleftarrows \text{endo,endo} \qquad (2.22)$$

the value of $\Delta G^{\neq}_{273\text{ K}}$ lies between 189.1 (for R = H) and 12.5 (for R = i-Pr) kcal/mole. This reduction in $\Delta G^{\neq}$ with the size of the substituent R is explained as being due to its destablizing effect on the allyl compound. In order to establish whether these lithium compounds are of the IP_c or IP_s type, NMR spectroscopy was used to see what changes occur when the metal was substituted and when a certain amount of hexamethylphosphoramide was introduced into the THF solution. The result showed the participation of IP_c of the lithium derivatives in the transformations (2.22) to be possible only for the cases R = H or CH_3.

The dependence of the type of ion pair on the nature of the substituent is seen somewhat differently in compounds of the benzyl type; see Section 2.2.3.2.

Information on the NMR spectra of analogous cationic active sites is incomparably less detailed. The instability of carbocations practically excludes the possibility of studying them in systems which are sufficiently close to those used in polymerization. Only under extreme conditions which favor the stability of such agents is it possible to occasionally obtain sufficiently sharp NMR spectra. Of the materials which are of interest from the point of view of the present discussion, it is possible to mention the substituted alkenyl cations obtained by Siehle and Mayr [38], according to the reaction

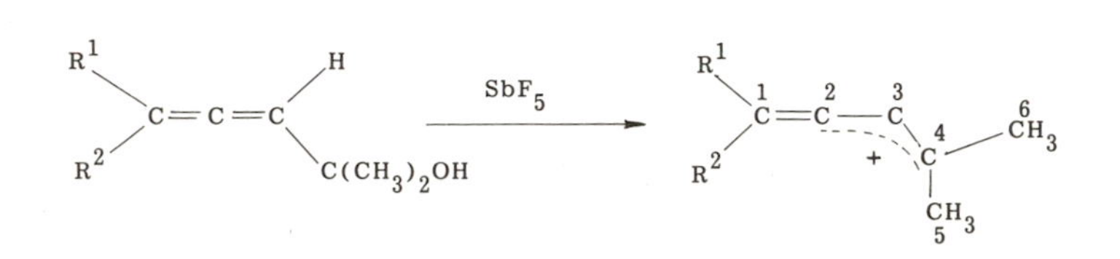

by the molecular beam method and studied in mixed solvents SO_2ClF–SO_2F_2 (2:1) at −120°C; here R^1 and R^2 are H or CH_3. The conclusion as to the delocalized nature of the bonds in the π-allyl groups is based on the inequality of the ^{13}C-NMR signals of the C^5 and C^6 atoms and on the maximum downfield chemical shifts for the C^2 and C^4 atoms. The corresponding values are given below for the simplest of the cations studied ($R^1 = R^2 = H$):

	C^1	C^2	C^3	C^4	C^5	C^6
Chemical shift (ppm)	79.01	241.88	113.68	261.58	32.81	36.91

For the common cationic active sites there is a certain amount of data for oxonium, sulfonium, and other agents with a positive charge in the heteroatom; their ^1H-NMR spectra are given in the review by Penczek et al. [21].

2.2.3.2. Derivatives of Unsaturated Monomers and Their Models

Fraenkel et al. [39] used ^{1}H- and ^{13}C-NMR to study benzyl derivatives, which model anionic active sites of the styrene series. They studied the structural effects which are caused by an α-substituent in a compound of the benzyllithium type and also the presence of electron donors (ED) in the coordination sphere of the lithium. The materials chosen were benzyllithium and the adduct of α-methylstyrene with t-butyllithium (MSt–BuLi) with tertiary amines or diethyl ether as electron donors.

In the absence of electron donors both compounds are contact ion pairs. A difference between them is found when an ED is introduced into the hydrocarbon solutions. The formation of corresponding complexes is not reflected in the type of ion pair in the case of benzyllithium, but it causes a transition of IP_c to IP_s for the MSt–BuLi adduct. Such a conclusion is based on the coincidence of the chemical shifts in the ^{13}C-NMR spectra of various MSt–BuLi–ED systems while analogous phenomena are absent for benzyllithium. The NMR spectra of its complexes exhibit noticeable sensitivity to the nature of the ED. The following magnitudes of the chemical shifts (in ppm), measured at 25°C, in the molecule

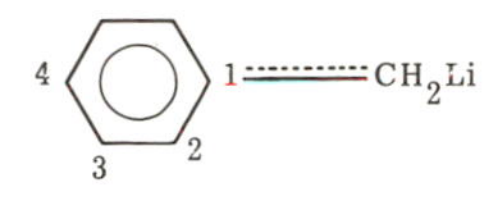

give some idea of this:

	C	C_1	C_2	C_3	C_4
Diethyl ether (DEE)	30.8	152.7	116.3	130.2	109.7
DEE + tetramethyl-ethylenediamine	35.4	157.5	116.6	129.2	107.7
DEE + 1,2-di-piperidylethane	34.7	158.9	118.0	128.7	109.3

The ^{13}C-NMR spectra of α-methylstyryl model anionic active sites have been examined from a somewhat different point of view in the recent work of Firat and Bywater [40]. This deals with the $PhC(CH_3)_2Mt$ compounds (2.IV), in which Mt is Li or K, and their 1:1 adducts with monomer (2.V); they were considered as single- and two-unit models of MSt active sites respectively. Both these and other agents are delocalized structures in THF. The restriction of the rotation about the Ph···C_α bond was possible in the case of adducts (2.V) where the inequality of o,o' and m,m' signals has been established. From the comparatively low sensitivity of the structures to the reduction in temperature to −70° and −80°C, the authors drew the conclusion that the migration of the counterions from one position relative to the plane of the π-system to another is rapid.

The question of the distribution of the electron density within the carbanion occupies a significant place in the same work. In particular, it is shown that the chemical shifts of the C_α-atoms in the compounds studied differ considerably (THF, −20°C):

Compound	Mt	Chemical shift (ppm)
(2.IV)	Li	53.0
	K	65.4
(2.V)	Li	64.7
	K	71.2

(It must be noted that the often used relationships between chemical shifts and the charges of the corresponding atoms reflect semi-quantitatively the distribution of the electron density in the molecule. The absolute values of the charges are neither observable experimentally nor subject to precise determination by theory; see Section 2.3.)

This result, which is the consequence of the different environment of the C_α-atoms in the compounds studied, is of interest as an illustration of the effect which is caused by the transition of the end group from $-C(CH_3)_2Mt$ to $-CH(CH_3)Mt$. Of course, the real α-methylstyryl active sites can only be modelled using the features given for the agents (2.V).

The structural parameters obtained by NMR methods for a series of other model RM_mMt compounds, in which M is a nonpolar or polar monomer and m is either 1 or 2, are given in Chapters 4 and 5. Here we mention only the work of Vancea and Bywater [41, 42], which concentrates on methacrylate models, whose structure coincides with or is close to the compounds studied by IR spectroscopy (see Section 2.2.2). These include the products $C_2H_5(CH_3)(COOCH_3)Mt$ [(2.VI)] and (2.II), where R = CH_3 and Mt is an alkali metal.

The ^{13}C-NMR signals, which are recorded in tetrahydrofuran solutions of these compounds, indicate the strong restriction of rotation about the C_α–carbonyl group bond, i.e., the enol-character of the end group. In the case of the compound with one monomer unit the existence of equilibrium of the type (2.23) remains undecided, since in the temperature interval from −60° to +20°C the distribution between the two isomers

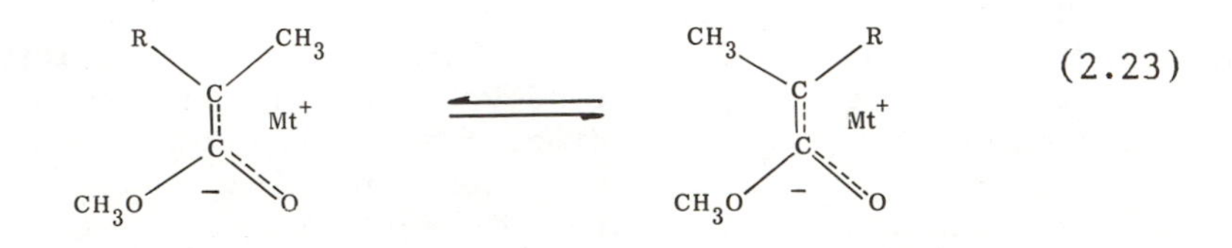

(2.23)

varies very little, remaining close to 1:1 (±10%). The effect of the nature of the counterion is reflected only in the charge on the C atom which increases with the size of the counterion.

The interpretation of the spectra of the two-monomer-unit model (2.II) is complicated by the additional isomeric effects. These are distinguished by a more definite tendency to mutual transitions (2.23). This follows from the sensitivity of the ratio between the isomers (which is clearest in the case of the lithium counterion) to changes in temperature. As the authors emphasized, precise assignment of the signals does not seem possible.

In particular, attempts to isolate intramolecular complex formation (established using IR spectroscopy; see Section 2.2.2) were unsuccessful. It proved impossible to assign the signals observed to the actual geometry of the material studied. The authors confined themselves to the statement that in the case of two-unit models the position of equilibrium (2.23) depends on the nature of the counterions. Thus, for the potassium derivative one of the isomers has a distinct superiority (1:4) whereas for its lithium analog a ratio of 1:2 has been established (data for THF at −20°C [42]). Apparently, this clear, although "noncorrelated," difference is related to the structural effects which are known for the polymerization of methacrylates, which takes place under the action of M_nK and M_nLi chains in a THF medium (see Section 5.1.2).

2.2.3.3. Ziegler–Natta Catalysis

Spectroscopic studies in this region have for a long time been confined to ESR spectra of the homogeneous $Cp_2TiEtCl–EtAlCl_2$ system which provides evidence of the transition Ti(IV) → Ti(III). These results, obtained by Henrici-Olivé and Olive [43], were used to establish the valence state of titanium in the active complex in the polymerization of ethylene. Only recently the same system together with its related systems $Cp_2TiEtCl–Et_3Al$ and $Cp_2Ti(CH_3)Cl–EtAlCl_2$ were studied in detail by Fink and Rottler in the presence of ethylene using ^{13}C-NMR [44, 45]. From the great quantity of material given in the cited works we will dwell on the results relating to the $CpTiEtCl–EtAlCl_2$ system. The interaction of these components at 220°K in the ratio 1:2 produces the most considerable chemical shift of the C atom of the methylene group which is bound to titanium; this shift is from 64.5 in the original compound to 84.6 ppm in the composition of the complex. The signal from the C atom of the methyl group of the $TiCH_2CH_3$ fragment, however, does not change (23.1 ppm in the original and the final products). The chemical shift found may be regarded as evidence for the reduction in the electron density in the titanium atom. As the authors note [44] (referring to [46]), the notion of "electron pressure" from the aluminum component of the complex to the titanium ought to be the reason for the oppositely directed chemical shift of the C atom of the methylene group in the alkyltitanium fragment.

In one of the works cited [44], the changes in the ^{13}C-NMR spectra of the same complexes observed on the introduction of ethylene to the corresponding hydrocarbon solutions have been analyzed in detail. These experiments were aimed at discovering the active site-monomer complexes. On the basis of the signals which are well known for certain ethylene complexes with transition metal compounds (for example, $Cp_2Nb(C_2H_4)C_2H_5$: 29.3 and 27.6 ppm) and C-atoms of methylene groups of polyethylene (30.7 ppm) the authors came to the conclusion that the indications of the formation of intermediate complexes of the monomer with the above agents were absent.

The ^{13}C signal for the free monomer C_2H_4 (about 123 ppm) suffers practically no changes in the ratio of 1:1 to each of the original components of the titanium—aluminum complex. This is shown in separate experiments at temperatures of 200-260°K. In the polymerization conditions where a chemical shift related to the active site-monomer complex might be expected, only a sharp signal 30.7 ppm which is characteristic of polyethylene is found. Such a negative result was also obtained by Fink and Rottler for the other systems which they investigated.

The interpretation of this may be different. It is simplest to propose, as the authors themselves do [44], that the concentration of the complexes lies beyond the sensitivity of the method used. On the other hand, it must be noted that the intermediate formation of π-complexes with the monomer during the polymerization of α-olefins using Ziegler–Natta catalysts does not find support among all researchers. However, these concepts are not in accord with any experimental data; see [47]. Finally, since there is a multiplicity of catalytic complexes of this type, it is permissible to assume that the mechanisms of their action are not all the same. It is to be borne in mind that the stage at which the active site forms complexes with the monomer, which is particularly probable for heterogeneous systems (which have not in general been studied experimentally from this point of view), may be less significant and may be absent completely in homogeneous Ziegler–Natta systems.

The significant shift of the C_α and C_γ signals on the alkyltitanium fragment which is observed in the initial stage of polymerization in $Cp_2TiEtAlCl$—$EtAlCl_2$—ethylene systems is of interest as an additional base for the mechanism by which the monomer is inserted into the active bond of complexes of the Ziegler–Natta type. Any changes in the alkyl group on the aluminum are, in this instance, absent [44]. These effects are the first direct experimental evidence that the reaction proceeds via the carbon—transition metal bonds in binary complexes of this type.

In concluding the present section, we emphasize that the fragmentary nature of the presentation is a simple consequence of the approach which we have chosen. We have intentionally transferred a considerable part of the other results, which relate to this field, to the main chapters.

2.3. QUANTUM-CHEMICAL STUDIES

Among those concerned with research into ionic polymerization there are extreme views concerning the value of theoretical calculations. These range from ignoring such results completely to overestimating their value considerably. This situation is connected with the fact that quantum-chemical calculations on real components of ionic systems can only be carried out to a limited extent. In the overwhelming majority of cases they have to be substituted by simplified models. The very assumptions themselves are open to discussion with respect to the worth of theoretical data. At the same time, correlations between the theoretical features of simple models and experimental values are often found. This inspires confidence in the validity of using the quantum-chemical approach to the solution of certain problems even when the necessity of resorting to primitive analogies of actual materials used is forced upon the researcher. A cautious if not negative attitude to theoretical data is caused by a fundamental dependence of quantitative features on the actual quantum-chemical method.

The bulk of present knowledge on the quantum-chemical features of the compounds and systems under discussion has been obtained using semi-empirical methods. The preference which they enjoy (we emphasize that we are concerned with materials which are directly related to ionic polymerization processes) is only partially the result of the great accessibility of the necessary computer programs. No less pertinent is the increase in calculation time by one to two orders of magnitude with nonempirical methods. It follows then that the application of the latter leads to the necessity of turning to the simplest models, particularly so if the material studied is a multicomponent system.

Thus in attempting to evaluate the value of quantum-chemical systems, it is necessary to take into consideration both the advantages and disadvantages of the methods employed and the degree to which the models chosen have been simplified. We will deal here with only the second of these questions.

2.3.1. The Problem of Modelling in the Study of Ionic Systems Using Quantum-Chemical Methods

The choice of models which are related to ionic polymerization processes is becoming a problem which does not require special attention in every actual situation. In the study of individual compounds of simple structure, for example, the simplest initiators, the necessity to construct models may, in general, not arise. Although it is possible to find in this field important unsolved problems, research directed to the elucidation of the mechanisms of formation of macromolecules is concerned more at present with more complex materials, mainly two- and three-component systems. However, to proceed without deviating from the real components of such systems is rarely possible. More often one is forced to turn to simplified models of each component. This essentially converts modelling into an independent methodological problem. Systematic studies of this type which are known in the field of quantum chemistry are not concerned with the materials which are of interest to us. Of course, the methodological elements are contained in many theoretical studies of ionic agents. However, they usually do not go beyond the bounds of narrow comparisons of certain parameters in determining the correctness of the selection of the given model.

One of the particular tasks of modelling must be the choice of materials of the same type which differ in complexity; their number must be sufficiently large to show the dependence of the parameters evaluated on the degree of simplification of the structures chosen. The combination of the limiting permissible simplicity and the validity of its use in solving a particular problem is optimal from this point of view.

We put forward several examples of the choice of models in the study of the conformation of molecules without, for the moment, dealing with their electronic structure.

The conformational effects in ionic systems are often regarded as responsible for the intramolecular interactions in growing chains of polar molecules. The question is concerned mainly with the transformation of the linear active sites (AS_{ℓ}) to cycles of intramolecular counterion complexes (AS_c) with the polar groups of "their own" macromolecules. The reactivity and stereospecificity of the active sites is usually associated with these phenomena, as is their tendency to unimolecular deactivation. From this arises the interest in comparing the stability of the two forms of active sites, AS_{ℓ} and AS_c. The first question in posing quantum-chemical problems (and this is the same for both carbon—carbon chains $-CH_2-CHX-$ or heteroatom chain macromolecules) concerns the choice of the number of links n in the substance M_nY. Naturally, in this case n must be greater than or equal to 2, although other problems are possible in the study of ionic active sites for which n = 1. Turning to the structures of the type AS_{ℓ} and AS_c, the choice of n depends upon what is required, whether it is only the difference in the relative advantage of the structures being compared or the dependence of the tendency to cyclization on the number of links. Plainly the value n = 2 is sufficient only for the first of these problems.

It is not difficult, however, to foresee that for two-unit models which involve polar substituents (or heteroatoms) and counterions and are capable of donor—acceptor interactions, the structures of type AS_c will always be superior to the linear active sites from the point of view of energy. Hence the solving of such problems becomes reasonable only if it is not an end in itself. Problems connected with the comparison of quantitative effects may be beyond the limits of such calculations. Another condition for which the use of a two-unit model is absolutely correct is the study of the connection between the stability of the structures and a) the nature of the counterion and b) the nature of the polar fragment.

In spite of their comparative simplicity, such special studies have not been carried out. At the same time, individual studies are known which concentrate on models of the type M_nY, in which n varies but M and Y remain the same. These include the following anionic active sites:

M	Y	n	Ref.
$-CH_2-CH(CN)-$	Li	1-3	48
$-CH_2CH_2O-$	Na	1-6	49

In both cases the increase of the relative advantage of cyclic structures with increase in n, which results from the tendency of the counterions to interact with certain donor fragments in the compounds, has been established. This conclusion is in agreement with some experimental characteristics of related real agents [28, 29, 50].

The results of the theoretical calculations cited may be taken as an illustration of the superiority of the state of real M_n^*Y chains in nonpolar media in the form of AS_c. In electron donor media the most probable event is the displacement of the equilibrium $AS_\ell \rightleftarrows AS_c$ towards the linear form $M_n^*Y \cdot mD$ where D is the molecule of the polar solvent, which is competing with the intramolecular donor groups of M_n^*Y. To confirm this using data for models which are sufficiently similar to the nitrile and ethylene oxide derivatives quoted above is not possible since, as yet, such data are not available.

A more detailed discussion requires the modelling of the composition of the ionic active sites in a polar medium. In the case of anionic reagents, molecules of H_2O and NH_3 are often used in addition to model structures of the type $AS \cdot 2D$. When attempting to evaluate to what extent the calculations of such structures may be regarded as characteristic for anionic active sites in polar media, it is necessary to proceed from the basic problem of calculating both the nature of the donors used and the number of such molecules in the material studied. Water and NH_3 are chosen as the simplest models of oxygen- and nitrogen-containing electron donors. The consequences of substitution by ethers and amines are illustrated by a comparison of the data for the $CH_3Li \cdot H_2O$ and $CH_3Li \cdot (CH_3)_2O$ complexes which are obtained using the CNDO/2 method. As shown in [51], the second of these complexes is approximately twice as stable as the first. The magnitudes of the electron density transferred by the corresponding donors to the CH_3Li molecule are in approximately the same ratio. At the same time, the nature of the changes suffered by the acceptors in each of these ststems coincides. Hence, conclusions about the tendency of the original organolithium compound to change during complex formation may be obtained from a consideration of the binary system with the simplest oxygen-containing donor. Such a level of modelling, which is sufficient for the solution of particular problems, is unacceptable for studies which require the calculation of both the degree of perturbation of the reagents and the role of their spatial structure. The latter circumstance acquires especially great significance for systems in which the final result depends upon the coordination saturation of the central atom. In these cases both the number and the structure of the ligands is vital. The scale of the screening effect which they exert may depend upon the second of these factors to an even greater extent.

The question of the correctness of the modelling of active sites of structure $M_nY \cdot 2D$ in polar media is quite often debatable. Such an approach is acceptable only when the consequences of solvation of the active sites by the majority of the molecules of the polar solvent is ignored. The structure of $M_nY \cdot 2D$ essentially models the stoichiometric complexes of the contact ion pairs in nonpolar media. Hence the informativeness of the calculations cannot be discussed out of context.

The desire to select materials which give results relatively easily when calculations are carried out have so far resulted in the exclusion of structures which are closer to real systems, i.e., $M_nY \cdot mD$, where $m > 2$. A way out of these restrictions is possible in some of the simplest cases, for example, the hydration of the lithium ion. The results of such calculations and the general examination of the influence of the solvent evaluated using quantum chemistry is contained in the review by Abronin et al. [52]. Of course the use of modest means in the modelling of large materials is unavoidable, but such limitations must be carefully considered when the results of such calculations are interpreted. Such caution is seldom met with.

2.3.2. Basic Approaches and Examples of Their Application

In spite of the various limitations, quantum-chemical calculations of single-component materials and of complex models of systems often lead to useful information. In some cases the results of such calculations prove to be the only criteria upon which the choice between alternative solutions to questions associated with the characteristics of ionic agents or with the mechanism of their actions in polymerization processes can be based. Nevetheless, the chief criterion regarding the value of the theoretical characteristics is their correlation with experimental data. Among the examples of the discovery of such a correlation may be named the structural effects which are typical of organometallic compounds which contain alkyl, aryl, and polar groups. Spectroscopic study of such agents indicate the donor—acceptor interactions between such substituents and the metal sites [24, 30, 50, 51, 53]. Data from a series of quantum-chemical studies lead to analogous conclusions. In particular this is shown in model compounds of the general type $CH_3CH(Li)R$, in which R is $-COOCH_3$, $-CN$, $-NO_2$, $-CH=CH_2$, or $-C_6H_5$ [53], and for butenyllithium derivatives of various structures [54].

The structural characteristics of the terminal links (TL) of polybutadienyllithium (PBuLi) and their corresponding model compounds are an interesting example of the correlation between experimental and theoretical values for more complex substances. NMR data relating to nonpolar media, i.e., to conditions in which, over the range studied, PBuLi is in the form of a tetramer, shows the advantage of the trans forms of TL [23, 24]. The lack of correspondence between these data and the microstructure of polybutadiene formed in the same medium is regarded as a consequence of the association of PBuLi. According to such an interpretation the equilibrium between the isomeric forms of TL must be displaced during the transition of individual chains in the aggregates to trans structures. This suggestion was later confirmed by the results of calculations on monomeric, dimeric, and tetrameric forms of crotyllithium used as models for the final unit in PBuLi [55, 56]. The differences between the energy values of the cis and trans forms for monomer and tetramer were 12 kcal/mole (in favor of cis) and 8 kcal/mole (in favor of trans) respectively.

From the point of view of methodology, there is yet another point connected with the results of [55] that deserves attention. The author used as a model of a tetramer the aggregate (crotyllithium)$_3$·methyllithium [(2.VII)]. Nevertheless, the given simplification did not prevent the appearance of the expected effect. In particular, in this case the choice of the original geometry of the model aggregate (2.VII) was based on the structure of the allyllithium tetramer, which was calculated by the same method of CNDO with the optimization of the wide selection of variation of structural parameters. In its turn structure (2.VII) was optimized on the basis of "internal" variables relating to the C_4H_7Li components. Unfortunately, the complete optimization necessary for correct conclusions of a fundamental nature were not always obtained. Often data on the completeness of optimization are generally absent and this does not permit the actual value of calculated data to be judged.

Nonempirical calculations during the period up to the mid-seventies were very often carried out without optimizing the geometry. The situation with semiempirical calculations changed much earlier in favor of complete optimization at least of the fundamental parameters. Of course, the full optimization of substances composed of many atoms in which a considerable number of the parameters are varied, and which leads to an enormous expenditure of computer time even on fast computers, is not needed. Essentially, the criterion for the correctness of calculations is the degree of optimization of the structures of such substances. Conclusions which go beyond the limits of the set of varied parameters are inadmissible. For example, conclusions based on calculations on flat structures which model anionic active sites of the styrene series are speculative. According to many experimental and theoretical studies, the metal atom in such agents is situated above the plane of the aromatic nucleus.

The objective value of quantum-chemical data generally becomes significant when results which demonstrate the tendency of certain parameters to change (energetic, geometric, or electronic) in a series of given compounds or multicomponent systems are examined. Although there are studies which concentrate upon an individual substance, we address ourselves at first to studies which contain a series of compounds.

2.3.2.1. Series of Compounds

When studying a series of compounds, the values which change, including, for example, the total energy (E), the charges (q) on the atoms, the distance (r) between the atoms, etc., are compared. The magnitude of ΔE, which is the quantitative expression of the energy effects (corresponding to thermochemical parameters), depends on the method used to calculate it. The theoretical characteristics which are closest to the true ones are obtained by nonempirical calculations. Of the semiempirical methods the best results in this respect are obtained using MINDO/1,2,3, MNDO, and SINDO. The results of calculations on the distribution of the electron density in real structures (which there is no need to exchange for models) show how useful comparative quantum-chemical characteristics can be. Such substances may be, for example, unsaturated monomers. The electron density distribution obtained for many of these enable monomers to be clearly differentiated according to selectivity with respect to the cationic and anionic active sites and allow construction of activity series which agree satisfactorily with the behavior of these monomers in copolymerization processes. The following data for vinyl methyl ether (VME) and vinyl acetate (VA) deserve mention. The electronic characteristics of the vinyl groups of these monomers calculated using the CNDO/2 method practically coincide. This is shown by the data for the π- and σ-components of the total charges of the C_β and C_α atoms (in fractions of e^-) [51].

	C_β^π	C_β^σ	C_α^π	C_α^σ
VME	−0.11	0.01	0.04	0.13
VA	−0.10	0.02	0.04	0.13

At the same time only VME exhibits the ability for cationic polymerization in accordance with the charge on the C_β atom. The passivity of the VA molecule to the cationic active sites must be ascribed to the high negative π-charge on the oxygen atom of the carbonyl group (−0.34). The ether oxygen atoms have quite high (but similar) negative charges in both monomers (about −0.2); however, the π-components of the charges in each of these cases are positive. Apparently, the reason for the passivating effects in cationic systems (analogous to that suggested for VA) may be the π-electron density on the heteroatom of the monomer.

For the present discussion the results quoted are important mainly because they illustrate the possibility of obtaining calculated data which are in good agreement with known experimental facts or with simple logical conclusions. We note that the cause of passivity of VA in cationic systems mentioned above had been made earlier by Plesch on the basis of general considerations [57]. In this sense the coincidence of the electronic structure of the vinyl groups in VME and VA is a more noteworthy consequence of the calculated data than in the high electronegativity of the carbonyl group. The total charges on the C atoms of the simplest alkyl cations calculated by the MINDO/2 method may be treated in the same way. The differences found agree qualitatively with the inductive effects common to the corresponding substituents attached to the cationic center [58].

	CH_3^+	$CH_3CH_2^+$	$(CH_3)_2CH^+$	$(CH_3)_3C^+$
qC^+, e	0.58	0.45	0.43	0.39

The unexpected results which sometimes come from theoretical calculations of a series of compounds are of much more interest if the known phenomena are to be approached in a new way. Calculations on the stability of butyllithium isomers (CNDO/2 method) give the following series:

$$\underset{2.VIII}{\text{n-BuLi}} > \underset{2.IX}{\text{sec-BuLi}} > \underset{2.X}{\text{t-BuLi}}$$

It is difficult to call such a sequence self-evident. The sharp inclination of electron deficiency of the lithium atom to be compensated enables structure (2.X) to be identified as the most stable. The opposite effect which is found in calculations is due to the optimum geometry of the compound studied. The symmetrical distribution of the lithium atom relative to the three CH_3 groups in t-butyllithium not only hinder the effective interaction of this atom with each of these but also weakens the C_α–Li bond. On the other hand, the saturation of the electron deficiency of lithium is much more complete in the n-derivatives. This is shown by the comparison of the values of the bond order (P) and the actual valence of the lithium atom V_{Li}.

Structure	C_α–Li		V_{Li}
	r, Å	P	
(2.VIII)	1.99	0.80	1.67
(2.X)	2.13	0.31	1.66

These data can be used for a more precise definition of the difference in the initiating activity of the (2.VIII) to (2.X) compounds. In the polymerization of nonpolar monomers in hydrocarbon media, n-butyllithium is considerably inferior in this respect to its isomers. Such a difference is usually explained by the smaller stability of the aggregates of the (2.IX) and (2.X) compounds (i.e., the greater equilibrium concentration of the corresponding monomeric forms which participate in initiation) as compared with the aggregates of n-butyllithium. Obviously it is necessary to also consider the difference in the relative activities of the monomeric forms of the C_4H_9Li isomers themselves. The rank order of stability of these compounds which is shown above will be reversed when we look at their rank order for reactivity.* Such a conclusion is the more appropriate since the substances used for the theoretical calculations were real initiators and not simplified models. Of course for some compounds an inverse relation between stability and reactivity does not generally hold for all possible transformations. This question is dealt with further in Chapter 4.

2.3.2.2. Individual Compounds

In theoretical studies of individual compounds, characteristics which are inaccessible experimentally are of the greatest value. For ionic agents which are fulfilling the function of active sites in initiation and growth, such characteristics are the optimum conformation of the compounds studied and the distribution of the electron density over the whole molecule. The possibility of obtaining the geometry of the ionic agents using X-ray structural analysis, which is limited for these comparatively rare materials (for example, certain RMt compounds and their complexes), does not give information which is especially useful in establishing the conformation of active sites in solution, i.e., in the state which is particularly important for the study of the mechanism of polymeriza-

*The idea that the activity of organolithium initiators depends not only upon the stability of the associates but also upon the properties of the monomeric forms was given earlier in [59].

tion in homogeneous ionic systems. As for using NMR data for the evaluation of electron density, they do not enable the charge characteristics in heteroatoms to be obtained. Furthermore, the NMR spectra of ionic agents in the literature relate almost exclusively to solutions of high concentration. These conditions favor aggregation which causes change from the common electronic structure to monomeric forms. The pulsed mode used in combination with amplifying techniques which enable NMR spectra of dilute solutions to be obtained have so far not been widely applied to the study of the materials discussed here.

Knowledge of the total charge does not always permit as full an interpretation of the phenomena as that which becomes possible when data on the fractions q^{π} and q^{σ} are available. There are also other examples, apart from the electronic characteristics of vinyl monomers, which show the difference in the value of the corresponding total and differentiated values. The linear oxonium ions $R\overbrace{-OCHX}^{+}$ [(2.XI)] may be taken as models of active sites in the cationic polymerization of vinyl ethers, formaldehyde, and cyclic acetals. It follows from the calculations of methoxymethyl [R = CH_3, X = H; (2.XII)] and methoxypropyl [R = CH_3, X = CH_3CH; (2.XIII)] cations that the positive charge is concentrated practically completely on the C atoms of the —O—CH systems. The oxonium nature of the cations is revealed only when the corresponding π-constituents are used. In both cases they are distributed comparatively evenly between the C and O atoms with a slight preference for the oxygen (CNDO/2) [60]:

	q_C	q_O
(2.XII)	0.43	0.58
(2.XIII)	0.44	0.48

The question of the correlation of these values with experimental characteristics does not arise (in general such data are not available), but the behavior of the growing chains of the type (2.XI) is easier to correlate with an oxonium than with a purely carbenium ion (see Chapter 4).

The choice of the optimum geometry for a given substance using minimization of the calculated energy for certain structural variants is essentially the comparison of the advantages of different conformations. Here the basic interest, apart from the optimum geometry itself, is the magnitude of ΔE which is a measure of the energy advantage of the structure found, as compared with some other. We have already met with such situations in discussions on the compounds (2.VIII) to (2.X); each of these may, from some point of view, be considered separately from the series as a whole. The same principle is found also in the problems associated with other structural differences which are possible for a given compound. These include its existence in the forms of cis and trans isomers, the π-form and the σ-form, etc. In any other such problem the corresponding value of ΔE has special significance; particularly in semi-empirical calculations the difference in the magnitude of E usually deserves attention in the conditions $\Delta E \geq 10$ kcal/mole. In particular, the "π-σ problem" of the structure of alkene organometallic compounds, which has been the subject of long-term experimental study over a number of years (see Section 2.2), is also studied using quantum-chemical methods. The theoretical characteristics of these derivatives are among a number of examples which are regarded rather negatively by experimentalists. It must be borne in mind that the view expressed by Schlosser et al. [34], in a paper on allyl derivatives of various metals (AllMt), is that calculated features obtained for isolated molecules do not correspond with the true structures of real compounds. The authors believe that the π-form of AllMt compounds has a higher coordination saturation of the Mt atom as compared with the σ-form (this seems natural) and assume that the molecules of the polar solvent must give rise to π-σ transitions. Calculations carried out for $C_mH_{2n-1}Mt$ compounds lead, depending upon the method used, to different results. For example, according to the data of [61] the introduction of an electron donor into the coordination sphere of a metal increases the π-nature of the unsaturated system (crotyllithium—H_2, CNDO method). On the other hand, in [62] the opposite result was obtained for the allyllithium—H_2O system (a nonempirical method with quite a limited based was used). The advantage of the second method in this case is less obvious, since the choice of basis in nonempirical calculations is reflected in the charge characteristics to a considerable degree, as was shown in particular by Clark et al. [63] for example, for allyllithium (data for a symmetrical structure):

Basis	q, $\bar{e}$			Population density		
	C_α	C_β	Li	C_α—C_β	C_α—Li	C_β—Li
STO-3G	−0.197	0.013	0.110	0.983	0.310	−0.66
STO-4-31G	−0.548	0.146	0.439	0.801	0.300	0.030

Turning to isolated compounds and their models, we note that even though we may have emphasized the great value of the comparative parameters in some of the above examples this is by no means necessary for each of the substances studied. In each individual case different approaches are possible for the evaluation of the calculated data. For example, we have already mentioned a number of models of lithium active sites in connection with the generality of the conformation, which results from the electron deficiency of the lithium atom (page 51). At the same time, each of the models relates to a definite process of anionic polymerization and hence is interesting in itself. In this sense, the information on the geometry and the electron structure of a given material is useful, irrespective of its less favorable conformers or analogs.

2.3.2.3. Multicomponent Systems

There is a considerable volume of work which has been carried out in the theoretical studies of binary and more complex systems, which is concerned in the overwhelming majority of cases with reversible donor—acceptor interactions, i.e., these or other forms of the complexes. These states are often regarded as the precursors of the irreversible elementary acts of initiation, growth, etc., and are used for the interpretation of the mechanism of the formation of macromolecules. The direct path of the explanation for the formation of details of the processes of ionic polymerization on the basis of theoretical data, namely calculations of the potential surface of the reactions studied, is still practically inaccessible. Rigorous solutions of such problems are obtained only for certain systems, which lie outside the area under discussion. For polymerization reactions only approximate evaluations are known which relate to a number of particular cases. General conclusions on these are as yet premature.

Of the reversible multicomponent systems calculated we single out substances for study which are free from polymerizing monomer and systems which include monomer as one of its components. The reason for such a subdivision is the fact that substances of the first type are models of complexes or aggregates whose physicochemical parameters lend themselves to experimental study. Consequently an attempt to correlate the theoretical and experimental results is possible. Such a possibility is practically completely absent for complex structures which contain monomer and ionic sites (initiator or a model of the growing chain). In research systems of this type experimental methods may, in the best case, be used for kinetic measurements based on the independent physicochemical characteristics of the original reagent and the final products. Preliminary intermediate stages, i.e., the formation of AS·M complexes, have, except in the rarest cases, not so far been found. In fact, earlier concepts about these intermediates were speculative but the modern concepts are based on the results of theoretical calculations.

Associates (dimers, trimers, and tetramers) of organometallic compounds and alkoxides of metals, their various donor—acceptor complexes, and certain carbene agents are among the calculated "non-monomer" systems. The majority of these results published by the mid-seventies have been cited in [51]. Of the more recent data, we will concentrate on systems which include alkene derivatives or metals, bearing in mind, in particular, the "π-σ" problem which we have already mentioned.

Allyllithium and its complexes with different electron donors were used as alkenyl components in these systems. In calculations carried out with a sufficient degree of optimization, the superiority of the π-structure [61] is revealed. The results of calculations

TABLE 2.6. Charges on the Atoms (in fractions of electron) and Order of Bonds in π-Type Alkenyllithium Compounds

Compound	Method	q_{Li}	q_C	q_C	q_C	P_{Li-C}[a]	P_{Li-C}[a]	P_{Li-C}[a]	Ref.
AllLi	ab initio[b]	0.159	−0.0197	−0.063	−0.197	0.189	0.0128	0.189	62
AllLi	ab initio[c]	0.439	−0.548	0.146	−0.598	0.300	0.030	0.300	63
AllLi	CNDO/2	0.12	−0.20	0.16	−0.20	0.37[d]	0.34[d]	0.37[d]	64
AllLi	CNDO	0.465	−0.481	0.282	−0.481	0.323	0.156	0.323	61
$(AllLi)_2$	CNDO	0.424	−0.466	0.258	−0.466	0.157[e]	0.111[e]	0.157[e]	61
$(AllLi)_4$	CNDO	0.443	−0.502	0.218	−0.502	0.107[e]	0.063[e]	0.107[e]	61
$AllLi \cdot 2H_2O$	ab initio[b]	0.323[f]	−0.223	−0.043	−0.223	0.171	0.018	0.171	62
$AllLi \cdot 2H_2O$	CNDO	0.306	−0.474	0.282	−0.474	0.303	0.129	0.303	61
$AllLi \cdot (CH_3)_2O$	MINDO	−0.179	−0.184	0.197	−0.184	-	-	-	65
$AllLi \cdot 2NH_3$	CNDO/2	−0.23	−0.21	0.17	−0.21	0.34[d]	0.29[d]	0.34[d]	64
CrtLi[g]	CNDO	0.452	−0.505	0.236	−0.355	0.330	0.151	0.181	61
$(CrLi)_2$[g]	CNDO	0.399	−0.484	0.219	−0.353	0.148	0.112	0.186	61
$(CrtLi)_3 \cdot CH_3Li$[h]	CNDO	0.410	−0.478	0.197	−0.383	0.079	0.064	0.071	61
$(CrtLi)_3 \cdot CH_3Li$[i]	CNDO	0.407	−0.543	0.180	−0.208	0.138	0.055	0.067	61
$CrtLi \cdot 2(CH_2)_2O$	CNDO	0.274	−0.498	0.242	−0.446	0.264	0.114	0.171	61

[a]In the case of nonempirical calculations the figures given are those for total population density between the atoms.
[b]Based on STO-3G.
[c]Based on 4-31G.
[d]Population density.
[e]For the interaction with the nearest lithium atom.
[f]Charge on the $Li(H_2O)_2$ fragment.
[g]Data for cis-crotyllithium.
[h]Data for the π-form of trans-crotyllithium.
[i]Data for the σ-form of trans-crotyllithium.

on the associated forms AllLi, carried out using semiempirical methods, lead to an analogous conclusion [61, 64].

Of the other consequences of the formation of complexes we should mention changes in the electronic characteristics of the original alkenyllithium, which include the order of the C–Li bonds. A number of examples are given in Table 2.6. Structures given in this table are usually taken as models of active sites for the polymerization of butadiene using lithium initiators in polar and nonpolar media. The necessity for well-defined criteria in the construction of such models has been noted on page 51. It is appropriate to pose the question as to whether any of the calculated parameters agree with those which have been established experimentally for the alkenyllithiums $RCH_2CH{=}CHCH_2Li$, which are used in such studies as models of polybutadienyllithium. In the present case, the value for $(q_{C_\alpha} - q_{C_\gamma})$ may be a parameter used for comparison. According to the NMR spectra, this difference is reduced on going from a nonpolar to a polar medium (see Section 2.1). The results calculated for crotyllithium and its complex with ethylene oxide lead to the same conclusion, but in the qualitative respect this effect is not great here. If an attempt is made to evaluate the magnitude of $(q_{C_\alpha} - q_{C_\gamma})$ from experimental data for t-C_4H_9–C_4H_6–Li compounds, the resulting values are 0.57 and 0.29 in benzene and THF, respectively.

The cause of the considerable difference in the scale of these values which are obtained both experimentally and theoretically is difficult to determine simply. It may be regarded either as a feature of the quantum-chemical method used, or as an imperfection in the modelling of a polar medium by the two ligands of the electron donor, it may be due to the poor basicity of the ligands chosen. The role of the latter circumstance is shown by the noticeable difference in the fraction of transferred charge on one and the same acceptor when the nature of the donor is changed [51].

On the other hand, the correlation used for the evaluation of the charges from NMR signals may be called into question, the more so since this evaluation is based on data relating to aromatic compounds. The correctness of the use of such an approach to alkenyl compounds while preserving the same numerical ratio of chemical shift:electron density is by no means obvious. Conclusions as to the reasons for the variation of microstructure of the polymers with changes of the charge characteristics of diene "living chains" when the medium is changed may also be made on the basis of spatial concepts; see Chapter 5.

Of the binary and tertiary systems which model "nonmonomer" cationic agents, only the complexes BF_3 with HF, CH_3F, and H_2O have been evaluated. The ionic forms $H^+BF_4^-$ (2.XIV), $CH_3^+BF_4$ (2.XV), and

$H_3O^+BF_4^-$ (2.XVI) [51] were used as initial structures. In the oxonium derivatives (which incidentally is the first example of quantum-chemical calculations being carried out on a series of such compounds) a higher total positive charge was found on the cation in comparison with carbene compounds and they proved to be concentrated completely on the H atoms. The charge characteristics of all these structures are given below (CNDO/2 method):

$H^{+0.52}$ $C^{+0.22}$ (2.XIV) $H^{+0.09}$ $O^{-0.26}$ (2.XV) $H^{+0.41}$ (2.XVI)

The range of calculations on oxonium cations was considerably broadened recently by the introduction of systems based on cyclic monomers. Some of the data relating to them are given below.

A qualitatively close result has been obtained for trimethyloxonium and calculated by the same method which illustrates the distribution of the electron density within the following fragment (see [21]):

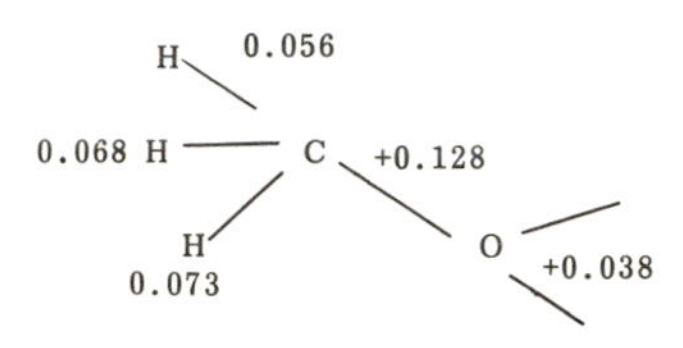

We now turn to multicomponent systems which include polymerizing monomers. Any of these may be regarded as a model of an intermediate stage, the formation of which precedes the insertion of the initiator into the growing chain. In the first theoretical studies of such materials such a differentiation was not often shown. Moreover, the molecule of the simplest ionic initiator, for example, methyllithium, was sometimes taken as a model of an active growth site. In its turn, the ethylene molecule often fulfilled the function of a model of an unsaturated monomer which was capable of polymerizations of any type, and in certain circumstances the ethylene itself was modelled by only its carbon backbone.

For the present discussion it is expedient to deal with systems which are closer to real materials. We will deal first with studies in which the calculations do not go beyond the bounds of the parameters of electronic and structural perturbations which are the result of reversible donor—acceptor interactions between the reagents. Even within these limits extremely varied information may be obtained. Its most interesting elements from the point of view of the mechanism of polymerization are the changes in the electronic structure of the monomer and its orientation with respect to the active bond of the ionic agent. It is precisely these elements which are usually taken as the features of the preconditioning of the monomer before its subsequent irreversible interaction with the active site. From the usual point of view, these can be considered separately.

Such an interpretation of the physical meaning of the parameters of a complex material is applied mainly to systems which are of the ion pair—unsaturated monomer type to which the majority of the multicomponent models of the common type considered belong. A somewhat different sense must be given to the characteristics of systems where the active agent is a free ion and the polymerizing monomer is a heterocyclic compound, particularly if the free ion is cationic. The peculiarity of such systems lies in the fact that the donor—acceptor interactions between components largely approximate to an irreversible preconditioning of the formation of an active site rather than to a reversible complex formation. In fact the material considered here is a product of the reaction

$$Ct^{+} + X\big) \longrightarrow Ct{-}X^{+}\big) \qquad (2.24)$$

where Ct^{+} is the attacking cation and X is the heteroatom of the cyclic monomer.

Calculations have been carried out in this field on a series of unsubstituted cyclic oxides using a methyl cation for Ct. The results obtained by the CNDO/2 method with full optimization of the systems [67, 68] are interesting in many respects. Here we will only deal with some of the electronic and geometric parameters which have been established in the works cited. The first of these deserves attention because the final products of the events given by Eq. (2.24), which here are oxonium ions, carry a positive charge which is concentrated almost completely on the three C_{α} atoms and on the H atoms attached to them; on the oxygen atom there is a small negative charge and only in the case of ethylene oxide is there an extremely small positive charge. We note that we have already met a similar phenomenon in the first example of $(H_3O)^{+}$ and $(CH_3)_3O^{+}$. This effect at first sight stands in considerable contradiction with the generally accepted notions on the natures and characteristics of oxonium compounds, according to which the cationic center in them is an oxygen atom. From our point of view too much should not be read into this contradiction. It is sufficient to regard the extremely reduced electron density as the important feature of oxygen atoms in oxonium derivatives. This is much lower than that of the original oxygen compounds. In particular, according to the data from the CNDO/2 method, in the unsubstituted cyclic oxides it is about −0.2 to −0.3 e^{-}. In their oxonium derivatives this value lies between −0.7 and +0.01, as it is shown in the following examples [67]:

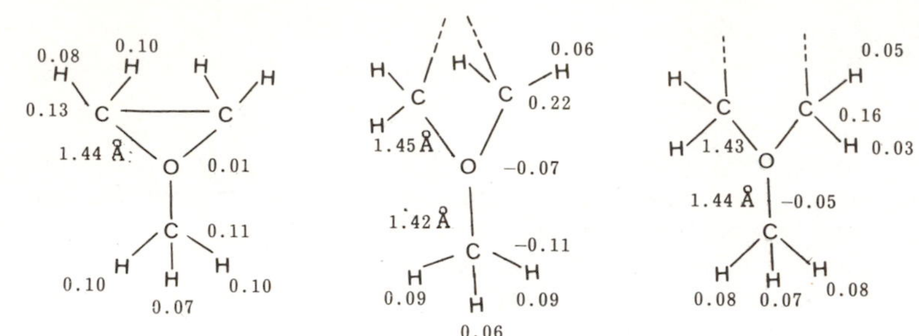
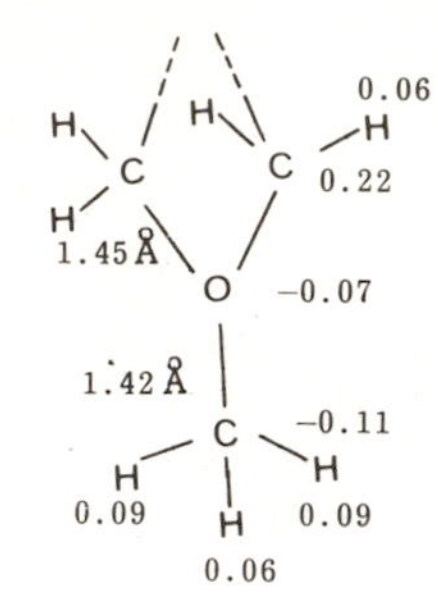
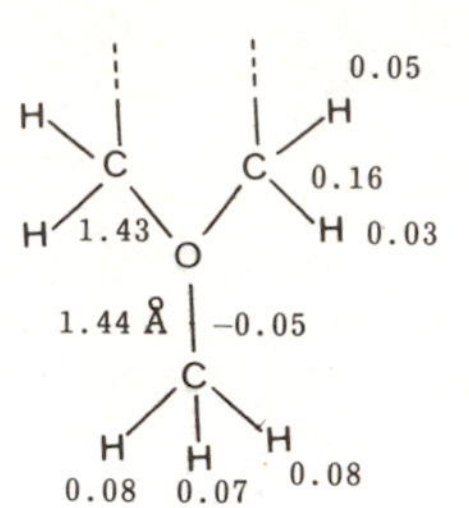

Active sites

ethylene oxide   oxacyclobutane   tetrahydrofuran

The affinity of the oxygen atoms of these structures for the electron is somewhat higher than in the C_α atoms. This follows from the contributions from the atomic orbitals to the energy of the lower vacant molecular orbitals. Hence the usual representation of oxonium centers in the form of symbols $\overset{+}{\mathrm{O}}$ is very disputable. A more accurate representation would be as "zero charged" with respect to the oxygen atom, i.e., $\overset{0}{\mathrm{O}}$; such symbols would reflect the electron deficiency of the oxonium ion more accurately.

Turning to the geometric parameters, we note only the circumstance that the approach of the C atom of the attacking cation to the oxygen atom of the monomer leads to all the C–O bonds in the end structure being of approximately equal length.

On the basis of the genesis of the oxonium ions [Eq. (2.24)], we regard the structures considered as multicomponent systems. This being so, there is no necessity for interaction between the monomers and the ionic agents which leads to the formation of products of the type AS·M. The data given in [67] relating to complexes of unsubstituted cyclic oxides with HF and CH_3Li compounds, which are models of cationic and anionic active sites respectively, correspond to this situation. Of the systems of the first type we will limit ourselves to the mention of the electronic characteristics of model oxonium ions of oxacyclobutane and THF. Again these indicate that the positive charge is here concentrated in the environment of the oxygen atom, just as it is in the situations which we have already examined;

The mutual orientation of the components in the optimized binary systems, which include unsaturated monomer, is of interest in connection with the mechanism of stereoregulation in the processes of formation of macromolecules. Data obtained by the CNDO/2 method for the CrtLi–butadiene and $(CrtLi)_2$–butadiene systems [69] permit the informativeness of the theoretical study of this question to be judged. In the first case the greatest advantage of the attack of the C_α atom of the model active site with the C atom of the methyl group of the monomer (C_1), which manifests itself in the approach of the above two atoms, has been established. As the authors emphasize, the factor which causes such an orientation is the degree of unsaturation of the C_α atom in crotyllithium. Its actual valence (see page 53) is 3.66. In the butadiene complex this increases to 3.90.

In the case of the butadiene complex with a model active site, which is in a dimer form, such a spatial selectivity is absent. Here, the order of the C_α–C_1 and C_γ–C_1 bonds not only differs little from one another, but also is extremely small in absolute terms (about 0.02).

The sum total of these data agree well with the concepts on the mechanism of polymerization of butadiene using organolithium compounds in nonpolar media, which are based on experimental results. In particular, it applies to the conclusion on the great stereospecificity of monomeric active sites in the sense of their tendency to form 1,4 links during the polymerization of butadiene. For more detail see Chapters 4 and 5.

The as yet small number of calculations for the stage at which monomer is inserted into the active bond of the ionic agent are of special interest. One of the more detailed studies of this type models the attachment of ethylene to a Ziegler–Natta catalyst site as in Eq. (2.25):

$$(CH_3)_2Al\langle^{Cl}_{Cl}\rangle Ti(CH_3)Cl_2 \xrightarrow{C_2H_4} (CH_3)_2Al\langle^{Cl}_{Cl}\rangle Ti(Pr)Cl_2 \qquad (2.25)$$

Novaro et al. [70], using a nonempirical method, note the comparative paucity of the basic choice and the incompleteness of the optimization of the systems studied. However, the sum total of the results obtained, as yet the only ones in this field (with the exception of some semiempirical calculations; see later), deserves some attention. This total amounts to the following.

Reaction (2.25) proceeds in 12 "steps," beginning with nonacting reagents, to the formation of a complex between them, and the subsequent act of insertion. Changes in the distance (Å) between the titanium and the C atoms, in the alkyl group of the original complex (Ti–CH_3), and in the monomer (Ti–C_1 and Ti–C_2), and also between the C atoms of the monomer itself, give some idea of the geometry of some of the intermediate and final states:

Stage	Ti–CH_3	Ti–C_1	Ti–C_2	C_1–C_2
4	2.15	2.59	2.59	1.383
6	2.15	2.40	2.43	1.406
7	2.26	2.24	2.39	1.328
9	2.48	2.14	2.60	1.473
12	2.81	2.15	3.03	1.54

The corresponding changes in the energy of the system are shown in Fig. 2.8. As can be seen, complex formation proceeds without activation energy, but the transfer of monomer from the structure of the complex to the growing chain is characterized by a magnitude of ΔE of about 15 kcal/mole. In [70] data are given of a series of experimental studies according to which the activation energy involved in the attachment of ethylene to the titanium–aluminum complexes is 8-16 kcal/mole. Nevertheless, such a correspondence between the theoretical and experimental values is not in itself an argument in favor of the unreserved conclusions which are drawn from the results of such calculations. Primarily, the experimental values obtained for the energy of activation are total values and include the features of both events of interaction of the monomer with the catalytic complex, and their breakdown is not possible. On the basis of the indirect results (polymerization kinetics and the composition of the copolymers) it is often considered that complex formation is a definite two-stage growth event. A similar conclusion on the role of this stage has been drawn in another theoretical study carried out by Armstrong et al. [71] in which the CNDO approximation was used. As this has been elucidated in some detail in works of a general nature we will limit ourselves to mentioning that the system $TiCl_4$–CH_3AlCl_3–C_2H_4 was chosen for study and that the conclusion of the necessity of energy expenditure solely on the formation of the initial complex with the monomer was based on the idea of the reorientation of the original trigonal bipyramid in the octahedral complex.

In attempting to evaluate the physical sense of these and other conclusions, we emphasize that for heterogeneous systems the idea of the necessity of the expenditure of energy on a detailed reorientation of the complex cannot be dismissed. The bulk of the experimental values which have already been given are a good example of this. However, it does not follow from this that an analogous result is natural in calculations on isolated systems of this type which model to a great extent the interaction of reagents in homogeneous conditions. Hence in this case it is difficult to agree with the interpretation based on semiempirical calculations, the more so since it leads to the conclusion that the monomer can be inserted without the expense of additional energy. It is interesting that the authors of [70], who used a nonempirical method, carried out a semiempirical calculation for comparison with their results using the same CNDO method which was used by Armstrong et al. [71].

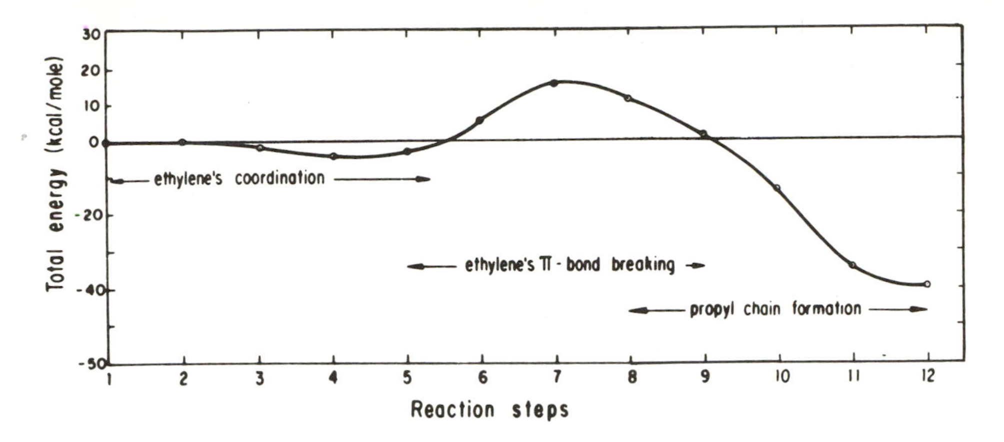

Fig. 2.8. Changes in the total energy when ethylene interacts with the titanium–aluminum catalytic complex [70].

The result proved to be fundamentally different from that shown in Fig. 2.8 and close to that obtained by [71].

We note finally that Novaro et al. [70] analyzed in great detail the possible role of the consequences of the simplifications which they made. In particular, the choice of one of the probable, but not the only, reaction paths, and of the somewhat limited basis, etc., was analyzed. The result of this analysis agrees qualitatively with the assumed mechanism of reaction (2.25).

The commonly used semiempirical methods (CNDO, MINDO) are of little use for the analysis of the reaction path and so researchers are forced to calculate the electronic structure of the transition states, the geometrical structure of which is based on some considerations which lack the necessary rigor.

In particular, an attempt to model the transition state in AS–M systems was undertaken by Miertush et al. [72, 73]: for example, the interaction of styrene with benzyl derivatives of alkali metals. The authors, using a modified Pariser–Parr–Pople method in π-approximation, neglected other works on the geometry of such ion pairs and used a flat structure for them; this method is unsuitable for the calculations of nonplanar structures.

A number of calculations on the dependence of the energy of nonplanar $PhCH_2Mt$–styrene systems (where Mt is Li, Na, or Cs) on the degree of approach of the reagents were carried out, essentially, without changes in the mutual orientation while varying the single parameter r:

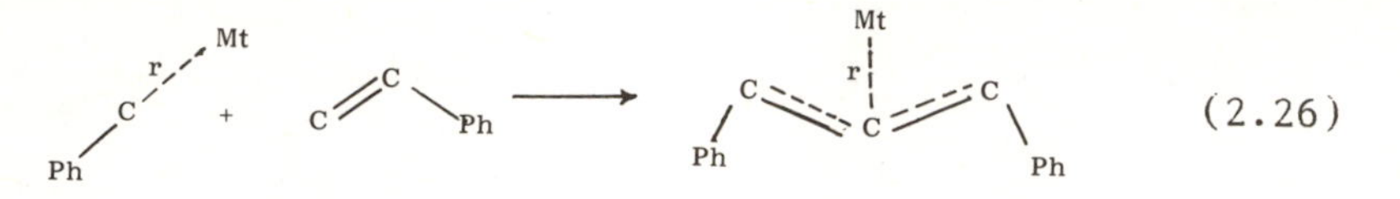

(2.26)

The final structure was taken as the activated complex, and the energy of its formation ($E_{(2.XVII)} - E_{AS} - E_M$) for the energy of activation of the stage of formation of the adduct

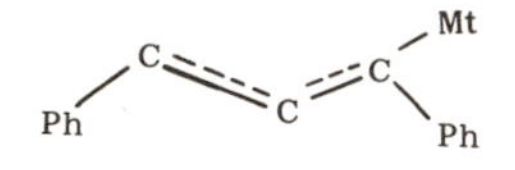

which as yet has not been examined further. The basic problem amounts to the evaluation of the difference in energy of event (2.26) as a function of the nature of the counterion and the polarity of the medium. The polarity is calculated by including the solvation energy calculated according to the formula

$$E_{solv} = \frac{1}{2} \sum_{\nu} \sum_{\mu} Q_\mu Q_\nu \nu_{\mu\nu} \, (1 - 1/\varepsilon)$$

in which Q_μ and Q_ν are charges on the μ and ν atoms of the dissolved molecules (calculated quantum-chemically), $\nu_{\mu\nu}$ is the integral of the electron repulsion, and ε is the permittivity of the medium. On the E vs r curves for $\varepsilon = 1$ distinct minima are obtained which correspond to the values of r from 3.5 (Li) to 4 (Cs); for $\varepsilon = 4$ the minima are much less well defined and lie in the region 5.5-6 Å. In the original active site the corresponding distances are 2.2-2.5 and 4-5 Å. These distances may be taken as approximating the experimental values for IP_c (for $\varepsilon = 1$) and for IP_s (at $\varepsilon = 4$).

Digressing from the absolute values of the energy characteristics (which differ considerably from the true ones in the majority of calculations) it is possible to ascertain the correctness of the differences which are caused by the changes in the nature of the counterion in the reactive medium. The data given in the graphs of [72] lead to the following values of the difference in energy of some of the systems studied (in electron volts):

($E_{RLi \cdot M}$, nonpolar medium) − ($E_{RLi \cdot M}$, polar medium) = −3.70
($E_{RCs \cdot M}$, nonpolar medium) − ($E_{RCs \cdot M}$, polar medium) = −2.70
($E_{RLi \cdot M}$, nonpolar medium) − ($E_{RCs \cdot M}$, nonpolar medium) = 0.35
($E_{RLi \cdot M}$, polar medium) − ($E_{RCs \cdot M}$, polar medium) = 2.00

In spite of the extreme simplicity of the model chosen, these values of ΔE correlate qualitatively with some conclusions of kinetic studies of the corresponding real systems. In particular, it deals with the difference in the energy of activation of growth in contact and separated ion pairs. The structures considered are at best only very distant models of the transition state and it is not possible to ascribe a physical meaning to the energy differences found (as was done by Miertus et al. [73]); yet it is impossible to deny these values a certain significance.

2.3.3. Summary

The material of the present section enables certain concepts to be formulated about the objective value of the results of theoretical calculations for the study of the nature of ionic agents and the mechanism of the reactions which take place under their action. We emphasize nevertheless the special role of those theoretical data about which there can be no doubts and which are inaccessible by experimental methods. Essentially, calculations which agree well with experiment but do not lead to conclusions which are new in principle have a mainly methodological value and are a confirmation of the validity of the chosen method of approach to the solution of a problem. Of course, the necessity of such studies is beyond argument, just as is the presence of control experiments in experimental work provided that in the setting of the problem itself the solution is not hidden, i.e., does not yield a "tuning" for obtaining a certain result by artificially limiting the number of parameters varied. The effects found from calculations or assumed earlier only on the basis of indirect data, but not observed directly, must be evaluated differently. Without turning to concrete examples (see the previous subsections) we note again that the results obtained from the study of nonmonomer systems may be treated with the greatest confidence. It is necessary to take into account the fact that the "optimal" mutual orientation of the monomer and the active site in the intermediate complex is not necessarily the most favorable for the act of inserting the monomer into the structure of the product of the irreversible reaction. It is very rare that a judgement on the advantage of this event can be made with any certainty from the geometry of the optimal structure of such complexes. Among calculations which have been carried out on systems of the general active site–monomer type, it is possible to find a great variety of mutual orientations. These vary from the quite close approach of the atoms of each reagent between which the covalent bond must be formed in the product resulting from the insertion (the butadiene–crotyllithium system [69]) to the comparatively large distances between the analogous atoms of the components of the complex (the acrylonitrile–methyllithium system [51]). In cases of the latter type, it is impossible to avoid noting the necessity of the displacement of the monomer from the optimum position for complex formation to one which is less favorable from the point of view of the stability of the complex, but actually more favorable for introduction into the structure of the active site. It is probable that the energy expenditure on such a displacement approximates in magnitude the energy of activation. It seems to us that a special study of this problem, which so far has not been carried out, is necessary.

REFERENCES

1. M. Szwarc, Carbanions, Living Polymers and Electron Transfer Processes, Interscience, New York (1968).
2. M. Szwarc (editor), Ions and Ion Pairs in Organic Reactions, Vol. 2, Interscience, New York (1974).
3. Ch. B. Tsvetanov, V. N. Zgonnik, I. Panayotov, and B. L. Erusalimskii, "Langkettige alkalimetallorganische verbindungen vom Type: $RCH_2CH(CN)M$: Herstellung, spektrosckopische und elektrochemische Charackterisierung," Ann. Chem., 763, 545-548 (1973).
4. L. V. Vinogradova, V. N. Zgonnik, N. I. Nikolaev, and Ch. B. Tsvetanov, "Electric conductivity of polybutadienyl- and poly-isoprenyllithium in tetrahydrofuran and dimethoxyethane," Eur. Polym. J., 15, 545-550 (1979).
5. I. V. Berlinova, I. M. Panayotov, and Ch. B. Tsvetanov, "Conductivity studies on living polymers with an α-oxide terminal unit in tetrahydrofuran," Eur. Polym. J., 12, 485-488 (1976).
6. L. V. Vinogradova, N. I. Nikolaev, V. N. Zgonnik, B. L. Erusalimskii, G. V. Sinitsina, Ch. B. Tsvetanov, and I. M. Panayotov, "Changes in the electrochemical characteristics and the UV spectra of polydienyllithium chains on storage in polar media," Eur. Polym. J., 17, 517-520 (1981).
7. A. Persoons, "Field dispersion effects and chemical relaxation in electrolyte solutions of low polarity," J. Phys. Chem., 78, 1210-1217 (1978).
8. A. Persoons and M. Van Beylen, "The dynamics of electric field effects in ion-pairing processes," Pure Appl. Chem., 51, 887-900 (1979).
9. T. E. Hogen-Esch and J. Smid, "Studies of contact and solvent-separated ion pairs of carbanions," J. Am. Chem. Soc., 88, 307-318 (1966).
10. R. M. Fuoss, "Non-coulomb variation of ion pairing in polar solvents," J. Am. Chem. Soc., 100, 5576-5577 (1978).
11. A Gandini and H. Cheradamé, "Cationic polymerization. Initiation with alkenyl monomers," Adv. Polym. Sci., 34/35, 1-289 (1979).
12. D. W. Grattan and P. H. Plesch, "Ionization of aluminum halides in alkyl halides," J. Chem. Soc. Dalton Trans., 1734-1744 (1977).
13. D. W. Grattan and P. H. Plesch, "The initiation of polymerization by aluminum halides," Makromol. Chem., 181, 751-775 (1980).
14. M. Chmeliř, M. Marek, and O. Wichterle, "Polymerization of isobutylene catalyzed by aluminum tribromide," J. Polym. Sci., C16, 833-839 (1967).
15. M. Chmeliř and M. Marek, "Influence of some Friedel—Crafts halides on the polymerization of isobutylene catalyzed by aluminum bromide," J. Polym. Sci., C23, 223-229 (1968).

16. P. Lopour and M. Marek, "Polymerisation des Isobutylenes durch zweikomponenten Katalysatorsysteme die Aluminiumhalogenid als eine der Kompenente enthalten," Makromol. Chem., 134, 23-31 (1970).
17. A. Ledwith and D. C. Sherrington, "Stable organic salts: Ion-pair equilibria and their use in cationic polymerization," Adv. Polym. Sci., 19, 1-56 (1975).
18. K. Matyjaszewski, S. Slomkowski, and S. Penczek, "Kinetics and mechanism of the cationic polymerization of tetrahydrofuran in solution: THF—CH_2Cl_2 and THF—CH_2Cl_2/CH_3NO_2 systems," J. Polym. Sci., Polym. Chem. Ed., 17, 2413-2422 (1979).
19. K. Matyjaszewski, S. Slomkowski, and S. Penczek, "Kinetics and mechanism of the cationic polymerization of tetrahydrofuran in solution: THF—CH_3NO_2 system," J. Polym. Sci., Polym. Chem. Ed., 17, 69-80 (1979).
20. K. Brzezińska, K. Matyjaszewski, and S. Penczek, "Macroion pairs and macroions in the kinetics of polymerization of oxepane," Makromol. Chem., 179, 2387-2395 (1978).
21. S. Penczek, P. Kubisa, and K. Matyjaszewski, "Cationic ring-opening polymerization," Adv. Polym. Sci., 37, 1-149 (1980).
22. D. W. Grattan and P. H. Plesch, "Binary ionogenic equilibria," Electroanal. Chem., 103, 81-94 (1979).
23. M. Szwarc (editor), Ions and Ion Pairs in Organic Reactions, Interscience, New York (1972).
24. V. N. Zgonnik, E. Yu. Melenevskaya, and B. L. Erusalimskii, "The study of active centers in anionic polymerization using spectroscopic and quantum-chemical methods," Usp. Khim., 47, 1479-1503 (1978).
25. S. Bywater, "Spectroscopic studies on the nature of the active centers in anionic polymerization," J. Polym. Sci., 12, 549-553 (1980).
26. A. Andresen, "UV-Spektroskopische Untersuchungen an homogenen Ziegler—Natta Katalysatoren mit Methylalumoxan als Katalysator-komponente," Dissertation, Hamburg (1980).
27. V. M. Sergutin, V. N. Zgonnik, and K. K. Kalnin'sh, "The study of the spectrum of oligopentadienyllithium and its complexes with electron donors," Vysokomol. Soedin., A22, 415-421 (1980).
28. Ch. B. Tsvertanov, I. M. Panayotov, and B. L. Erusalimskii, "Investigation by means of IR spectroscopy of styrene oligomers containing acrylonitrile active ends," Eur. Polym. J., 10, 557-562 (1974).
29. Ch. B. Tsvetanov and I. M. Panayotov, "On the nature of the active centers in the initial stages of the methacrylonitrile anionic polymerization — I. Spectral studies," Eur. Polym. J., 11, 209-214 (1975).
30. L. Lochmann and J. Trekoval, "Esters of diacids and oligo (carboxylic acid)s (oligomers of methyl methacrylate) substituted in the α-position with an alkali metal. Their stability and IR spectra," Makromol. Chem., 183, 1361-1370 (1982).

31. V. N. Zgonnik, A. A. Davidyan, N. I. Nikolaev, and E. R. Dolinskaya, "On complex formation in polydiethyllithium chains with electron donors in the presence of monomer in a hydrocarbon medium," Vysokomol. Soedin., A25, 749-754 (1983).
32. S. Brownstein, S. Bywater, and D. J. Worsfold, "Allyl alkali metal compounds," J. Organomet. Chem., 199, 1-8 (1980).
33. M. Schlosser and M. Stahle, "Nicht-ebene Strukturen von Allyl und Pentadienylmetall-Verbindungen," Angew. Chem. Supppl., 198-208 (1982).
34. M. Schlosser and M. Stähle, "Magnesium, Lithium- and Kaliumverbindungen vom Allyl-Typ: π-order σ-Strukturen?" Angew. Chem., 92, 497-499 (1980).
35. M. Stähle and M. Schlosser, "Neue ^{13}C-spektroskopische Untersuchungen zur Struktur und Allylmetall-Verbindungen," J. Organomet. Chem., 220, 277-283 (1981).
36. P. West, J. I. Purmort, and S. V. McKinley, "The ionic character of allyllithium," J. Am. Chem. Soc., 90, 797-798 (1968).
37. G. Boche, K. Buckle, D. Martens, and D. R. Schneider, "Konformation und Rotationsbarriere bei 1,3-Diphenyllithium Verbindungen," Ann. Chem., 1135-11771 (1980).
38. H. U. Siehl and H. Mayr, "Stable vinyl cations. Direct spectroscopic observation of substituted vinyl-cations," J. Am. Chem. Soc., 104, 909-910 (1982).
39. G. Fraenkel, M. I. Geckle, A. Kaylo, and D. W. Estes, "Effects of ligands on ion-pairing behavior of benzylic lithium compounds," J. Organomet. Chem., 197, 249-259 (1980).
40. Y. Firat and S. Bywater, "A ^{13}C-NMR investigation of a dimer anion of α-methylstyrene," Eur. Polym. J., 18, 265-267 (1982).
41. L. Vancea and S. Bywater, "Carbon-13 nuclear magnetic resonance of anion pairs related to acrylate polymerization. 1. Monomeric models," Macromolecules, 14, 1321-1323 (1981).
42. L. Vancea and S. Bywater, "^{13}C-NMR studies on anion pairs related to acrylate polymerization. 2. Dimer models," Macromolecules, 14, 1776-1778 (1981).
43. G. Henrici-Olivé and S. Olivé, "Koordinative Polymerisation and löslichen Übergangsmetallkatalysatoren," Adv. Polym. Sci., 6, 421-472 (1969).
44. G. Fink and R. Rottler, "ethyleninsertion durch lösliche Ziegler Katalysatoren. Direckter Einblick durch reagierendes Ethylen-^{13}C mit Hilfe der ^{13}C-NMR Spektroskopie," Angew. Makromol. Chem., 94, 24-47 (1981).
45. G. Fink, R. Rottler, and C. G. Kreiter, "Die Primärkomplexbildung in löslishen Ziegler-katalysatorsystemen. Kinetische und thermodinamische baten durch ^{13}C-NMR-Spektroscopie," Angew. Macromol. Chem., 96, 1-20 (1981).
46. G. Olivé and S. Olivé, Polymerisation. Katalyse-Kinetik-Mechanismen, Verlag Chemie, Weinheim (1969).
47. G. Henrici-Olivé and S. Olivé, "Mechanism for Ziegler—Natta Catalysis," Chemtech., 746-752 (1981).

48. Ch. B. Tsvetanov, Yu. E. Eizner, and B. L. Erusalimskii, "Structure of terminal and penultimate units of a living chain of polyacrylonitrile with lithium counterion. Quantum chemical investigation," Eur. Polym. J., 16, 219-226 (1980).
49. P. A. Berlin, V. L. Lebedev, A. A. Bagatur'yants, and K. S. Kazanskii, "Quantum-chemical modeling of the active centers in the anionic polymerization of ethylene oxide," Vysokomol. Soedin., A22, 1600-1606 (1980).
50. K. S. Kazanskii, "Donor—acceptor and solvation interactions in anionic polymerization of some heterocycles," Pure Appl. Chem., 53, 1645-1661 (1981).
51. Yu. E. Eizner and B. L. Erusalimskii, The Electronic Aspect of Polymerization Reactions [in Russian], Nauka, Leningrad (1976).
52. I. A. Abronin, K. Ya. Burshtein, and G. M. Zhidomirov, "The quantum chemical determination of the effect of the solvent on the electronic structure and reactivity of the molecules," Zh. Strukt. Khim., 21, 145-164 (1980).
53. Yu. E. Eizner and B. L. Erusalimskii, "The electronic structure and geometry of the anionic centers in anionic polymerization of vinyl monomers," Eur. Polym. J., 12, 59-63 (1976).
54. B. L. Erusalimskii, N. V. Smirnova, N. S. Dmitrieva, and V. N. Zgonnik, "Quantenchemische Untersuchung von Anionisch Aktiven Zentren am Beispiel von Butyl- und Butenyl-Lithium-Verbindungen," Acta Polym., 31, 357-362 (1980).
55. G. B. Erusalimskii and V. A. Kormer, "Quantum-chemical study of the effects of the association phenomenon on the active site structure in butadiene polymerization reactions initiated by organolithium compounds," Eur. Polym. J., 16, 463-465 (1980).
56. G. B. Erusalimskii and V. A. Kormer, "A quantum-chemical study of the structure of the active centers and the mechanism of polymerization of 1,3-dienes under the action of organolithium compounds (on a sample of 1,3-butadiene)," Zh. Vses. Khim. Ova., 26, 266-272 (1981).
57. P. H. Plesch, "Cationic polymerization," Progr. High Polym., 21, 137-188 (1968).
58. N. Bodor, M. J. S. Dewar, and D. H. Lo, "Ground states of σ-bonded molecules. XVIII. An improved version of MINDO/2 and its application to carbonium ions and protonated cyclopropanes," J. Am. Chem. Soc., 94, 5303-5310 (1972).
59. H. L. Hsieh, "Kinetics of polymerization of butadiene, isoprene, and styrene with alkyllithiums. Part II. Rate of initiation," J. Polym. Sci., A3, 163-172 (1965).
60. Yu. E. Eizner and B. L. Erusalimskii, "The electron structure of the active centers of a linear oxonium ion," Vysokomol. Soedin., A12, 1614-1620 (1970).
61. G. B. Erusalimskii, "A quantum-chemical study of the nature of the active centers in the polymerization of butadiene under the action of organolithium compounds," Dissertation, Leningrad (1981).

62. A. Bongini, G. Cainelli, G. Cardillo, P. Palmieri, and A. Umani-Ronchi, "A theoretical study of the allyllithium ion pair," J. Organomet. Chem., 110, 1-6 (1976).
63. T. Clark, E. T. Jemmis, P. v. R. Schleyer, J. F. Pinckles, and J. A. Pople, "Ab initio structure of allyllithium," J. Organomet. Chem., 150, 1-6 (1978).
64. E. T. Tidwell and B. R. Russell, "Electronic structure and bonding of allyllithium," J. Organomet. Chem., 80, 175-183 (1974).
65. J. F. Sebastian, J. R. Grunwell, and B. Hsu, "Electronic structure and geometry of bis(dimethyl ether)allyllithium," J. Organomet. Chem., 78, C1-C3 (1974).
66. S. Bywater and D. J. Worsfold, "Charge distribution in disubstituted allyl-alkylmetal compounds by ^{13}C-NMR," J. Organomet. Chem., 159, 229-235 (1978).
67. N. M. Geller, Yu. E. Eizner, and V. A. Kropachev, "Change in electron structure of cyclic oxides during their interaction with electron acceptors. Quantum-chemical investigation," Acta Polym., 32, 144-149 (1981).
68. N. M. Geller, Yu. E. Eizner, and V. A. Kropachev, "Effect of substituents on the electronic structure of cyclic oxides with an electron acceptor. Quantum-chemical investigation," Acta Polym., 34, 584-588 (1983).
69. G. B. Erusalimskii and V. A. Kormer, "Quantum-chemical study of the effect of butadiene interaction with active sites on the polymer microstructure," Eur. Polym. J., 16, 467-470 (1980).
70. O. Novaro, E. Blaiston-Barojas, E. Clementi, G. Guinchi, and M. E. Ruiz-Vizcaya, "Theoretical study on a reaction of Ziegler—Natta type catalysis," J. Chem. Phys., 68, 2237-2351 (1978).
71. B. R. Armstrong, P. G. Perkins, and J. Stewart, "Theoretical investigation of Ziegler—Natta type catalysis. Part I. Soluble catalyst systems," J. Chem. Soc. Dalton Trans., 1972-1980 (1972).
72. S. Miertuš, O. Kysel, and P. Májek, "Quantum-chemical study of the reactivity in anionic polymerization. 1. The effect of the polarity of the medium and alkali-metal cations on the rate of propagation of reaction," Macromolecules, 12, 418-421 (1979).
73. S. Miertuš, O. Kysel, and P. Májek, "Quantum-chemical study of the reactivity in anionic polymerization. 2. Effect of electronic structure of monomer on the rate of propagation reaction," Macromolecules, 12, 421-432 (1979).

Chapter 3

General Questions on the Problem of Multicenteredness

The parallel or alternating course of polymerization under the action of various active sites in systems of the ionic type has been the subject of discussion in one way or another for about a quarter of a century. Without going into the history of the problem, we note that it is obvious that the phenomena which constitute the basis for the conclusions about the existence of agents which are distinguished for their kinetic and/or stereochemical features were discovered during anionic polymerization in polar media and in processes initiated by Ziegler-Natta catalysts. The number of actual systems, which in this respect proved to be analogous to those mentioned above, has increased considerably. This process continues today. Now it is considerably more difficult to find an example of an ionic system which may have been regarded as single centered, than to substantiate the fact that systems of this type have many active sites. On the other hand, the generality of the phenomenon which we examine in the present chapter has a formal character since the effect of the different active sites may result from very different causes. Apart from this feature concerning the genesis of the required species, there is also another important question, i.e., whether mutual transitions between individual species takes place or not. No less important for the final result is the relative activity of each of the sites which are functioning in a given system because this is one of the factors which determine their contribution to the total effect. Special attention is devoted to this point later on.

3.1. TYPES OF MULTICENTERED SYSTEMS AND THE FORMS OF THEIR CORRESPONDING ACTIVE SITES

A classification of such systems using these features is not, as it might at first appear, a purely academic problem. Apart from such a classification, which may be justified by its making the task of further exposition easier, we have in mind an analysis of polymerization processes which is based on the principle of the extent to which one functioning active site transforms into another.

This is all the more important since with the great frequency of mutual transformations, the discovery of the fact that such systems have many active sites may well prove to be beyond the bounds of experimental possibility. In similar situations, the effects observed may coincide qualitatively with those which are caused by individual active sites which are distinguished by a variety of functions, i.e., by the absence of a clearly expressed selectivity in the interaction with the monomer. The most significant examples of such active sites are the so-called "unbound" active sites, i.e., free ions which are similar in this respect to free radicals.

Such nonselectivity can be most simply illustrated by the structural characteristics of polydienes which are formed under the action of any of the possible unbound active sites. Naturally the processes in which the growing chains are bound to ion pairs or their close analogs can also lead to similar results. However, in such circumstances it is necessary to take account of the possibility of parallel action of the active sites of a related nature, but which are distinguished by different structural parameters.

The variety of possible systems which are distinguished in this respect reflect the limiting conditions which either include equilibria of the types AS-I $\rightleftarrows$ AS-II (3.1) (in which AS-I and AS-II are active sites which coexist in the system) or which are free from such reversible transformations. Naturally in real conditions the number of different coexisting active sites may be greater, but the cases of such a type known at present (see Chapters 4 and 5) show that in fixed experimental conditions far from all the active sites which are present make a significant contribution to the overall effect.

Turning to the ways in which the multicenteredness is determined, irrespective of the types of the active sites themselves, it is possible to state that the case with which a discovery is made is facilitated by the minimal frequency of mutual transitions (3.1). Nevertheless, even when the rate of such transitions is high, the possibility of the identification of the phenomenon considered is not always completely excluded. At the basis of such an identification may be both kinetic parameters of the polymerization processes as well as structural and molar-mass characteristics of the polymers formed. Concepts of the relative value of the data may acquire sense only with the choice of a definite aspect of examination from two possibilities. One of these is the maximum detailing of the mechanism of formation of the macromolecules, and is not connected directly with any applied questions; the second is the appraisal of results intended for use in the control of synthesizing polymers with a given structure and molar mass.

In adopting the first of these approaches we will not ignore those systems in which the effects of multicenteredness do not appear particularly sharp.

TABLE 3.1. Varieties of Multicenteredness

Origin	Initial compounds or components of the systems	Types of processes
Deaggregation of the associates	RMt and ROMt compounds	Anionic polymerization of unsaturated and heterocyclic monomers
Ionic dissociation	RMt and ROMt; carbenium, oxonium, sulfonium salts	Anionic polymerization; cationic polymerization
Formation of intramolecular complexes	Carbon chain growing chains with polar substituents; heterochain growing chains	Anionic polymerization of polar unsaturated monomers; polymerization of heterocyclic monomers
Formation of intermolecular complexes	Combinations of Lewis acids and bases; binary systems based on Lewis acids	Anionic and cationic polymerization; process initiated by Ziegler–Natta catalysts
Isomerization	Growing polydiene chains; growing polyolefin chains	Anionic polymerization; cationic polymerization

For all the systematic discussion of the number of questions which constitutes the subject of the present chapter, it seems expedient to group the various cases of multicenteredness together, proceeding on the basis of some general principle. Such an attempt is seen in Table 3.1 which contains the possible origins of the phenomenon being considered. Although we commence the discussion of each case as if it were independent, we must emphasize that in real situations they are at least partially interlocked. Subsequently we note that the existence of some of these particular situations shown in Table 3.1 is not excluded.

In the general consideration of these cases we confine ourselves to the small number of concrete examples, taking into account material from Chapters 4 and 5.

3.1.1. Dissociation of Aggregates

The influence of association which is especially strong in compounds RLi and ROLi is easy to establish by direct physicochemical measurement of some of the parameters of the corresponding solutions. The methods used for this enable the degree of dissociation of the monomeric forms of these compounds to be determined, sometimes in the form of average values. However, as a rule they are of no use for determining spontaneous dissociation, i.e., the presence in the system of tiny amounts of equilibrated dissociated forms. These can only be deduced indirectly from the kinetics of certain reactions with those compounds in the particular polymerization processes which they initiate. Such a situation may cast doubt upon the very existence of the spontaneous dissociation of the basic form of the compound. Its splitting may be ascribed to the "inducing" action of the co-reactant, i.e., in the above case to the monomer. Rigorous proof for such a proposition is not yet possible, but participation of the co-reactant in the dissociation of the compounds (at least as an additional enhancing factor) is considered to be absolutely necesssary.

The extremely small tendency of aggregates $(M_nLi)_p$ to spontaneous dissociation is illustrated by the approximate values of the equilibrium concentrations of polystyryl and polybutadienyllithium (PStLi and PBuLi) in hydrocarbon media at room temperature and at an original concentration of the above compounds of $1\cdot10^{-3}$ moles/liter. These values, which have been evaluated by Melenevskaya [1] using values of K_d which characterize the equlilibria

$$(PStLi)_2 \overset{K_d}{\rightleftharpoons} 2PStLi \qquad (3.2)$$

and

$$(PBuLi)_4 \overset{K'_d}{\rightleftharpoons} 2(PBuLi)_2 \overset{K''_d}{\rightleftharpoons} 4PBuLi \qquad (3.3)$$

are of the order of $8\cdot10^{-4}$ and $1\cdot10^{-5}$ moles/liter respectively. In turn the values of K_d obtained from kinetic studies [1, 2] are no more than tentative.

Naturally, it makes sense to examine the systems for which equilibria of the type (3.2) or (3.3) are typical. Only those cases in which the associated and dissociated forms make tangible contributions to the reactions are being considered. We will turn to this matter in Section 3.2 noting until then the coexistence of different reagents which are in principle capable of exhibiting a given activity.

More significant are the effects of deaggregation of components produced by the action of electron-donor compounds, the limit of which is the complete transformation of condensed systems $(M_nLi)_p$ into the monomeric forms of complexes of the general type $M_nLi \cdot mD$. Such a limiting result by no means coincides with the complete elimination of multicenteredness, which may, in certain circumstances, actually be obtained. We have in mind here the actions of bidentate electron donors when the ratios D/Li are nearly stoichiometric in hydrocarbon media. More often the transition from hydrocarbon media to systems which contain a small amount of electron-donor components causes the formation of complexes which differ stoichiometrically and in the degree of deaggregation of the original component. This circumstance deserves mention not only in connection with the use in some studies of mixed solvents of the hydrocarbon—THF type, etc. It is also of great significance for the much more widely occurring cases of polymerization of polar monomers in nonpolar media. Here the monomer may become the source of multicenteredness as a consequence of its participating in complex formation. Ideas on the probability of occurrence of such events which were formerly based on data which characterize the features of a certain process of the polymerization or the structure of the macromolecules are in agreement with the results of physicochemical studies of certain systems.

The information which characterizes systems using anionic reagents of the alkoxide type is less certain. As is known such initiators as lithium alkoxides are found in a strongly associated state in hydrocarbon and polar media, for each of which direct data of the existence of equilibria

$$(ROLi)_n \rightleftharpoons nROLi \tag{3.4}$$

is practically nonexistent. Growing chains which are formed under the action of such initiators in the polymerization of polar unsaturated monomers obviously cannot be described by the simplest possible formula, i.e., by ROM_nLi. Such a formula seems doubtful due to the low efficiency of initiation (usually on the order of 10^{-2}), combined with the absence of those reactions which are characteristic of systems in which metal alkyls participate. In particular, after the complete conversion of the monomer in ROLi—acrylonitrile systems, the bulk of the initiator remains unexpended. The sum of these facts leads to the hypothetical structures $ROM_nLi \cdot (ROLi)_{n-x}$ for the active growing chains in these cases. Accepting this it is possible to assume the distinctive multicenteredness which results from a certain set of values of x. From this point of view the coexisting active sites may be regarded as growing chains which differ in the degree of association of their counterions.

As for the growing chains M_nOMt, which are formed in the polymerization of epoxides, it is apparent that intramolecular complex formation is more characteristic than aggregation; see Section 3.1.4.

3.1.2. Ionic Dissociation

As to the experimental approach to the finding of effects which are described in the previous section and those to which we now turn here, the original problems are always contrary. In the first case, above all, it is necessary to state the fact of the existence of associates, i.e., of the main forms of the corresponding typical systems. On the other hand, in the second case, the form being sought is usually in incomparably smaller concentration than the main bulk of the material studied. The formal similarity between the phenomena being compared may be seen in that each of them requires a double evaluation, i.e., a statement of the coexistence of the various agents and a statement of the coparticipation of each of these in the reaction studied.

In systems in which the subject being studied is ionic dissociation, the question of coexistence is decided completely by using electrochemical methods which are capable of giving a sufficiently accurate evaluation of the constant K_{diss} which characterizes the equilibrium between ion pairs and free ions. The multicenteredness of the system which is the result of such equilibra has been studied in the greatest detail. The information in this field has been supplemented recently by the results given in the previous chapter; see Section 2.1.

3.1.3. Intermolecular Complex Formation

In the previous sections we have mentioned one of the particular cases of this type relating to anionic systems. We will deal with it here in somewhat greater detail and we will include in our examination only the events of complex formation which are not accompanied by ionic dissociation. Such conditions are created in hydrocarbon media at low concentrations of electron-donor compounds whose presence has practically no effect on the polarity of the medium.

The various forms of complexes which may be described by the formula $(RLi)_n \cdot mD$ are determined by the stoichiometry of the original components and the degree of dissociation of the organometallic compound. The complexes of composition $(RLi)_n \cdot D$ and $RLi \cdot xD$ are extreme structures which are formed when the corresponding relations $RLi/D > 1$ and $RLi/D < 1$ hold true. The first of these cases, in the formal sense the simplest, is far from being an example of a single-centered system since the excess of $(RLi)_n$ components continues to remain free for equilibria of the types (3.2) and (3.3). On the other hand, with the complete transition of the original compound $(RLi)_n$

to a complex which corresponds to the second of these limiting structures, the exclusion of the effect of multicenteredness is more real.

A complete concrete definition of complexes which are present in one or another system, which is relatively accessible when IR spectroscopy methods are used (see Section 2.2), is not as yet capable of giving exhaustive information on conditions which are usual in processes of polymerization. The existing information on this relates to the region of much higher concentrations than are used in such processes. Hence ideas of the active forms of the various complexes and of their relative contributions as participants in complex multistage equilibria are more often than not tentative. Complications which result from these circumstances become especially obvious when the high acceptor capability of the monomeric forms of the organometallic compounds are considered. Such a feature leads to the probability of the existence of monomeric forms RLi in the form of complexes with D-agents even in systems in which these reagents are found in insignificant concentrations.

Of the other anionic reagents the effect of multicenteredness becomes apparent to a much greater extent in the case of organomagnesium compounds. In comparatively simple systems which contain these compounds, the coexistence of RMgX and R_2Mg derivatives which is described by the Schlenk equilibrium is possible. The situation becomes noticeably more complicated when the usual Grignard reagents are used and, in particular, alkylalkoxymagnesium derivatives for which (considering Schlenk's equilibrium) the possibility arises of the formation of a large number of complexes which are structurally different while retaining the same stoichimetry (see Section 5.1.2.1).

The stoichiometric differences are typical of many cationic initiators which are complexes of Lewis acids and bases of the type $EX_n \cdot ROH$, $EX_n \cdot 2ROH$, etc., and which may be represented by the ionic structures $H^+[EX_nOR]^-$ and $H_2RO^+[EX_nOR]^-$ respectively. The fact of their existence is shown by the conductometric data obtained, in particular for the cases $EX_n = AlCl_3$; see Section 2.1. The multicenteredness of such systems is possible in molar ratios EX_n/ROH which lie between 1 and 2. In fact in such systems only the initiating agents differ and this may be reflected in the overall kinetics of the polymerization.

The formation of various complexes with the interaction of the components of Ziegler–Natta catalysts, which in the majority of cases are accompanied by changes in the valence state of the transition metal, is due to other reasons. As a consequence of this a nonuniformity in the composition of the reaction products may appear; this also occurs in catalytic complexes which are the active sites for growth. In these cases both stoichiometric and structural differences between coexisting active sites are possible, caused, amongst other factors, by the simultaneous presence of transition metal derivatives which are found in more than one valence state. This fact may be revealed using quantitative analytical methods. Of course, the multicenteredness of the systems of this type can also manifest itself as a structural feature which is formally similar to that mentioned above for organomagnesium compounds when the transition metal derivatives are uniform in valence. Regarding the "classical" cases of the multicenteredness of polymerization processes using heterogeneous Ziegler–Natta catalysts, they obviously combine both the valence and structural features of the corresponding active sites.

We do not include substances which are considered in the present section when we consider those intermolecular complexes of active sites with monomer molecules (i.e., of the interaction before entry of the monomer into the structure of the growing chain). This is because it is difficult to ascribe the formation of such complexes to parameters related to multicenteredness. In particular, the commonly held concept of the course of the growth reaction in Ziegler–Natta complexes via an intermediate stage in which they form complexes with the monomer applies to all the presently coexisting active sites without causing additional multicenteredness.

3.1.4. Intramolecular Complex Formation

There are various facts which indicate the inclination of the growing ionic chains to transfer from linear forms to pseudocyclic forms as a consequence of the interaction of the counterion (or the end atom) with the internal links (or heteroatoms). These include IR spectroscopy data (see Section 2.2), the results of quantum-chemical calculations (see Section 2.3), and the sequence distribution in the chains of certain macromolecules (see Chapter 5). In considering this phenomenon from the aspect which is of interest here, we emphasize that the variety of forms of the growing chains which coexist as a consequence of this cause is not necessarily connected with the great variety of the possible reactions by which macromolecules are formed.

The significance of this remark is that generally intramolecular cyclization has various consequences including the exclusion of the growth reaction. Such a result may be both reversible (the formation of 'dormant' active sites) or irreversible. Of the cases under consideration here, those in which the intramolecular pseudocyclic complexes formed retain the ability to grow, i.e., are modified active sites, deserve attention.

In contrast to the effects which we have touched upon in the previous sections of this chapter, confirmation of the coparticipation of intramolecular complexes as active components of multicentered systems has so far not been strictly substantiated. On the other hand the tendency of ionic growing chains to intramolecular interactions is so general (at least in homogeneous systems) that it is necessary to pay careful attention to this question. As far as we know, the only attempt to discuss this from the above point of view has been made by Erusalimskii [4].

Using an extremely simple representation, we have the following:

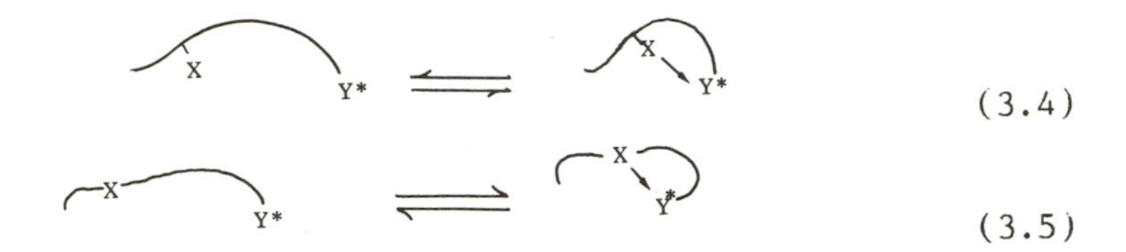

Here X is a side substituent in the monomeric chain or a heteroatom in the backbone of the chain and Y* is a counterion of either sign. We note that, in the first place, these schemes combined represent the interaction of the counterion with the X fragment of any of the groups of the growing chain and, secondly, that there may be any number of cyclic forms.

It seems plausible to assume that both linear and cyclic forms take part in growth reactions if their reactivities satisfy the inequality $AS_{\ell} > AS_c$, in inverse ratio of their concentrations. This is not the only possible situation because the above inequality is correct only when sufficiently stable AS_c forms are formed. In reality, depending upon the donor-acceptor properties of the reacting moieties of the growing chain, interactions which have inherent variations in efficiency are possible. With relatively weak interactions, differences both in the reactivity and the concentration of the coexisting AS_c and AS_{ℓ} forms may become so smoothed out that the phenomenon of multicenteredness acquires a purely formal character. However, the smoothing out of the individual parameters of each type of active site may be limited by the kinetic parameters not affecting the stereochemistry of the growth reactions.

When using schemes of the type (3.4) for the interpretation of the statistical distribution of the structural units of macromolecules (for bibliography see [5]), selective complex formation of the counterion with a chain unit is often proposed. A fairly good correlation between the detailed structure of the polymer and the Markov statistics of some order is often found. Deviations from such a correlation, which become the more noticeable the greater the structural variety chosen for evaluation (pentads and hexads), are not usually considered to be of serious importance. Nevertheless, the determination of an intramolecular complex of the counterion with one definite unit is a rather particular case, which occurs in especially favorable circumstances. These are due either to a special feature of the substance or simply occur by accident. Such a concept is easy to associate with the physicochemical characteristics of certain growing chains and their models, where the counterions have a tendency to simultaneously interact with certain polar substituents (or heteroatoms), i.e., a tendency to form "polyligand" complexes (PC) (see Sections 2.2 and 2.3).

It is natural that these PC should be diverse in the sense of their total number and the actual ligands which take part in their formation, particularly when the fluctuations which are caused by growth processes are taken into account. In changes of this type there exists the probability of the periodic transformation of PC with several relatively weak bonds X—Mt to single-ligand complexes (SC) with a single but much stronger bond of the same type. From our point of view, the irreversible reaction of the breaking of the multiple bonds in the polar group ($>C=O$, $-C \equiv N$), i.e., the termination event, should be taken as the main, if not sole, path to further transformation.

We will quote several facts which are in agreement with these propositions. Firstly we would remind the reader of the considerable role played by the formation of oligomers in the anionic polymerization of methacrylates and acrylonitrile (AN) in hydrocarbon media. In these conditions the above reaction is fundamental for the initiator where the nonactive oligomers are mainly cyclic trimers [6, 7] which are not difficult to associate with the intermediate formation of unstrained cyclic SC. Usually only a small fraction of the active sites which arise at the initiation stage overcome this trimeric threshold and are transformed into high-molecular-weight products. This situation has been known for a long time and indicates the reduced inclination of true growing chains to deactivation in comparison with low-molecular-weight active sites; this was previously ascribed to steric hindrance or to the stabilizing influence of such products, see for example [8]. The interpretation associated with the concepts of PC and SC seems more correct in that it takes into account the differences in the ability of short and long growing chains to form such forms as AS_c for which termination becomes the more favored reaction. The situation corresponding to the scheme which is generalized by the following concept must be borne in mind:

$$\text{Initiator} + n\text{M} \longrightarrow \text{AS}_1 \longrightarrow \text{AS}_c\text{–SC} \longrightarrow \text{termination} \qquad (3.6)$$

$$\text{AS}_1 \xrightarrow{m\text{M}} \text{AS}_l \rightleftharpoons \text{AS}'_c\text{–PC} \xrightarrow{\text{M}} \text{growth} \qquad (3.7)$$

$$\text{AS}'_c\text{–PC} \longrightarrow \text{AS}'_c\text{–SC} \longrightarrow \text{termination} \qquad (3.8)$$

TABLE 3.2. Efficiency of Initiation (F) for the Anionic Polymerization of Acrylonitrile in Toluene

Initiator	t, °C	F, %	Ref.
n-Butyllithium	−50	0.9	3
Tributyldimagnesium iodide (TMI)	−75	1.0	9
TMI—Dimethylformamide complex	−75	3.0	9
Polystyryllithium (PSLi)	−25	40.0	10
PSLi—Tetramethylethylenediamine complex	−50	55.0	10
Poly-2-vinylpyridylmagnesium iodide[a]	−50	60.0	11
Poly-4-vinylpyridyllithium	−50	95.0	11
Poly-2-vinylpyridyllithium	−50	100.0	11

[a]Obtained under the action of TMI.

Here n has an order of several monomer units and m may take any high value. Of course, these are tentative values, but for reaction (3.6), which may be regarded as the main termination reaction, the most probable value is n = 3. Chains which have overcome competition between oligomerization and polymerization and for which deactivation (3.8) has a limited significance due to the existence of active sites in the form of polyligand complexes, are designated by the symbol AS'. Transformations of the type (3.9), which are more often than not irreversible and which lead to termination, are of comparatively rare occurrence for high polymer chains:

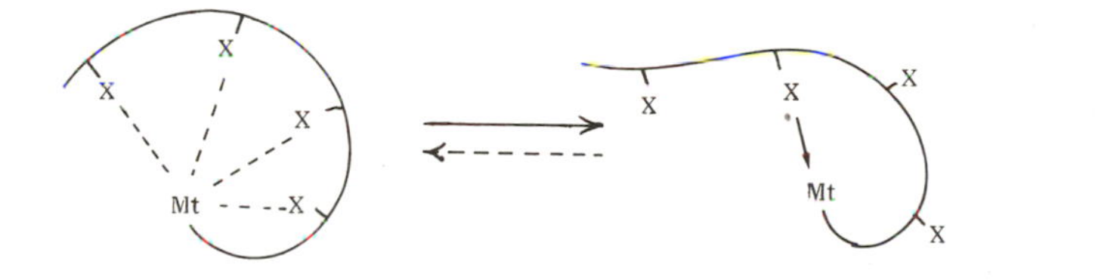

(3.9)

This already follows from the possibility of obtaining significant values for M in those polymerization conditions of polar monomers in which 90-95% of the initiator is expended on oligomerization. Another circumstance is the sharp suppression of reaction (3.6) when living chains of certain "foreign" monomers are used as initiators. The corresponding differences in the roles of such reactions are reflected by the values of the efficiency of initiation (F) established for AN polymerization processes which take place under the action of the usual and high polymer initiators; see Table 3.2.

The maximum suppression of reaction (3.6) in the case of living polyvinylpyridine chains is easily explained by participation of pyridine rings in the formation of AS_C—PC. This circumstance hinders the appearance of the stable forms AS_C—AC at the expense of AN links right at the initiating stage of the growth reaction. Obviously polystyrene chains behave in an analogous way but are inferior in the quantitative sense to polyvinylpyridine chains because of the significantly smaller donor activity of the phenyl groups.

As another fact which accords with the ideas under discussion here we adduce the anionic polymerization of ε-caprolactone which is characterized by the splitting off of macrocycles of various sizes in parallel with the growth reaction. It is possible to assume that the wide variety of these cycles established by Slomkovskii and Sosnovskii (up to about 20 units) [12] is caused by the periodic formation of fixed AS_C—SC structures along with As_C—PC forms of the growing chain.

The anionic and cationic polymerization processes of polar monomers are, in this respect, analogous. The similarity in forming intermediate AS_C forms of the poly- and single-ligand type must be borne in mind. However, the cationic active sites are distinguished by a feature which is more suited to discussion in the later chapters (see Sections 4.2 and 5.1.3).

3.1.5. Isomerization of Active Sites

One can name two types of processes in which the existence of multicenteredness which results from the isomerization of the active site is firmly established. These are the cationic polymerization of α-olefins and the anionic polymerization of dienes. In the first case the coexisting active sites may be derivatives which contain sec- and tert-carbocations, and in the second, geometric isomers of the end links of the growing chains. That these and other processes are formally of the same type can be seen from the fact that each of them may be examined from the point of view of competition between isomerization and growth, which determine the microstructure of the polymer being formed. But if this is the only criterion for cationic polymerization then, in the case of anionic diene active sites, by virtue of their great stability, it is possible to obtain direct physicochemical characteristics. We emphasize that the latter circumstance does not as yet permit the omission of debatable points which are connected with the structure and genesis of such active sites.

The most essential of these is the evaluation of the role of the intermediate which accompanies 1,4 (or 4,1) cis-trans transformations of the end units of polydiene chains. In contrast to the original and final products of such transformations, it has so far not been possible to determine the intermediate forms, i.e., terminal 1, 2 (or 3, 4) units, by means of spectroscopic methods. Only in certain cases are weak spectra found which are assigned by certain workers to end units of the vinyl type (see Section 5.2). It is sometimes proposed that these difficulties are the results of a significantly smaller concentration of the growing chains with end units of this structure and their high reactivity. Hence it is permissible to pose the question of whether systems which contain growing $\sim M{-}Mt$ chains, where M is the final diene unit, should be considered as two centered (bearing in mind the $\sim M_{cis}Mt$ and $\sim M_{trans}Mt$ structures) or as more complex. Of course the answer to this is important not so much for the determination of the number of forms for the active sites as for a firm conclusion regarding the origin of the formation of vinyl (or vinylidene) units. In spite of the point of view that the structure of diene polymers depends mainly upon the point of attack of the monomer molecule on the end unit, which is becoming more and more widely accepted, the possibility of another mechanism for the formation of the structure of the chains is not excluded. Naturally, a question arises in the interpretation of a phenomenon which concerns only the unassociated forms of growing polydiene chains; their cis-trans isomerization is more easily imagined as passing through a stage in which a "third" AS(C) arises:

$$\begin{array}{ccc} \underset{\text{(A)}}{\sim CH_2(R)C{=}CH{-}CH_2Mt} & \rightleftharpoons & \\ \Big\updownarrow & & \underset{\text{(C)}}{\sim CH_2CHMt{-}C(R){=}CH_2} \\ \underset{\text{(B)}}{\sim CH_2(R)C{=}CH{-}CH_2Mt} & \rightleftharpoons & \end{array} \tag{3.10}$$

Such a situation corresponds to the growing chains in polar media with any counterion and to nonpolar media for the highest alkali metals. In the case of the polymerization of dienes, which proceeds under the action of M_nLi chains in polar media, the appearance of 1,2 or 3,4 units in the macromolecules can be completely explained without the concept of the participation of active sites of type C; see Section 5.2.

We have dealt here only with one concrete example of multicenteredness as a function of isomerization transformations of the active sites, since they are of sufficient general interest. Certain other cases, which are formally related to the same type, are included in Chapter 5.

3.2. OVERALL AND INDIVIDUAL EFFECTS

The coexistence in the system of several types of active sites which have different characteristics still does not determine the significance of multicenteredness. The deciding feature is the degree of difference between the total effects which are typical of the given process of polymerization and the effects which are the results of features of each active site.

In using the word total we also include those complex effects which are not connected with multicenteredness. To this category is assigned any effect which is characterized by some relative or absolute parameter, for example, the experimental values of the rate constants and equilibrium constants between the active sites and the nonactive forms, i.e., "dormant" polymers. It would seem that in a given case of such a type the differentiation between mono- and polycentered systems is simple. However, such a conclusion is only correct if one can be completely confident that in such a system only one of the components of the equilibrium really fulfills the function of the active site. Consequently, with any other approach to the question of the causes which determine the total character of a certain experimental value, information both of the existence of the different agents and the ability of more than one of them to function as active sites is necessary. We considered it necessary to make this remark because of the possibility of the appearance of experimental effects which, with the absence of such information, might be mistakenly regarded as due to multicenteredness.

Thus in the most general case the availability of data such as that described in Section 3.1 is still not exhaustive in a way which is of interest to us. An independent stage of study is necessary even if it is directed to the elucidation of the qualitative coexistence of simultaneously (or alternately) functioning different active sites. The solution of such problems is possible only on the basis of data which characterize the kinetics of polymerization and certain parameters of the molecules formed.

3.2.1. The Kinetic Features of Polymerization Reactions

Conclusions on the possibility of the coexistence in the system considered of different active sites, which are based only upon the deviations from the most simple kinetic laws, do not have the force of final conclusions. Such deviations, in particular, nonintegral orders of reaction for active sites (or initiator) or the inconstancy of rate constants of the elementary process over some concentration range may be observed also in single-centered systems. The condition for this is the coexistence of the original initiator and corresponding products of its interaction with monomer basically as passive forms, which are capable of severely limited transition to active sites. The presence of multicenteredness in such cases may be established with the help of various methods which sometimes prove to be sufficient without going beyond the bounds of kinetic measurements which do not include obtaining the molar mass (MM) characteristics of the corresponding polymers. This is well illustrated by the results which have served as a basis for the discovery of parallel growth reactions on free anions and ion pairs or on growing organolithium chains which are found in monomeric or associated forms. It is opportune to note at this point that the second of the above cases has begun to be regarded only in comparatively recent times as an example of a multicentered system. Over a long period of studies on the polymerization of nonpolar monomers under the action of organolithium compounds in nonpolar media, these systems have been shown to be essentially single centered in which active monomeric forms of growing chains coexist with their completely passive associates. The gradual, although until now incomplete, transition to the other point of view has demanded the accumulation of a great deal of experimental material. This is still insufficient to obtain irrefutable quantitative data. So far the only attempt in this direction is the work of Melenevskaya [1], which was cited in Section 3.1., on the basis of which the growth rate constant corresponding to monomer, dimer, and tetramer forms of polybutadienyllithium chains were later evaluated. Their order was found to be 1, 0.001, and 0.0001 liter $mole^{-1} \cdot sec^{-1}$ respectively (data for 20°C) [13]. Naturally the relatively large contributions of the associated forms must be considered only in the region of sufficiently high concentrations of the growing chains. In particular when the concentration is of the order of 0.01 moles/liter this contribution approaches 30%. However, in the concentration region which is more usual for real polymerization processes, the kinetic contribution of the associates remains appreciable and this also affects the structure of the macromolecules which are formed (see Section 5.2). Thus the given systems may be regarded as three-centered systems and are described by Eq. (3.3).

The above quantitative evaluations must be treated with certain caution because the constant used in the calculation of the particular constants of the complex equilibrium (3.3) were found indirectly and the errors could not be determined precisely. This, however, does not affect the order of the differences in the growth rate constants of the active sites compared or the fact of their coexistence. The changes in the order of reaction in active sites with concentration over a wide range favor the latter.

In spite of the reservation made at the beginning of the present section, orders of reaction and experimental values of the constants which exhibit "creeping," i.e., which depend upon the concentration interval, are the most usual primary indicators of multicenteredness. It is sometimes possible to establish the true nature of such effects by the artificial elimination of the multicenteredness, i.e., by creating conditions which exclude the possibility of the existence in the system of growing chains in several active forms. The most usual of such examples consists of the suppression of the dissociation of ion pairs by buffer compounds. Nevertheless, although its use has proved its worth many times, it leads in certain cases to results for which there is no simple interpretation. This may be determined by different conditions, for example, the participation of the buffer in the modification or deactivation of the active sites which is assumed in particular for methyl methacrylate—THF—sodium initiator [14]. There is insufficient precision in this respect also for those systems in which, apart from contact ion pairs, there is some portion of separated ion pairs. The discovery of the latter using purely kinetic methods is difficult in such circumstances. In order to determine the type of ion pair, the electrochemical characteristics of the "nonbuffered" system which allow the interionic distance to be evaluated are sometimes used; see Section 2.1.

If, however, all the known cases in which for one reason or another a buffer has been added are taken into consideration, then it is possible to assert that the results obtained in anionic systems are more informative than those obtained in cationic systems. We have in mind not just lower accuracy of the experimental kinetic data of cationic polymerizations in the presence of buffer (see Section 2.1). Sometimes this indeterminacy also applies to the nature of active agents which in cationic systems may be both ionic and covalent.

The multicenteredness of certain cationic systems is the result of the transition of the ionic site of the growing chains to the covalent state (M_n-Y); this is shown in the simplest form by the scheme

$$M_n^+ , Y \begin{array}{l} \rightleftharpoons M_n^+ + Y \\ \rightleftharpoons M_n\text{-}Y \end{array} \tag{3.11}$$

Similar equilibria which are possible for processes which are initiated by protic acids, the anions of which are able to form ester groups by interaction with the end atoms of the growing chains, are sometimes found on analyzing data which characterizes the kinetics of the process and certain parameters of the polymers. The fact of "ionic—covalent coexistence" itself has been successfully established by a combination of kinetic and spectroscopic methods of which the cationic polymerization of styrene and p-methoxystyrene in a CH_3SO_3H—dichloroethane system is an example [15].

Of course, such a combination is necessary in the majority of other cases in which, apart from the ascertainment of multicenteredness, the aim is to establish the nature of the active sites. The range of active reagents which is possible in principle in any system can usually be foreseen with sufficient certainty. For the above cationic processes, however, such assumptions may easily prove to be mistaken.

One especially valuable method associated with the kinetics of the process and which is used in establishing the effect of multicenteredness, but which requires the evaluation of the molar mass of the molecule, may be approached from a similar point of view. Some remarks on this are given below.

3.2.2. Molecular-Mass Distribution in Polymers

Methods for the study of the kinetics of very rapid polymerizations have not as yet been widely investigated. It is still possible to name polymerization reactions which have been studied for many years, but which have not yet been described from the kinetic point of view because of the high reaction rates. Among these are the cationic polymerization of isobutylene and the homogeneous anionic polymerization of acrylonitrile. In such cases the molecular-mass parameters of the polymers formed are of especially important significance and the molecular-mass distribution (MMD) of the macromolecule becomes the only criterion for assigning the system as a mono- or polycentered type. However, a definite choice in this respect is possible only when the MMD is not unimodal. For reasons which do not require explanation, the unimodality of the MMD as yet does not indicate anything. Therefore a more general method for elucidating this problem is to obtain data on the MMD of polymers in combination with other parameters which are characteristic of the system studied.

Because of the analogy between the independent purely kinetic and MM characteristics it has been implied (Section 3.2.1) that they may reveal the coexistence but not the nature of different active sites (because of the lack of any other data). Thus a discussion on the nature of active sites which are responsible for the multimodal MD only becomes possible when it is found that the MMD is sensitive to the presence of additional passivating or reducing agents (for example, buffer compounds in kinetic measurements).

Without dealing at length with the relative role of multicenteredness (see Section 3.3), we note that in studying it extensively the question of the preference for using kinetic or MM parameters may not arise. Nevertheless, there are cases in which the simple solution of such a problem on the basis of kinetic data alone, without MM data or the structural characteristics of the polymers, is, in general, excluded. The existence of independent active sites, the equilibrium between which is either totally absent or is of limited significance, may be sufficient for this. Radiation polymerization of styrene is somewhat remote from the field under discussion, but is nevertheless an extremely curious example of this. This growth may proceed by both cationic and radical mechanisms in parallel and the resultant polymers have a clear bimodal MMD. The cationic component of the total process is suppressed by small quantities of water and leads to a sharp reduction of bimodality [16]. Nevertheless, the presence of water does not affect the kinetic order of this process.

The variety of growing chains, which is the result of intermolecular complex formation, may influence the stoichiometry and the structure of the active sites without affecting their nature (Ziegler–Natta catalysts or certain organometallic initiators); this makes them difficult to differentiate kinetically.

The sensitivity of the MMD of polymers to the nature of the reaction medium, which has already been noted in earlier works on ionic polymerization, acquires a particular character in certain cationic systems. An obvious example of this is polymerization of styrene under the action of various protic acids, in particular trifluoromethylsulfonic acid [17]. The effects as discovered in various media are illustrated by the following characteristics of polymers obtained at 0°C:

	Solvent		
	Benzene	Dichloromethane	Nitrobenzene
Type of MMD	unimodal	bimodal	unimodal
M	10^3	$10^3, 5\cdot10^4$	$5\cdot10^4$

Together with data contained in the work quoted above on the dependence of the MMD on the presence of a buffer in each of the three cases, these results constitute the basis of the conclusions on the nature of the coexisting active sites (see Section 4.1).

Polymodality of polymers can be accompanied by the nonuniformity of their fractional composition. The corresponding structural effects deserve some attention outside their dependence upon the MMD.

3.2.3. Structural Features of Macromolecules

From a general consideration of this question, which forms the theme of Chapter 5, we will isolate only points connected with the problem of multicenteredness.

Just as with other approaches, the actual problem in this case amounts to an attempt to break down the total effect into its components; more precisely, into a comparison of the microstructure of the separate fractions of the polymer which have been isolated by one means or another. Naturally, the coexistence of active sites which are distinguished by their stereospecificity does not by itself guarantee that such a problem can be solved in this way. Apart from the absence of rapid mutual transformations of the type (3.1), the existence of such structural differences in macromolecules which lead to the possibility of the selective isolation of individual fractions even comparatively close in molecular mass is necessary. We have given this very little consideration because it is probable that, if only the structural features of the macromolecules are chosen as criteria, the conclusions about the nature of the system studied (in the sense of whether it is a mono- or polycentered type) could be incorrect. It will be recalled that the first facts which indicated the coexistence of active sites of different stereospecificity were established in cases in which at least one of the active sites was distinguished by a high stereospecificity (the polymerization of styrene and propylene under the action of Ziegler–Natta catalysts). As a consequence of this there is a marked difference in the solubility of components of the reaction products, i.e., stereoregular and atactic macromolecules; see Section 5.1.1. Analogous conditions are sometimes created in other systems, in particular, in certain cases of polymerization of α-substituted acrylates under the action of anionic initiators. For example, isotactic and syndiotactic polymers are formed in parallel for which selected solvents can be found (see Section 5.1.2). At the same time it is not difficult to give examples of active sites which are fundamentally different in their reactivity but do not exhibit great differences in the formation of the structure of the polymer chains which are formed under their action. Typical in this respect is the behavior of separated ion pairs and free ions in the anionic polymerization of styrene.

Foreseeing a more detailed discussion of this question in Section 3.3., we note that for the cases where active sites of different stereospecificity coexist, the macromolecules which are typically formed in parallel differ in their degree of atacticity rather than in their stereoregularity. In such a situation the possibility of dividing the polymer into fractions of different structure may prove complex, even sometimes impossible. The quite noticeable difference in the MMD of such molecules may facilitate a solution. On the other hand such a difference is not necessary when differentiating polymers which are essentially distinguished by structural characteristics and for which thin-layer chromatography is especially suitable [18, 19].

The significance of the various circumstances touched upon in Section 3.2 depends directly upon whether the problem of multicenteredness comes into the range of problems which are to be explained within the terms of the given study. The absence of such a special aim may provide grounds for casual but far from always correct conclusions.

3.3. THE ROLE OF MULTICENTEREDNESS IN REAL PROCESSES

The solution to the problem of the role of multicenteredness of a given process amounts to the evaluation of the contributions of the reactions which occur on each active site, taking into account the nature and degree of difference in the behavior of the active sites and the frequency of their mutual transformations.

We will examine the types of systems whose active sites are differentiated in the following ways: by reactivity and stereospecificity (type A); only (or mainly) by reactivity (type B); and only (or mainly) by stereospecificity (type C). For each of these variations additional distinguishing features are possible which detail differences in the frequency of exchange (see Eq. 3.1). It is necessary to single out cases in which transformations are completely absent; they can be regarded as very rare transitions from one type of active site to another.

The reality of variations A and B can be easily illustrated by a number of examples, some of which are given below:

Type	Systems			Frequency of exhange
	initiator	monomer	reactive medium	
A	Catalysts	Olefins	Nonpolar	Extremely low
B	Sodium and potassium organic compounds	Dienes	Polar	Extremely high

Obviously some processes of polymerization of methacrylates which are initiated by organomagnesium compounds are of type C.

The number of possible individual cases grows when the termination reactions which accompany many processes of ionic polymerization are taken into account. In the present context it is especially important when there are significant differences in the termination rate constants for the coexisting active sites. The mechanism of these reactions is also important provided that they are of the mono- or bimolecular type. On the other hand, the termination reactions which influence the total effect qualitatively do not affect all the features of the polymers. It must be borne in mind that changes in polydispersity which result from termination cannot be the cause of either the appearance or the disappearance of the polymodality of the MMD. This circumstance which simplifies the interpretation of the experimental facts when the MMD of polymers formed in the system being studied is polymodal does not lead to a simple conclusion when the MMD is unimodal. In particular, the considerable polydispersity of polymers is in principle equally probable both for a multicentered nonterminated system in which exchanges between the active sites are not uncommon (when there are significant differences in their reactivity and comparability of their contributions) and for single-centered systems in which polymerization is accompanied by termination reactions and/or chain transfer.

In the absence of side reactions a high frequency of exchanges between all the coexisting active sites may ensure the synthesis of macromolecules which are homogeneous with respect to all parameters, i.e., may mask the fact of multicenteredness.

The concept of frequency of exchange acquires definition in the estimation of a number of other factors, the most important of which is the duration of the experiment which, in its turn, is connected with the rate of polymerization in the conditions chosen. Consequently the frequency of exchange affects the degree to which multicenteredness of the system is manifested. We note that the causes of the observed effects have not been explained simply in all known cases of this type. This refers for instance to the conversion dependence of the polymer structure which is sometimes noted and which may be ascribed either to the different stability of the coexisting active sites or their modification which takes place gradually during the course of polymerization. The products of side reactions which are typical for many anionic systems including polar monomers may fulfill the function of modifiers.

Another more distinctive effect is possibly connected with multicenteredness. The question arises as to the differences in polydispersity of polystyrene formed in identical conditions (in a hydrocarbon medium at 60°C with equal concentrations of initiators) under the action of sec-butyllithium and polybutadienyllithium (PBuLi) with an average degree of polymerization of about 500 [20]. In the first case the resulting polystyrene is characterized by a ratio M_w/M_n = 1.05. In the second case this value is 1.4 although in the initial PBuLi it did not exceed 1.05. From the point of view formulated in [20] the difference found is to be attributed to the low frequency of exchange between the monomeric and the associated forms of PBuLi, each of which initiates the polymerization of styrene. The obstacles to exchange which are typical of high-molecular-weight initiators must be borne in mind.

It is possible to agree with such an interpretation only if it is assumed that the final result depends upon the nature of the chain (or the final group of the high-molecular-weight initiator). The following circumstances makes it necessary to consider this feature. The use of "living" butadiene—styrene block copolymers for the synthesis of a three-block copolymer leads to the formation of monodisperse alternating butadiene blocks with M_w/M_n = 1.05.

This curious effect, in its turn, has not yet received an unqualified explanation but that it is connected with the fact that the frequency of exchanges is a function of the above feature cannot be excluded.

The most valuable information concerning the connnection between the conditions under which polymerization takes place and the degree to which multicenteredness may be manifested or smoothed out can be obtained by the studying of processes of type A. The most suitable conditions for this are homogeneous conditions in which the mutual transformation of the active sites are regulated within the limits required for studying the consequences of changes in the number of such transformations.

For a more detailed discussion of this problem we will adhere to those possible events which are covered in a general form by Eq. (3.1) and which are connected with reversible complex formation; such events are often stages which are capable of determining the mutual superposition of effects which are shown in Table 3.1 as independent. Such secondary consequences of complex formation are manifested particularly clearly in systems which include organometallic growing chains and independent electron donors (ED). They may consist of making the nature of the multicenteredness both more complex and more simple, up to the transformation of the initial system to a practically single-centered one. In particular, the complete transformation of the associated active sites (which are represented in the initial system by forms which differ according to their degree of association) to complexes of their monomeric forms and the transition from contact ion pairs to free ions with a high degree of dissociation (of the order of tens of percent) are examples of this. When the difference in reactivity of the two latter agents is sufficiently great the contribution of the ion pairs may be neglected, even if their concentration is considerably in excess of that of the free ions.

Both the corresponding limiting and intermediate cases which are differentiated by the nature and number of individual active sites are of considerable interest but we will deal only with systems which are more convenient sources of information. From our point of view hydrocarbon solutions of organolithium compounds(OLC) which include nonpolar monomers and strong electron donors with the ratio OLC/ED ≤ 1 fall into this category.

The state of the initial OLC when there are no ED in some systems is described by Eqs. (3.2) and (3.3). For the general case such a situation may be expressed by the system of equilibria which included the forms $(M_nLi)_m$, $(M_nLi)_{m-x}$, and M_nLi when $1 \leq x \leq m-1$. If an equimolar quantity of a strong ED (for example, tetramethylethylenediamine [TMED], glyme, etc.) causes the complete transition of components to complexes of the form M_nLi—ED, then, if the ratio

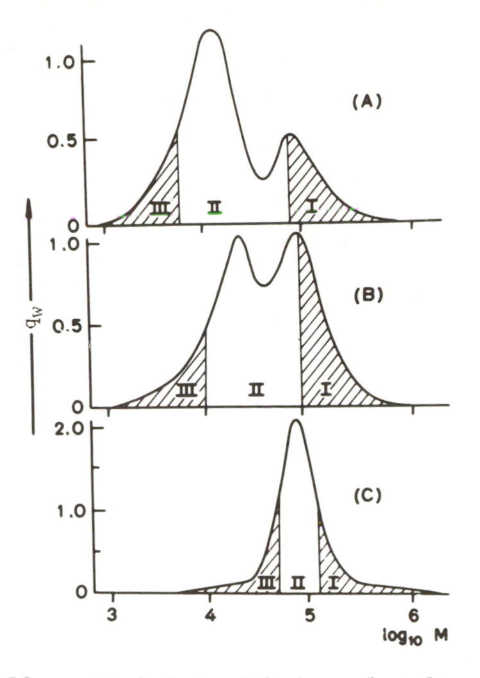

Fig. 3.1. The differential logarithmic molecular mass distribution for polyisopropene (gpc) formed under the action of tetramethylethylenediamine–oligoisoprenyllithium (TMED–OILi) systems with molar ratio TMED:OILi = 0.01 [23]. Temperature of polymerization, –30°C. Concentration, moles/liter: isoprene, 3.07 (A), 1.08 (B), 1.60 (C); OILi, $1.4\cdot10^{-3}$ (A), $1.1\cdot10^{-3}$ (B), $2.5\cdot10^{-3}$ (C). Conversion, %: A) 10; B) 35; C) 80.

OLC/ED < 1, a set of complexes is possible which differ in their degree of association and stoichiometry from $(M_nLi)_m\cdot ED$ to $M_nLi\cdot xED$. At the same time there also remains some fraction of the original OLC. The ratio between these and other reagents can be easily controlled by the relative concentration of ED.

Without concerning ourselves for the moment with the question of the actual forms of the "free" OLC and its complexes we note that the total contribution of both types of growing chain can be controlled in real polymerization processes. Such regulation is especially favored by the great difference in the rate of polymerization in a given M_nLi–M–nonpolar solvent system which contains no ED and when ED is present in quantities close to the stoichiometric OLC/ED ratio. Monomers which are suitable from this point of view are butadiene and isoprene, the polymerization of which under the action of OLC is accelerated considerably by the introduction of small amounts of bidentate ED [21, 22]. This circumstance ensures the comparable participation of the growing chains and their complexes which are free from ED in the growth reaction, practically without changes in the nature of the reactive medium. Thus the presence of very small amounts of TMED in an M_nLi–diene–hydrocarbon medium creates the conditions which allow the dependence of the fractional composition of the polymers formed on the frequency of exchanges to be traced. Studies on the use of such a method for the evaluation of the reactivity of active sites are more appropriately considered in Chapter 4. Here we note only the influence of the chosen conditions and the nature of the reagents on the frequency of mutual transformations of the coexisting active sites. In this case the results obtained at −30°C and a conversion of 10% are particularly distinct. The greater degree of conversion gradually smooths out the effects which result from the nature of each type of active site as a consequence of the increase of the total number of exchanges (Fig. 3.1). Increasing the temperature, which produces an increased rate of exchange of the type given by Eq. (3.1), leads to a similar averaging out of the nature of the effects of the different active sites. In any attempt to define precisely the possible type of active site in cases similar to those above (i.e., for which $OLC/ED \gg 1$), it must be borne in mind that, in numerical terms, the least active components of $(M_nLi)_m$ continue to remain the main form of active site. The corresponding equilibrium concentration of the monomeric form M_nLi (even when the equilibrium shifts under the action of ED) is clearly not sufficient to consume all the ED produced. This leads to the suggestion that the function of the chief acceptor with respect to ED in these conditions is fulfilled by the associated growing chains. Of course the considerable increase in the dissociation constant of the product $(M_nLi)_m \cdot ED$ compared with its initial value may be a secondary effect in this case (i.e, the splitting off of the monomeric form $M_nLi \cdot ED$ or M_nLi).

If, however, donorless systems of the type (3.3) are compared with those containing a subcatalytic quantity of ED then the main difference between them, irrespective of the nature of the active site, is that the frequency of exchanges is practically immeasurable (i.e., is extremely high) in the first case and is, in principle, capable of evaluation in the second. It will be recalled that the obligatory entry of the monomeric forms M_nLi into the structure of the component after each individual growth event is often taken for system (3.3) [24]. The frequency of exchange in such systems is insensitive to appreciable influences of this type. Only control of the relative contribution of the different active sites, from the practical exclusion from participation of the associated forms in the extremely low concentration region to their dominant role when the concentration of active sites is a maximum, is possible; see Section 3.1.1.

The relation between experimental conditions and the nature of the reagents on the one hand and the frequency of exchanges [Eq. (3.1)] on the other deserves special study. However, among the "natural" multicentered systems hardly any can be found which are suitable. For this reason the above artificial systems may be regarded not simply as curious particular cases but as examples useful for the general approach to the study of the problem. Here, apart from the method already used, similar means of modifying some fraction of the active sites while keeping the rest unaltered must be borne in mind. Information which is of interest in such cases may be obtained at comparable participation of the original and modified active sites in the processes being studied, i.e., by the choice of the corresponding amount of the modifier.

In particular, anionic systems with subcatalytic quantities of crown ethers or cryptands which are capable of converting some of the original contact ion pairs into separated ion pairs are suitable for studies of this type. Also suitable are cationic systems with subcatalytic quantities of salts which contain counterions similar to the basic counterions of the growing chains. The suitability of the latter for this purpose can be seen from the data of [25], which describe the influence of the salts of various protic acids on polymerization in a styrene–$HClO_4$–dichloroethane system. The results of such studies, which are useful primarily for the precise definition of the mechanism of polymerization and the evaluation of the reactivity of the active agents, are also interesting in quite a different respect. The question arises as to the possibility of ensuring the same total effect with different methods of varying the structural and molecular mass parameters of the polymers. The question of to what degree such a variation may affect the macroscopic properties of polymers deserves attention. As yet it is only possible to foresee that the degree of such an effect must, apart from all other variables, depend essentially upon the nature of the monomer itself.

Turning to the main subject of the present discussion we give attention to the necessity of developing further MMD methods of analysis for polymers which are formed in multicentered systems; this has been recently emphasized by Shamanin [26]. The results relating to systems with partially modified active sites have considerable long-term significance in this respect. Data on the conversion path of the MMD which is accompanied by the gradual disappearance of multimodality may provide a basis for useful qualitative evaluations even when absolute growth rate constants for coexisting active sites are absent. As follows from Section 3.3., the role of multicenteredness may be examined from different points of view and does not yield to a short and simple treatment. Also, essentially, the volume of data necessary for well founded conclusions from any possible aspect is as yet insufficient.

3.4. MULTICENTEREDNESS IN COPOLYMERIZATION PROCESSES

In a certain sense copolymerization processes were the first to lead to the problem of multicenteredness. Of course, primarily there are active sites which differ in the nature of their end units and in the majority of cases it proves to be sufficient to take into account only differences of this type. Nevertheless, there are systems such as M_1–M_2 in which the differences in the penultimate links of the growing chains have to be considered when the final links are identical. In the framework of our considerations those in which these or other specific features appear on chains of the general type $m_1m_2{}^*Y$ and $m_2m_1{}^*Y$ (where Y is the counterion) must also be treated as multicentered copolymerization processes even if, according to the general classification, they belong to one and the same type (to contact ion pairs, for example). As yet the existence of such cases can be illustrated by only a small number of facts (see Chapter 5) but it is to a considerable extent the result of the absence of directed attempts to discover corresponding effects in suitable examples.

We would like here to draw more attention to another aspect of multicenteredness and its application to the processes of ionic copolymerization. The question arises as to additional changes which the introduction of a second monomer produces in the multicenteredness which is inherent in homopolymerization. Having in mind a short but comprehensive discussion of this question, we will adhere to the format laid down in Section 3.1.

3.4.1. The Aggregated State of Active Sites

At first, processes in which growing comonomer chains differ in their degree of association and/or deaggregation constants (K_{deag}) should be of special interest. Under this condition the difference in the end units may give rise, in copolymerization, to deviations from the state which is typical for each homopolymer. From the formal point of view a simple and graphic example in this respect is the copolymerization of styrene with butadiene in a hydrocarbon medium under the action of OLC. Homoassociates of corresponding growing chains, i.e., dimers of polystyryllithium and tetramers of polybutadienyllithium, should, in an M_1–M_2 binary system, undergo at least partial modification, forming mixed associates. The magnitudes of K_{deag} which are characteristic of these independently of the stoichiometry and degree of association of the new aggregates must differ from the magnitudes which are typical for the original formations. However, it is not difficult to see that similar phenomena must also be expected when the degrees of association typical of the growing chains of each individual system corresponding to the monomers M_1 and M_2 coincide.

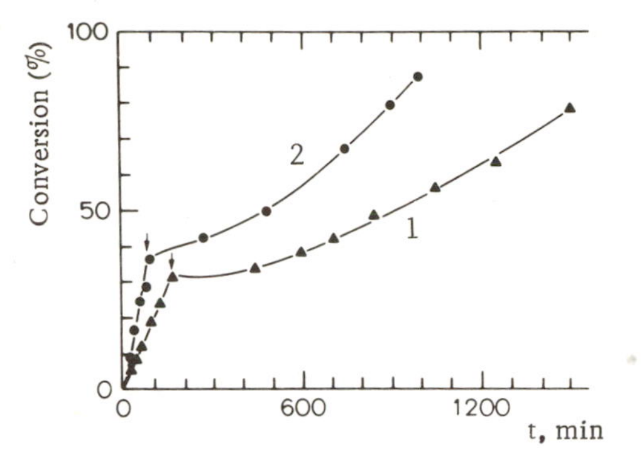

Fig. 3.2. The effect of n-butyllithium on polymerization in the 2,3-dimethylbutadiene (DMB)—oligo(2,3-dimethylbutadienyl)lithium (DMBLi)—hexane system [28]. Temperature of polymerization, 60°C. Concentration, moles/liter: DMB, 0.96 (1), 0.95 (2); ODMBLi, 0.0049 (1), 0.00129 (2); BuLi, 0.47 (1), 0.063 (2). Arrows show the point of introduction of BuLi.

Concerning the role of such effects on the kinetics of polymerization in real processes it is possible only to present a few assumptions which are based on the existence of the action of mixed associates on the homopolymerization kinetics. This may be seen in the disappearance of the induction period when a quantity of some "living" chains are introduced before the polymerization commences (sec-butyllithium—isoprene system [27]) or in a sharp reduction in the rate of polymerization when a low-molecular-weight initiator is introduced into a process initiated by "seeding" which has already commenced (an oligo-2,3-dimethylbutadienyllithium—2,3-dimethylbutadiene system [28], see Fig. 3.2). These facts accord with the following stability series of the aggregates discussed, at least when $x > p + q$:

$$(RLi)_x > pRLi \cdot qM_nLi > (M_nLi)_y.$$

Naturally, actual magnitudes of K_{deag} for mixed associates must be determined by both their nature and stoichiometry with which the different probabilities of the splitting off of each different type of monomeric form OLC from the associate can be connected.

Among binary systems $(M_nLi)_p$—$(M'_mLi)_q$ it seems expedient to differentiate cases corresponding to the conditions $p = q$ and $p \neq q$ based on comparable magnitudes of n and m (or simply sufficiently large magnitudes of these quantities). For the first of these an intermediate value of K_{deag} is probable as compared to the values which are typical for the initial homoaggregates. A situation which is possible for the second case is less obvious. Nevertheless, it is more probable that of all the values compared the lowest value of K_{deag} is that for mixed components.

3.4.2. Ion Pairs and Free Ions

The most obvious consequence of the transition from homopolymerization to an M_1–M_2 system in which conditions for the dissociation of ion pairs to free ions exist is the possibility of the buffering effect of one growing chain upon another. For practically complete suppression of dissociation a difference in the values of K_{diss} of the ion pairs of 2-3 orders of magnitude is required. This is approximately the case for K_{diss} for polystyryl- and polybutadienyl- (or polyisoprenyl-) lithium chains in THF. Here the order of these values is 10^{-7} and 10^{-10} liters/mole at 25°C. Thus from this it follows that the contribution of the free ions of the dienyl type in the copolymerization of dienes with styrene must be less in comparison with their contributions in the homopolymerization of the same dienes, especially in the region of increased styrene–diene ratios. Some reduction in the contribution of the free ions is permissible in circumstances in which the difference between the values of K_{diss} for growing chains corresponding to M_1 and M_2 monomers is not too great. Refraining from attempts to evaluate limits of possible changes in such systems, we emphasize that when studying the mechanism of ionic copolymerization these effects should not be ignored. In particular, it is perfectly possible that the variation in the ratio M_1–M_2, which effects the relative contributions to the growth of the ions and ion pairs in chains for which the value of K_{diss} is low, influences the corresponding parameters r_1 and r_2. In these circumstances the question of the physical meaning of these parameters is complex.

3.4.3. Complex Formation with External Reagents

The great sensitivity of many ionic copolymerization processes to the nature of the reaction medium is due in a greater extent to the result of changes in the coordination saturation of the growing chain counterions while the initial type of ion pairs are preserved than to the formation of free ions. This follows from the fact that the effects which are very similar in scale can be obtained by two paths. These are the transition from a nonpolar to a polar medium and the introduction into the polar medium of small quantities of modifying reagents. Especially obvious in this respect are the sharp changes in the values of r_1 and r_2 and the presence of small quantities of electron donors which are well known for anionic copolymerization.

Of particular interest for the present discussion are processes in which the modification of active sites, which results from complex formation with independent electron donors, produces effects which are specific for M_1–M_2 binary systems. These include phenomena which are connected with competing complex formation (with the participation of both monomers and an independent ED) and with the incomplete modification of the active sites. We will deal with the latter case, which consists essentially of the combination of features which are present in the initial and completely modified systems.

For this purpose we draw upon one of the most distinctive systems of the general type M_1–M_2–OLC–nonpolar medium, namely the copolymerization of styrene with butadiene for which (bearing in mind the nature of the remaining components) the concept of copolymerization itself is applied, essentially, only when the styrene–butadiene ratio is extremely high. When styrene is present in great excess a random copolymer may be obtained which contains comparable fractions of both monomers [2]. On the other hand an analogous result is also obtained when $M_1/M_2 = 1$, provided a strong electron donor is present with a concentration of the same order of magnitude as that of OLC [29]. The practically complete equalization of the selectivity of both types of growing chains (i.e., $\sim m_1Li \cdot D$ and $\sim m_2Li \cdot D$) with respect to the comonomers, which is caused by the transition of the counterion Li to Li·D, enables one to foresee the parallel formation of macromolecules which differ considerably when the concentration of donor is reduced and the transformations such as described by Eq. (3.1) only seldom take place. Among the macromolecules formed must be butadiene homopolymers, copolymers of styrene with butadiene (corresponding to chains with the first and second of the above counterions respectively), and hybrid products, i.e., block and multiblock copolymers which correspond to single and multiple exchanges [Eq. (3.1)].

The results obtained for the styrene–butadiene system using subcatalytic quantities of TMED (0.2 and 0.01 with respect to OLC) and by carrying out low-temperature polymerization to low conversion indicate that it is possible to create conditions which limit such exchanges [30]. In both these cases copolymer was found together with the formation of a certain fraction of polybutadiene which was free from styrene. However, "hybrid" multiblock products were determined for only the smallest TMED/OLC ratios (for more detail see Chapter 4). A qualitatively similar difference in the character of the final products (which is caused by changes in the number of "jumps" of the molecule of the electron donor from one growing chain to another) may also be ensured when the TMED/OLC ratio remains constant while varying either the temperature or the conversion. A change in the nature of the electron donor is yet another possibility.

Systematic study of the consequences of such a variation can yield extremely valuable information which can be used for the elucidation of a number of problems connected with the mechanism of these processes. Furthermore, we find here an unprecedented approach to the study of the effect of the distribution in the macromolecule of sequences which correspond to monomers M_1 and M_2 on the macroscopic properties of the corresponding polymer products.

3.4.4. Intramolecular Effects

In M_1–M_2 binary systems a feature of the intramolecular events of the type described by Eq. (3.4) may manifest itself in an increase in the number of variants of the active site due to the participation of parts of each unit of the growing chains in complex formation:

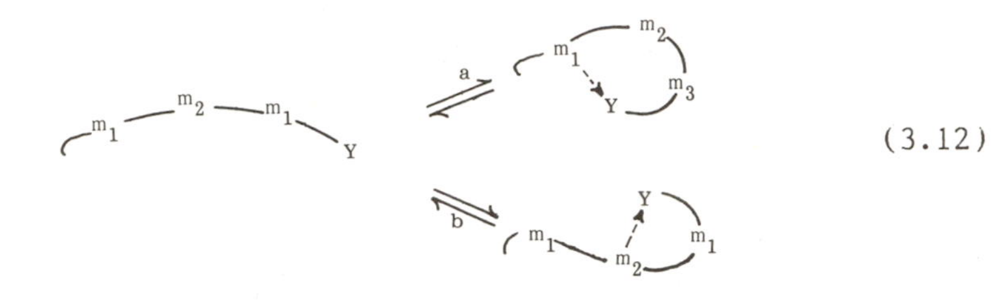

(3.12)

As far as we know there has not as yet been a research program directed especially to the study of this question, but it is not difficult to determine the possible consequences of the events described by Eq. (3.12), which are particularly real for the anionic copolymerization of polar molecules in nonpolar media.

In the first place, when the composition of the growing chains, which depends upon the original ratio M_1–M_2, is varied, the stability of the complexes formed when the counterion, Y, interacts with the polar group of units corresponding to monomers M_1 and M_2 may alter its significance as a factor determining which of the directions (a or b) is taken. Secondly, Eq. (3.12), being sensitive to the presence of polar molecules because of the probability of the reactions

$$\sim m \cdots Y \xrightarrow{M} \sim m — Y \cdot M \qquad (3.13)$$

must (because of the differences in the electron donor properties of M_1 and M_2), in its turn, depend upon the composition of the reaction mixture.

The most probable consequence for the final product of the latter transformation is the transition of the monomer from the coordination sphere of the counterion to the growing chain. Nevertheless, the reversibility of complex formation should not be excluded completely, especially if the geometries of the complexes which arise as a result of Eq. (2.13) and the direct interaction of the monomer with the linear form of the active sites are not assumed to be identical.

All these circumstances may to a greater or lesser degree complicate the relationship between the ratio M_1–M_2 and the structure of the copolymer, depending mainly upon the nature of the components of the system.

3.4.5. Isomerization

We now turn to processes which were cited as examples of multicenteredness, which result from the isomerization of the end unit of the growing chain (see Section 3.1.5). Considering the analogous effects in M_1–M_2 systems, two cases may be distinguished, namely the ability of only one type of end group to isomerize (for example, that of monomer M_1) and the possibility, in principle, of such events for both types of final link. For brevity we will deal only with the first case.

The first consequence of the transition from a "homo-system" to a binary system consists in the different correlation between isomeric final links of the chain with different penultimate links. That such a dependence might be very significant is shown, in particular, by data on the $RCH{=}CHCH_2Mt$ compounds in which the fraction of cis and trans isomers for a given Mt is determined, in a particular medium, by the nature of the substituent r (see Table 5.6). If in this case the active sites which contain end units of different geometrical structure differ from one another to a significant extent and in their reactivity (which can occur, see Section 5.2.2), then the nature of the penultimate group acquires an independent significance as a factor influencing the behavior of active sites of the type $\sim m_1 m_1 Y$ and $\sim m_2 m_1 Y$. This circumstance may affect the structure of the units of the macromolecule, and possibly the complete structure of the copolymer.

It must be borne in mind that for $\sim m_1 Y$ chains four kinetically distinct growth reactions are possible. These are conveniently given, using, for example, butadiene for M_1 (counterions omitted):

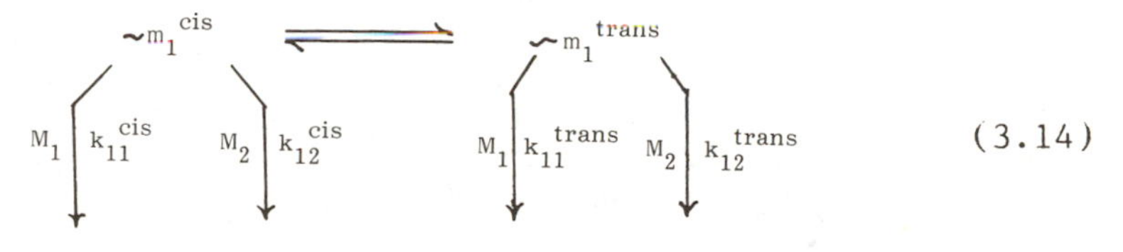

(3.14)

Since the ratio cis–trans in the chain depends in its turn upon the nature of the penultimate group it follows that the total fractions of cis and trans conformations for all $\sim m_i m_i Y$ and $\sim m_2 m_i Y$ chains must be some function of M_1–M_2.

Such a situation is comparatively close to that in which the penultimate link has a direct influence upon the reactivity of the active sites which are participating in the copolymerization but are free from changes like those considered in the present subsection. We emphasize that we are discussing the possibility of a different kind of effect which may have a considerably greater significance in cases such as the copolymerization of butadiene (or isoprene) with styrene under the action of OLC.

More precise conclusions require data on the difference in the ratio of the cis and trans conformations of the end units of $M_n(C_4H_6)_2Li$ and $M_nCH_2CH(Ph)C_4H_6Li$ chains which are as yet unavailable. Nevertheless, data describing the hydrolysates of such chains (see Section 5.2.2) indicate the possibility of corresponding structural effects. This circumstance somewhat increases confidence that in copolymerization processes of the kind we are considering the isomerization events are among the real factors which determine the values of r_1 and r_2 found experimentally. We now turn to the considerable divergence in the values which have been obtained by various authors for systems of the general type OLC–butadiene (M_1)–styrene–nonpolar solvent. The values of r_1 (from 3 to 23.7) and r_2 (from 0.004 to 0.1) account for the comparatively narrow range of temperature (from 20° to 50°C) (for bibliography see [2]). Without going into details (these can be found in [32]), we note that variations in the temperature and the solvent (both aliphatic and aromatic hydrocarbon solvents are used) may play some part in these differences. However, these changes too may be significant not simply of themselves but as variables which, in their turn, influence the correlation between different types of coexisting active sites and/or the contributions of active sites which differ in their degree of association and the geometry of the end unit. From our point of view the very fact of additional multicenteredness is of special importance because it is sensitive to the ratio M_1–M_2 independently of whether changes in the temperature or the nature of the solvent affects the degree and nature of such sensitivity.

If in the systems considered the possible effects are evaluated on the basis of scheme (3.13), then it is necessary to take into account certain other circumstances as well as those mentioned above. One of these is the dependence on the nature of the substituent which is attached to the end unit: not only on the geometrical structure of this unit but also on the rate of its cis–trans isomerization. This follows from the data relating to compounds n-$BuCH_2C(CH_3)$=$CHCH_2Li$ and t-$BuCH_2C(CH_3)$=$CHCH_2Li$ (see Section 5.2.2).

Differences in the reactivity of the cis and trans conformations of dienyllithium end groups whose penultimate links are different are also not excluded.

Another essential point concerns the competition between growth and isomerization, a consequence of which for the structure of diene polymers has been the subject of a number of studies; see Erusalimskii et al. [5]. When applied to copolymerization, the connections between this competition and the structure of the macromolecules formed become important and in the case under discussion here may be determined not only by the ratio M_1/M_2 but also by their absolute concentrations. Such a feature must also be considered to be one of the causes of the divergence between the magnitudes r_1 and r_2 obtained by different authors for precisely the same systems.

We have dealt in more detail with the copolymerization of butadiene with styrene under the action of OLC because it is possible to examine this process by using data characterizing both the relative fractions of the geometric isomers in the end units of living chains and their dependence upon at least several factors. Analogous considerations can justifiably be applied to other copolymerization processes. For example, for the cationic diene—olefin systems, scheme (3.13) in principle also covers these processes. We note that the detailed structural analysis of isobutylene copolymers with butadiene and pentadiene which have been carried out recently give some grounds for regarding our hypothesis as being justified (see Section 5.2.1). It is natural to group cationic copolymerization with the processes under dicussion here even if it includes only one of the monomers which are capable of bringing about the isomerization of the final links. The results of studies on the homopolymerization of such monomers show a considerable dependence of the fraction of isomerized links in the macromolecules on the structure of the monomers. This leads to the assumption that changes are possible in the relative roles of the events of isomerization which are typical of the final links m_1 during transition of the active site $\sim m_1m_1Y$ to $\sim m_2m_1Y$. Thus, regarding processes in which diene monomers participate, the situation can be expressed by the functional relation $m_1/m_2 = f([M_1]\cdot[M_2])$, which differs from the usual condition $m_1m_2 = f([M_1]/[M_2])$. The coexistence of two structural forms of the end group whose reactivity depends upon the nature of the penultimate group makes it necessary to consider four varieties of active site of the general type $\sim m_1Y$. It follows from this that there are up to ten separate growth events in such systems even if for active sites with final links m_2 such varieties do not exist. For the sake of simplicity the possibility that reactivities of the active sites $\sim m_2m_2Y$ and $\sim m_1m_2Y$ are different is not considered. The complexity of the connection between the composition of the reaction mixture and the structure of the copolymer produced by these circumstances is intensified by the existence of Eq. (3.13) and the aforementioned competition between isomerization and growth.

Thus, it is possible to distinguish variations which affect only (or mainly) the physical meaning of the copolymerization constant (see Sections 3.4.3 and 3.4.4) among the varieties of multicenteredness examined in this section and cases which cause the breakdown of the usual interrelationships which are used in regulating the composition of the copolymer. The example of the latter type, which we discussed in Section 3.4.5, is not unique in this respect. The influence of the monomer concentration and not only as in the ratio M_1–M_2 can be seen in processes in which the reversible transformations of the active site compete with the growth reactions. This refers in particular to systems which contain active sites in the form of associated and disassociated forms, when these and others participate in growth reactions and when there are differences in their selectivity with respect to the comonomers. Of course, in observing this situation another circumstance also becomes important, namely the total concentration of the growing chains themselves, which determines the relative amounts of the various types of coexisting active sites.

Besides the various causes which lead to the appearance of multicenteredness as a consequence of the inherent nature of ionic reagents (see Table 3.1), the effects which are produced by the action of uncontrolled admixtures or of the side products of polymerization deserve mention. These are given by the equations

$$\mathrm{AS + X \rightleftarrows As{\cdot}X} \tag{3.14}$$

$$\mathrm{As{\cdot}X + (AS)' \rightleftarrows AS + (AS)'{\cdot}X} \tag{3.15}$$

in which X is an electron donor or an electron acceptor and AS is the growing chain or an anionic initiator. Events given by Eqs. (3.14) and (3.15) correspond formally to the situations in which the presence of reagents, X, in the sphere of action is already known. It is possible, for example, to assume that the phenomena discovered when D–$M_nLi \ll 1$ (see Sections 3.3 and 3.4.3) model the masked multicenteredness which results from the presence of admixtures. Small amounts of active reagents, X, are capable of leading to a large number of different active sites which react in parallel.

On the other hand, the formation of side products which are capable of carrying out the function of X agents need by no means be the cause of multicenteredness. In particular, the well-known splitting off of the ROMt alkoxides in the polymerization of acrylates and alkylacrylates under the action of R'Mt[6] compounds can lead (depending upon the nature of the actual reagents) both to ratios X–$AS \gg 1$ and also to the reduced concentration of the metal alkoxide. Naturally, in the first case active sites of the form $M_nMt{\cdot}ROMt$ may be practically the only ones present.

REFERENCES

1. E. Yu. Melenevskaya, "The structure and reactivity of organolithium active centers in the polymerization of butadiene and styrene," Dissertation, Leningrad (1980).
2. R. Ohlinger and F. Bandermann, "Kinetics of the propagation reaction of butadiene—styrene copolymerization with organolithium compounds," Makromol. Chem., 181, 1935-1947 (1980).
3. B. L. Erusalimskii and A. V. Novoselova, "Mechanismus der durch Lithiuminitiatoren angeregten Polymerization von Acrylnitril," Faserforsch. Textiltech., 26, 293-300 (1975).
4. B. L. Erusalimskii, "Das Mehrzentrenproblem in ionischen Polymerizationsprozessen," Acta Polym., 34, 667-673 (1983).
5. B. L. Erusalimskii, I. G. Krasnosel'skaya, V. N. Krasulina, A. V. Novoselova, and E. V. Zhashtsherinskii, "Mechanisms of the side reactions in the anionic systems of low efficiency of initiation," Eur. Polym. J., 6, 1391-1396 (1970).
6. L. Reich and A. Schindler, Polymerization by Organometallic Compounds, Interscience, New York (1966).
7. B. L. Erusalimskii, E. Yu. Melenevskaya, and V. N. Zgonnik, "Modern views on the stereospecificity of anionic active centers," Acta Polym., 32, 183-195 (1981).
8. B. L. Erusalimskii, "Uber einige Besonderheiten der anionischen Polymerization polarer Monomerer," Plaste Kautsch., 15, 788-792 (1968).
9. B. L. Erusalimskii and I. G. Krasnoselskaya, "Zum Mechanismus der Aktivierung der anionischen Polymerization polarer Monomerer durch Lewis-Basen," Makromol. Chem., 123, 80-90 (1969).
10. I. G. Krasnoselskaya, E. S. Gankina, B. G. Belen'kii, and B. L. Erusalimskii, "The polymerization of acrylonitrile under the action of polystyryllithium," Vysokomol. Soedin., A19, 999-1003 (1977).
11. W. Berger, Ch. Steinbrecher, H. J. Adler, I. G. Krasnosel'skaya (Krasnoselskaya), G. V. Lyubimova, and B. L. Erusalimskii, "Die anionisch initiierte Blockcopolymerisation von Vinylpyridinen mit Acrylnitril," Acta Polym., 34, 396-398 (1983).
12. S. Slomkovskii and S. Sosnovskii, "The kinetics of macrocycle formation in the polymerization with ε-caprolactams," V. Int. Mikrosymp. Fortschr. Ionenpolym., Prague (1982), prepr. 70.
13. V. V. Shamanin, E. Yu. Melenevskaya, and V. N. Zgonnik, "Influence of the concentration of growing chains on the polymerization rate and the microstructure of the polymer formed in the polybutadienyllithium—butadiene—aliphatic hydrocarbon system," Acta Polym., 33, 175-181 (1982).
14. B. J. Schmitt and G. V. Schulz, "Uber zwei Formen des Initiators Na-Naphtalin und die Bestimmung der 'lebenden' Kettenenden in der anionischen Polymerisation," Makromol. Chem., 121, 184-204 (1969).
15. M. Sawamoto and H. Higashimura, "Stopped-flow study of the cationic polymerization of p-methoxystyrene. Evidence for the multiplicity of the propagation species," Macromolecules, 11, 501-504 (1978).

16. R. T. M. Huang and J. F. Westlank, "Molecular-weight distribution in radiation-induced polymerization. III. γ-Ray induced polymerization of styrene at low temperatures," J. Polym. Sci., A1, 8, 49-61 (1970).
17. M. Sawamoto, T. Masuda, and T. Higashimura, "Cationic polymerization of styrene by protic acids and their derivatives. 2. Two propagating species in the polymerization by CF_3SO_3H," Makromol. Chem., 177, 2995-3007 (1976).
18. B. G. Belenkii and E. S. Gankina, "Thin-layer chromatography of polymers," J. Chromatogr., Chromatogr. Rev., 21, 13-90 (1977).
19. G. Glöckner, Polymercharacterisierung durch Flüssigkeitschromatographie, VEB Deutscher Verlag d. Wiss., Berlin (1980).
20. V. V. Nesterov, V. D. Krasikov, V. N. Zgonnik, E. Yu. Melenevskaya, I. V. Kosheleva, and B. G. Belen'kii, "Studies of block copolymers based on polystyrene and polybutadiene using gel permeation chromatography and ozonolysis," Vysokomol. Soedin., A25, 2568-2574 (1983).
21. B. L. Erusalimskii, "Unresolved problems in ionic polymerization," in: Advances in Ionic Polymerization [in Russian], Z. Jedlinsky (editor), Warsaw (1975), pp. 9-23.
22. A. F. Halasa, D. N. Schulz, D. P. Tate, and V. D. Mochel, "Organolithium catalysis of olefin and diene polymerization," Adv. Organomet. Chem., 18, 55-97 (1980).
23. A. Davidyan (Davidjan), N. I. Nikolaev, V. N. Zgonnik (Sgonnik), B. G. Belen'kii (Belenkii), V. V. Nesterov, V. D. Krasikov (Krasikow), and B. L. Erusalimskii (Erussalimsky), "Subkatalytische Effekte im System Isopren—Oligoisoprenyllithium—N,N,N'N'-Tetramethylethylendiamin. 2. Umsatzabhangigkeiten der Molekulargewichtsverteilung und Mikrostruktur der Polymere," Makromol. Chem., 179, 2155-2160 (1978).
24. W. Gebert, J. Hinz, and H. Sinn, "Umlagerungen bei der durch Lithiumbutyl initiierten Polyreaktion der Diene Isopren und Butadien," Makromol. Chem., 144, 97-115 (1971).
25. O. F. Olaj, H. Rehmann, and J. W. Breitenbach, "Der Gegenion-Effekt beir der durch Perchlorsaüre initiierten Polymerization von Styrol," Monatsh. Chem., 110, 1029-1043 (1979).
26. V. V. Shamanin, "Molecular-weight distribution analysis of the mechanism of addition polymerization," Acta Polym., 31, 353-356 (1980).
27. S. Bywater and J. E. L. Roovers, "Polymerisation of isoprene by sec-butyllithium in hexane," Macromolecules, 8, 251-254 (1975).
28. N. V. Smirnova (Smirnowa), V. N. Zgonnik (Sgonnik), K. K. Kalnin'sh, and B. L. Erusalimskii (Erussalimsky), "Über einige Besonderheiten der Polymerisation von 2,3-dimethylbutadien in durch lithiumorganische Verbindungen initiierten Prozessen," Makromol. Chem., 178, 773-790 (1977).
29. V. N. Zgonnik, N. I. Nikolaev, E. Yu. Shadrina, and L. V. Nikonova, "Copolymerization of butadiene with styrene on butyllithium complexes with tetramethylethylenediamine and 2,3-dimethoxybutane," Vysokomol. Soedin., B15, 684-686 (1973).

30. B. L. Erusalimskii (Erussalimsky), B. G. Belen'kii (Belenkii), A. A. Davidyan (Davidjan), V. D. Krasikov, V. V. Nesterov, N. I. Nikolaev, V. N. Zgonnik (Sgonnik), and V. M. Sergutin, "Subcatalytic effects in the anionic polymerization processes," 27th Int. Symp. on Macromolecules, Strasburg (1981), Vol. 1, pp. 182-185.
31. M. Schlosser and J. Hartmann, "2-Alkenyl anions and their surprising endo preference. Facile and extreme stereocontrol over carbon-carbon linking reactions with allyl-type organometallics," J. Am. Chem. Soc., 98, 4674-4676 (1976).
32. R. Ohlinger, "Kinetische Untersuchungen der mit Lithium-organischen Wachstumskatalysatoren initiierten Copolymerisation von Butadien und Styrol mit dem der Darstellung von statistischen Copolymeren mit bestimmter Butadien-Styrol-Zusammensetzung," Dissertation, Hamburg (1974).

Chapter 4

The Reactivity of Active Sites and Monomers in Homogeneous Ionic Systems

The problem of the reactivity of initiators of growing chains, and in particular of monomers, is treated from one or another point of view in all the general works and in a considerable number of original papers on ionic polymerization. In this chapter an attempt is made to determine the progress which has been made in modern concepts in this field.

The usual criteria used to evaluate the relative activity of monomers in growing chains are the reverse copolymerization constants (1/r) and the propagation rate constants (k_p). Apart from a correct choice of limits within which the values of the above required quantities may be obtained, the researcher must determine the connection between the position in the activity series of each component studied and its physicochemical characteristics. The problem has only been solved satisfactorily for comparatively small groups of monomers. When the number of monomers is increased any attempt to find similar correlations is usually impeded by a number of factors. The most important of these are (1) the inapplicability to this system of the usual methods used to calculate the copolymerization constant, i.e., the incorrectness of the values obtained for 1/r (for example, as a consequence of the multicenteredness; see Section 3.4), and (2) the fact that the reaction proceeds in two stages:

$$M_n^*Y + M \underset{}{\overset{K_c}{\rightleftarrows}} M_n^*Y \cdot M \underset{}{\overset{k_{in}}{\rightleftarrows}} M_{n+1}^*Y \qquad (4.1)$$

Here K_c is the constant of complex formation and k_{in} is the rate constant for the introduction of the complexed monomers into the growing chain M_n^*Y. In such a mechanism the necessary condition for a rigorous evaluation of the reactivity of the reagents studied no longer holds, i.e., the reference reaction is no longer single-staged.

The above circumstances may be considered to be both independent and interconnected. This is determined by the actual nature of the system studied. Dealing with the two-staged mechanism [Eq. (4.1)], we note that this affects the physical meaning of the values of $1/r$ and k_p used as criteria of reactivity. Nevertheless, the overall constants of the general type $K_c \cdot k_{in}$ may be sometimes used as provisional indices of the reactivity of growing chains, because generally a more rigorous approach is not possible. So far the question has had to be limited to that of the advisability of evaluating the relative activity of the coreagents at each stage of reaction (4.1); only isolated attempts have been made to evaluate this. One can only assume that some of the deviations from the known laws could be explained if detailed information on the above stages were available. This also applies to the fact that the study of the reactivity of active sites and monomers does not lead to the discovery of strict laws.

Many of the complications which arise in the interpretation of effects relating to monomers are the result of the distinctive features of the ionic active sites. We will therefore deal first with those questions which concern the relative activity of growing ionic chains.

4.1. THE RELATIVE REACTIVITY OF GROWING CHAINS

In order to cover a sufficiently wide range of ionic systems, apart from the absolute constants (k_p), it is necessary to appeal to both total and relative magnitudes which, if carefully handled, become quite valuable. Information on the values of k_p which are known for ionic polymerization is insufficient for an attempt to discuss the problem covered by the systems described briefly in Chapter 1. Such a situation, taking into account the discrepancies between data from different authors, considerably complicates the most important aspects of the problem. We have in mind the search for reactivity indices, a task which goes well beyond merely establishing a connection between the kinetic activity of the active sites and their physicochemical characteristics.

In principle any feature of the electronic structure of the molecule which can be correlated with the reactivity of a certain group of compounds in a standard reaction may be taken as a reactivity index (RI). The possibility of finding such values as indices of the donor and acceptor properties of a number of compounds is clearly shown for a large number of unsaturated and heterocyclic monomers. In the case of active sites in ionic polymerization it is difficult even when the necessary quantitative indices, i.e., the value of k_p for the growing chains, is present. This is particularly clearly illustrated by a comparison of ion pairs and free ions. On the basis of the data calculated for the various compounds available at present, it is difficult to regard the sign of the charge of the terminal atoms of the active site as significant. Within the

TABLE 4.1. Principles of the Variation of M_n^*Y Compounds

Case	Components of the system	
	variables	constants
1	M_n^*	Y, reactive medium
2	Y	M_n^*, reactive medium
3	Reactive medium	M_n^*, Y
4	Reagent added (catalytic quantities)	M_n^*, Y, reactive medium

limits of the ion pairs themselves it is still more difficult to find a use for the charge characteristics as direct reactivity indices. Nevertheless, systematic research in this field sometimes leads to encouraging results (see Section 2.3).

The preliminary information necessary for the study of this problem generally amounts to a knowledge of the kinetic features of the interaction of the standard monomer with different growing chains. The multitude of forms of growing chains possible for each different type of ion pair (see Chapter 3) leads to a great variety of possible solutions for such problems. The possibility of using other principles to produce different varieties of growing chains which we have discussed earlier for examples of anionic systems [1] must be borne in mind. These same principles may be applicable to any ionic system as may be seen in Table 4.1. Here the symbol M_n^*Y is taken as a general symbol for the active site in which M_n^* is strictly the growing chain proper with a charge of either sign and Y is the component with a charge of the opposite sign, i.e., in the case of true ion pairs it is a counterion.

The differentiation of these alternatives from the point of view of preference is not necessary. It needs only to be asserted that data on the consequences of modifying the nature of the growing chains (case 1) yields much less information than the data on the influence of the state and the environment of active sites of a given nature (cases 2-4) on their reactivity. The majority of the known reactivity series for these or for other growing chains is constructed using data for the initial compound M_n^*Y with a constant component M_n^*. We will initially examine the results of just this type.

4.1.1. The Role of the State and Environment of Growing M_n^*Y Chains

The interpretation of the connection between the reactivity of the M_n^*Y compounds and their state is comparatively simple providing the changes are caused by the transition from the weakly polarized initial agents to free M_n^* ions. Also a relatively clear picture is obtained regarding those states which are intermediate between the above limiting forms. Nevertheless, concepts in this field cannot be regarded as exhaustive as is found in attempts at generalizations which go bevond the confines of the small number of related systems.

We emphasize that the individual examination of such parameters as the state and the environment of the M_n^*Y compounds is no more than an artificial device which is sometimes convenient in the formal sense. The usual method for describing the above changes in the form of the system of equations

$$\underset{\text{polar compound}}{M_n^*Y} \rightleftarrows \underset{\text{contact ion pair}}{M_n^*, Y} \rightleftarrows \underset{\text{separated ion pair}}{M_n^* \,||\, Y} \rightleftarrows \underset{\text{free ions}}{M_n^* + Y} \qquad (4.2)$$

reflects only the changes in the character of the connection between the oppositely charged components of the M_n^*Y compound without taking into consideration the modification of the components themselves. The displacements of the particular equations covered by scheme (4.2) when the nature of the reactive medium changes are caused by donor—acceptor interactions between the solvent molecules and at least one of these components.

The modification of the Y components is the most typical result of an interaction of such a type. With all the great variety of particular cases, such a modification can lead to the introduction of additional ligands into the coordination sphere of the Y component or to the substitution of some ligands by others. Either of these events converts component Y into Y', affecting the stability of the active bond, its spatial accessibility, and consequently the difference in the reactivity of the M_n^*Y and M_n^*Y' compounds. From the formal point of view such a modification corresponds to case 3 (Table 4.1). However, the consequences of varying the reaction medium often coincide with the changes which are caused by the modification of the Y component within the same medium (case 4) or even the substitution of one of the Y components with another (case 2). In spite of this the differentiation of cases 2-4 is sensible since the totality of the final effects is not found in all ionic systems. Furthermore, fundamentally different causes may underlie the external similarity of the results obtained when the parameters in the above cases are varied.

In approaching the examination of the consequences which result from each variation of Y in the active site M_n^*Y while keeping the M_n^* constant, we emphasize that the interpretation of the corresponding effects differs greatly in their degree of complexity. Case 4 possesses certain advantages in this respect, particularly when applied to nonpolar media. On the other hand, when an attempt is made to extract the required information from the large number of facts associated with the changes in reactivity of the active site caused by such a modification, information suitable for our purposes proves to be quite limited. Therefore in discussing this further we will not specially classify material according to the variations given in Table 4.1. The series which corresponds to the transition from relatively simple to more complex materials is more suited to our purpose. From this point of view, it is more convenient to begin with the alkali metal derivatives and primarily with the organolithium compounds. Their natural features and behavior in polymerization have been widely discussed in the literature for many years and are still receiving considerable attention today (see, for example, [2-10]). The view which is widely held by experimentalists concerning the mainly covalent character of the C—lithium bond has not been generally substantiated by theoretical work. This shortcoming has been particularly noted in the quantum-chemical studies of Streitwieser et al. [4]. Diagrams of the electron density distribution in monomeric forms of methyl-, ethyl-, and vinyllithium obtained by a nonempirical method with a wide variation of the basis led the authors to conclude that the C—lithium bond contained no covalent compounds. In particular, for methyllithium the total charge on the alkyl group was found to be greater than 0.8 e when the electron density is a minimum in the internuclear space 0.038 e (in atomic units^{-3}): the latter value is compared with that for the covalent C—H bond which has the value 0.275 e (atomic units^{-3}).

As has been emphasized in [4], the opposite conclusions as to the character of this bond are based on the charge indices of the atoms which are calculated according to Mulliken. This method for analyzing the electron-density distribution is not free from defects, the chief of which is as follows. The overlapping population density is taken as a measure of the covalence of the A—B bond. However, in fact the overlapping of the orbitals of atoms A and B is actually taken over the whole volume and not just in the internuclear space. The population density of any atomic orbital which is centered on atom A is formally assigned to that atom irrespective of the dimensions of the orbital. Hence, if there are diffuse orbitals, even if only on one of the atoms, the covalence of the bond may be markedly overestimated. At the same time the electron density, which is concentrated in the volume of the atom, also turns out to be overestimated.

Precisely such a situation is found in methyllithium in which the 2s and 2p orbitals cover practically the whole molecule. This led Streitwieser et al. [4] to conclude that the use of the Mulliken index in these circumstances is incorrect.

As a counterweight to the above, Graham et al. [7], who made a detailed analysis of [3], regarded the criterion that the minimum electron density falls in the internuclear space as insufficient ground for classifying the bond as either ionic or covalent. As a basis for this are adduced similar values calculated for ionic LiF and covalent Li–Li compounds (these are 0.080 and 0.013 e (atomic units^{-3}) respectively). Nevertheless, it is very significant that in drawing their conclusions as to the nature of the C–Li bond, Streitwieser et al. used, amongst other parameters, a value for the electron density found by integrating over the whole "atomic" volume. This value corresponded to the charge on the lithium atom in methyllithium of about +0.84 e. According to the calculations of Graham et al., the charge on the lithium atom in CH_3Li lies in the range +0.55 to +0.66 e.

A comparison of these results with the experimentally obtained characteristics is difficult since the latter relate either to organolithium associates or to their crystalline states. Moreover, in the usual conditions in which polymerization is carried out, the models are predominantly monomeric forms of M_nLi and for these nonempirical calculations are completely absent.

Theoretical evaluations of the influence of aggregation on the nature of the C–lithium bond have been carried out by, for example, Clark et al. [6] who chose the spin–spin splitting constant $J_{C,Li}$ as a parameter for comparison. The calculation of this value for the monomer, dimer, and tetramer forms of methyllithium (using the INDO method) led the authors to assume that the ionic nature of the bond increased on transition from the monomeric to the associated form. We emphasize that any extrapolation of the results relating to the simplest lithium alkyls to their alkene or benzyl analogs (i.e., to active sites of the diene or styrene series) is scarcely possible. Even the characteristics of the closest homologs of lithium alkyl derivatives sometimes exhibit features which are quite unforeseen. We give as an example the data of Graham et al. [8] which allows the values of $r_{C\text{-}Li}$ in CH_3Li and C_2H_5Li compounds and their symmetric dimers to be compared. As has been shown (using the PRDDO method), in the monomeric forms this value is greater for ethyllithium but in dimers the reverse is true.

Turning to the main topic of our discussion we will deal first with the results which are associated with such a modification of the active sites which does not affect the nature of the growing chain itself. This concerns any variation in the coordination sphere of the counterion, including changes in the degree of association of organolithium compounds. Thus, the monomeric forms of growing chains, the complexes which they form with electron donors

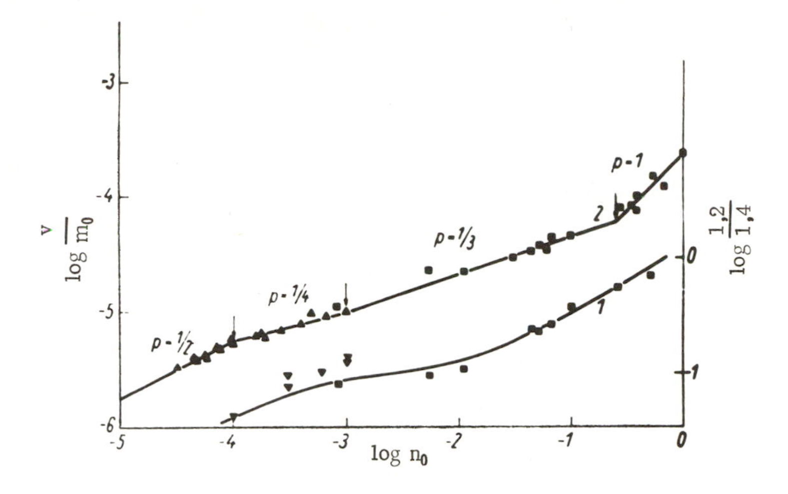

Fig. 4.1. The dependence of (1) the microstructure of polybutadiene and (2) the reduced initial growth rate on the concentration of the living chains (n_0) in polymerization in nonpolar media [11]. ■) Isooctane, 25°C; ▲) hexane, 20°C; ▼) heptane, −10°C.

and complexes of $(M_nLi)_p \cdot xD$ associates, where x ranges from 1 to p, come into consideration.

4.1.1.1. Organolithium Growing Chains and Their Associates

The generally accepted concept that the reactivity of the monomeric forms of M_nLi is greater than that of their associates has not for some time been subjected to any qualitative assessment. Now such a possibility arises for the polymerization of butadiene which is based on data for changes in the growth reaction rates ($V_p/[M]$) over a wide range of concentrations and the corresponding structural characteristics of polybutadiene; see Fig. 4.1. The satisfactory agreement between the experimental and theoretical data shown in the figure was obtained from a mathematical treatment of scheme (4.3) [11] in which K_1 and K_2 are the constants for the particular equilibria, and k_T, k_D, and k_M are respectively the rate constants for the growth on tetramer, dimer, and monomer forms of polybutadienyllithium (BuLi):

$$
\begin{array}{ccccc}
(M_nLi)_4 & \underset{}{\overset{K_1}{\rightleftharpoons}} & 2\,(M_nLi)_2 & \underset{}{\overset{K_2}{\rightleftharpoons}} & 4\,M_nLi \\
M \downarrow k_T & & M \downarrow k_D & & M \downarrow k_M \\
 & & \text{growth reactions} & &
\end{array}
\qquad (4.3)
$$

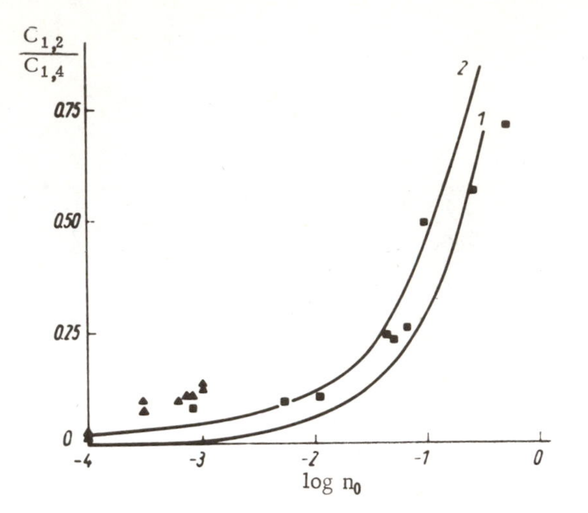

Fig. 2. Calculated curves for the dependence of the microstructure of polybutadiene on the concentration of living chains formed, based on the assumption that 1,2 units are formed, due to the tetrameric form (1) and due to the dimeric forms (2) of active sites [11].

According to Fig. 4.1a, the order of the reaction (n) for growing chains in different concentration ranges is given by the following magnitudes:

Total concentration PBuLi, moles/liter	$1 \cdot 10^{-5}$ to $5 \cdot 10^{-5}$	$1 \cdot 10^{-3}$ to $5 \cdot 10^{-2}$	$5 \cdot 10^{-2}$ to $5 \cdot 10^{-1}$
n	0.45	0.27	0.57

An analysis carried out taking into account the kinetic data and the concentration path of the microstructure of the polymer enables the values of k_1, $k_2^{0.5}$, k_M, and k_T to be obtained; these were used to find the value of k_D. The values calculated in this way are given below (data for 20°C, growth rate constants liter·mole^{-1}·sec^{-1}).

$k_2^{0.5} k_M$	k_D	k_T	K
$8 \cdot 10^4$ liter$^{0.5}$/mole$^{0.5}$sec	$2 \cdot 10^{-3}$	$4.5 \cdot 10^{-4}$	$2 \cdot 10^{-4}$ moles/liter

The correctness of the scheme (4.3) and the values obtained is shown by the fact that the conversion curves calculated for the concentration range of the active sites from $8 \cdot 10^{-4}$ to 0.5 moles/liter agree well with the experimental data (Fig. 4.2) and the theoretical curves showing the dependence of the microstructure of the polymer upon the concentration of living chains (Fig. 4.3).

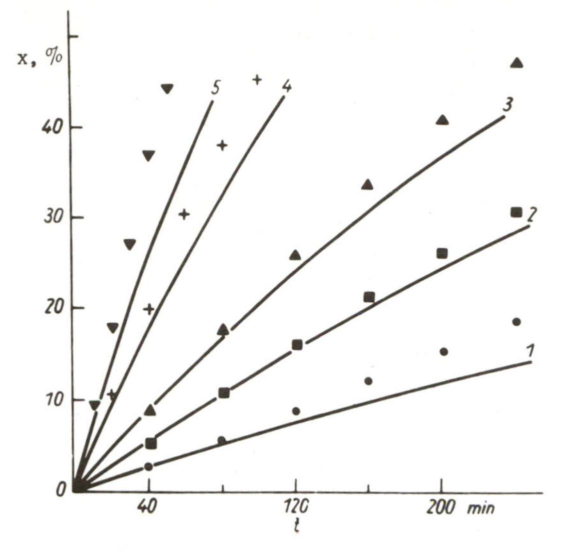

Fig. 4.3. The kinetics of polymerization of butadiene under the action of oligobutadienyllithium in isooctane at 25°C [11]. Initiator concentration, moles/liter: •) $8.3 \cdot 10^{-4}$, ■) $1.1 \cdot 10^{-2}$, ▲) $4.3 \cdot 10^{-2}$, +) $2.5 \cdot 10^{-1}$, ▼) $5.15 \cdot 10^{-1}$. Monomer concentrations for the same points are 1.75, 1.54, 1.69, 1.89, and 3.72 moles/liter, respectively. The curves are calculated using the constants from [11].

The data obtained by Zachmann et al. [12] showing concentration as a function of degree of association for polyisoprenyllithium in a hydrocarbon medium were used for the evaluation of k_M. This has led to the conclusion that the magnitude of K_2 for PBuLi cannot exceed 10^{-6} moles/liter. When this is so the value of k_M is 0.8 liters·mole^{-1}·sec^{-1} and this is its lower limit.

The most recent work on the association of "living" chains is that of Zachmann et al. [12] which deals with the heptane solutions of several samples of polyisoprenyllithium (PILi) and their corresponding deactivated polymers. The degree of association of (PILi) was carried out using a viscometric method. This was done by comparing the true molecular mass of the "living" chain $\bar{M}_{Li}$ (which was found independently) with a range of values of $\bar{M}$ taken for a number of polymers prepared at a known concentration. The degree of association was evaluated using the ratio $\bar{M}/\bar{M}_{Li}$ at the points of intersection of the curves (see Fig. 4.4).

Drawing upon low-angle-scattering data for the same solutions and results of the older kinetic studies, and taking the "star-shaped" character of the $(M_nLi)_m$ associates into consideration, the

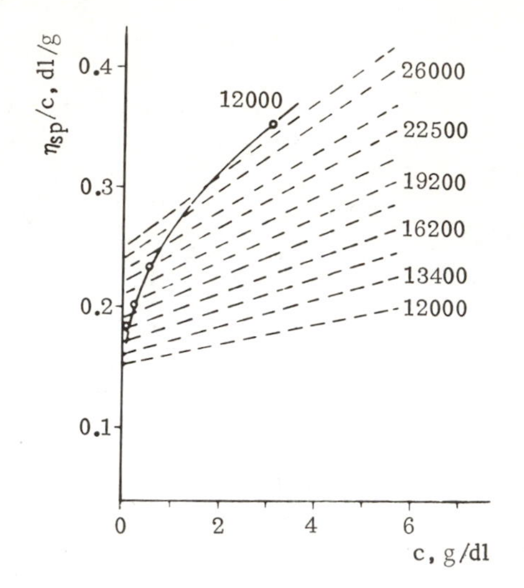

Fig. 4.4. Intrinsic viscosity of living chains (———) from data for the corresponding deactivated isoprene polymers (– – –) [12]. MM characteristics of polyisoprene are given on the curves.

authors take an average value of ≈3 for the degree of association at a concentration of 10^{-3} moles/liter for PILi. Extrapolation to a concentration of 10^{-2} moles/liter leads to a degree of dissociation for PILi of 4.

The correction which must be brought in when calculating the molecular mass of star-like polymers from viscometric data has been given by Bywater [13] for "three-rayed" macromolecules as 15%.

4.1.1.2. Complexes of Organolithium Active Sites in the Polymerization of Nonpolar Monomers

We turn now to the change in the reactivity of the monomeric forms M_nLi when various electron donors are introduced into the coordination sphere of the lithium atom. The use of data on the effect of small amounts of electron donor on the kinetics of the polymerization of nonpolar monomers in nonpolar media enables one to conclude that the transition M_nLi to $M_nLi \cdot D$, which is not accompanied by the formation of separated ion pairs, reduces the reactivity of the growing chains. The total kinetic effect which results from the presence of an electron donor may be a beneficial consequence of the increase in the number of active sites which are distinguished by increased reactivity as compared with the initial associates. The relative roles of these two factors which act in opposite directions depends upon the state of equilibria such as Eq. (4.3), i.e., for a given solvent and temperature, upon both the

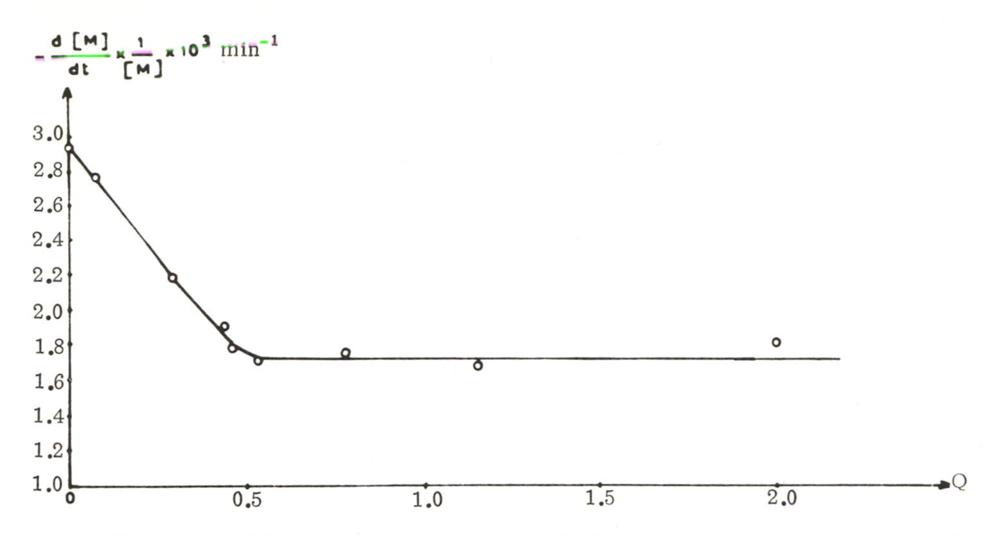

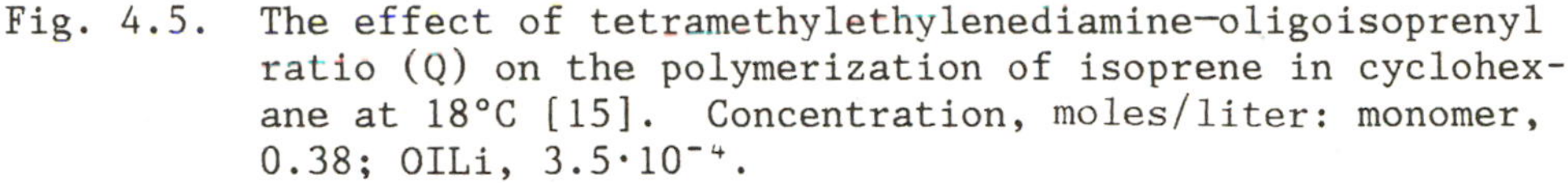
Fig. 4.5. The effect of tetramethylethylenediamine–oligoisoprenyl ratio (Q) on the polymerization of isoprene in cyclohexane at 18°C [15]. Concentration, moles/liter: monomer, 0.38; OILi, $3.5 \cdot 10^{-4}$.

temperature and the nature of M, and upon the concentration of $(M_nLi)_m$. In the case of the most stable associates of growing chains, of which polybutadienyllithium is an example, all that is known is the fact that the rate of polymerization increases under the action of ED. For the less stable associates the appearance of a negative overall kinetic effect when the concentration is sufficiently low is possible. This follows from the reduction in the rate of polymerization of 2,3-dimethylbutadiene in the presence of TMED with poly-2,3-dimethylbutadienyllithium at a concentration of $3.5 \cdot 10^{-4}$ moles/liter [14], and from the dependence of the nature of the change of rate of polymerization in the polyisoprenyllithium (PILi)–isoprene system under the action of TMED upon the PILi concentration [14] (see Fig. 4.5).

Thus the total sum of the action of ED on the kinetics of the polymerization of nonpolar monomers in the cases under discussion is determined by whether the reduction in the reactivity of the monomeric forms of the active site is compensated by the increase in the total number of simultaneously growing chains whose activity is higher than that of the $(M_nLi)_m$ reagents.

Some idea of the difference in the relative activity of the M_nLi and $M_nLi \cdot D$ chains is given by the following values which may be taken for the growth rate constants (k_M and k_M^D, respectively),

TABLE 4.2. The Effect of the Nature and Concentration of the Electron Donor (ED) on the Reaction Order (n) with Respect to Organolithium Reagents in the Polymerization of Dienes in a Hydrogen Medium (initiator concentration, 10^{-3} moles/liter)

Monomer	Initiator	ED		n^a	Ref.
		name	molar ratio K		
Styrene	Polystyryllithium	THF	200[b]	1	23
Butadiene	n-Butyllithium	TMED	4	1	18
Isoprene	Oligoisoprenyl-lithium	THF	4	0.25	24
		THF	100	1	24
		DME	1	0.90	21
		DME	4	0.94	21
		TMED	4	1	20

[a]In the absence of an electron donor the magnitude of n for the polymerization of butadiene and isoprene under the same conditions is 0.25.

[b]When the ratio of THF to polystyryllithium is reduced, n = 0.5.

which correspond to these and to other active sites [data for room temperature (20°C), liter·mole^{-1}·sec^{-1}]:

Monomer	K_M	k_M^D	
		TMED	DME
Butadiene	0.8 [11]	0.20 [18]	0.38 [19]
Isoprene	0.65 [16]	0.18 [20]	0.23 [21]
Styrene	0.50 [17]	0.13 [17]	

The magnitudes of k_M^D, in particular for dienes, are approximate; they are evaluated on the basis of kinetic data for the range of values of the ratio ED—lithium which results from the complete dissociation of the associates [1]. Nevertheless, the order of magnitude of the values is sufficiently reliable to conclude that they are several times smaller than the corresponding values of k_M. Such a conclusion, which follows from results given above, was first drawn much earlier by Morton [22] from results on the polymerization of isoprene in hexane and in diethyl ether.

Before discussing the question of the possible reasons for the differences in the activity of the growing chains being compared we will deal with the essential dependence of the degree of dissociation of the associates on the nature of the electron donor in these conditions.

It is usual to consider two parameters: the reaction order with respect to the growing chains (approximating it to unity) and the viscometric features of the solutions of "living" and deactivated polymers as indices of the degree of dissociation (as they approach one another). In the case of bidentate donors the first order is attained at a molar ratio ED–Li of about unity. When monodentate donors are used analogous effects are observed for much higher ratios (see Table 4.2). The reason for this amounts to the difference in the mechanism of complex formation, the limiting case of which is described by Eqs. (4.4) and (4.5):

$$(M_nLi)_m \xrightarrow{mD} (M_nLi)_m \cdot mD \qquad (4.4)$$

$$(M_nLi)_m \xrightarrow{mD} mM_nLi \cdot D \qquad (4.5)$$

Obviously the efficient dissociation to monomer of the associates is particularly favorable to the formation of bidentate complexes (D–R–D) or the biligand complexes which model them in the case of monodentate donors (D), i.e., in the case of structures (4.I) and (4.II), respectively:

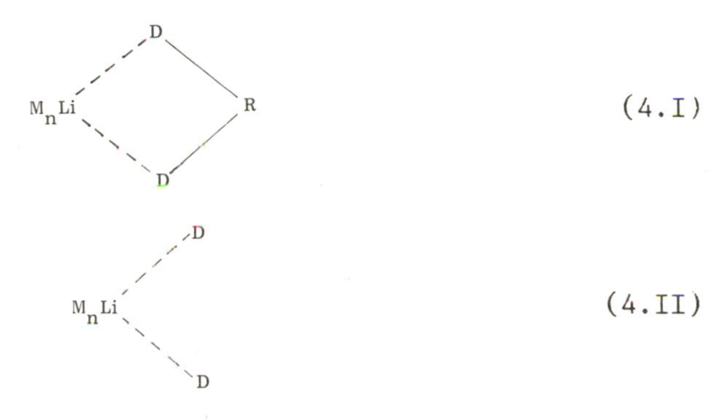

It is impossible to ascribe the incomparably greater ease of formation of biligand complexes of type (4.I) in comparison with those of (4.II) to the difference in the energy advantage of one or another set of events in complex formation. On the contrary, in the general case complexes (4.II) must be energetically favored. There are no particular close examples which might be used to illustrate this but for our purpose the results of theoretical calculations for the energy of complex formation (E_k) under the action of methyllithium with mono- and bidentate π-donors may be called upon:

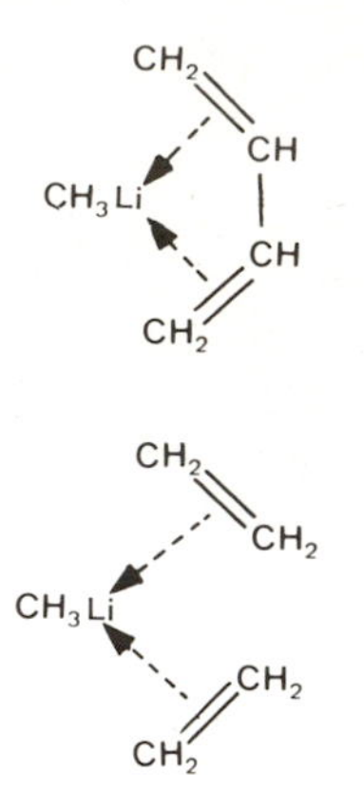

(4.III)

(4.IV)

According to data obtained by the CNDO/2 method [25], E_k (4.III) − E_k (4.IV) = 53 kcal/mole.

An analogy between structures (4.III, IV) and (4.II, I) is quite fortuitous. Nevertheless, the difference in the magnitudes of E_k for complexes (4.III) and (4.IV) are quite remarkable considering the known superiority of butadiene as a π-donor over ethylene. This superiority becomes particularly clear when the simplest equiligand complexes formed between these and other agents are compared, namely:

$$\begin{array}{c} CH_3Li \\ \uparrow \\ CH_2{=}CH_2 \end{array} \qquad (4.V)$$

$$\begin{array}{l} CH_3Li \\ \uparrow \\ CH_2{=}CH{-}CH{=}CH_2 \end{array} \qquad (4.VI)$$

From the results given in [25], it follows that the butadiene complex is more advantageous: E_k (4.VI) − E_k (4.V) = 45 kcal/mole. These data may be taken as evidence for the entropy advantage of the formation of the chelate complexes (4.I, IV) over biligand complexes (4.II, III).

Regarding structure (4.I) as a final form it scarcely follows that it can be assumed that the formation of analogous chelate complexes occurs at the initial stage of the interaction of associates with bidentate donors. In such circumstances the more correct path is from the tetramer form (T) of the polydienyllithium to the monomer form (m) via "quasichelate" complexes (4.VII) and (4.VIII), in which each heteroatom of the compound is connected with one lithium atom of the associate as shown below (D is the dimer form of the associate):

$$T \xrightarrow[a]{2\,DRD} R\begin{smallmatrix} D \\ D \end{smallmatrix}T\begin{smallmatrix} D \\ D \end{smallmatrix}R \xrightarrow{b} 2\, d\begin{smallmatrix} D \\ D \end{smallmatrix}R \xrightarrow{c} 2m\cdot DRD + 2m$$

(4.VII) (4.III) (4.IX) (4.6)

For products which are formed at stages a and b reversible reactions may be less significant due to the tendency of such "strained" structures to dissociation. The reversibility of stage c probably limits further transformation of the final products to complexes m·DRD·m and in particular (4.I).

The necessity for the presence of a large excess of monodentate electron donor for complete splitting of the associates may be the result of the considerably reduced tendency of complexes T·4D and d·2D to events of the type (4.6a) and (4.6b) [in comparison with structures (4.VII) and (4.VIII)]. It cannot be excluded that a significant shift of the equation towards dissociation requires an increase in the number of monodentate ligands, i.e., the formation of T·(4 + x)D and d·(2 + x)D complexes. The practically complete prevention of the association of complexes of monomeric forms of growing chains with monodentate electron donors obviously only becomes possible for structures of the type (4.II); they may become dominant only for high values of the ratio D/m.

Turning to the reasons for the lower values of k_M^D as compared with those of k_M, we note that this cannot be regarded as a natural consequence of complex formation. Events given by

$$\text{RMT} \xrightarrow{\text{D}} \text{RMT}\cdot\text{D} \tag{4.7}$$

are always accompanied by some stretching of the C—metal bond which, in the absence of circumstances which may lead to complications, must lead to an increase in the reactivity of the initial agent. Obviously, just such a situation exists in metallation [26]. A contradictory result observed during the anionic polymerization of nonpolar monomers demonstrates the dominant role of other factors, one of which is the screening of the active bond of the growing chain by D ligands and/or the reduction in the acceptor properties of the counterion which are essential for the growth reaction which proceeds according to scheme (4.1). On the basis of the differences in the reactivity of the agents compared in the metallation (RMT < RMT·D) and in the interaction with the monomer with the C—metal bond (RMT > RMT·D) then the second of these factors must be accorded preference. Stage (4.1a) may be regarded as the activation of the monomer which is caused by the redistribution of electron density to favor stage (4.1b). At the growth of $M_nLi\cdot D$ chains this effect must be weakened to a considerable extent. On the other hand any parallels of a similar nature with the metallation reaction are excluded here. It is possible to assume that the qualitative difference in the character of the dependence of the polymerization and metallation reactions on the presence of an electron donor is a consequence of just this circumstance.

The effect of the stretching of the C–metal bond on the introduction of the monomer into the structure of the growing chain, which is exceeded by other factors in the case of contact ion pairs, becomes decisive when these are transformed into separated ion pairs. We return to this question somewhat later (see Section 5.1.1.6).

4.1.1.3. Anionic Chains in the Polymerization of Polar Monomers

It is difficult to trace the changes in the reactivity of M_nMt chains under the action of electron donors using an analogous approach. This is made difficult by kinetic termination (whose rate is different for M_nMt chains and their complexes); such processes and the formation of side products are sometimes capable of fulfilling the role of electron donor with respect to the active site.

We will deal first with termination taking for simplicity the monomolecular mechanism of these reactions. The situation is then described by $k_t \neq k_t^D$ where k_t and k_t^D are the rate constants in Eqs. (4.8) and (4.9), respectively:

$$M_nMt \xrightarrow{k_t} \qquad (4.8)$$

termination

$$M_nMt \cdot D \xrightarrow{k_t^D} \qquad (4.9)$$

The polymerization processes considered are usually distinguished by a low initiating efficiency (F) and are very sensitive to the presence of electron donors where the most typical condition is $F < F^D$; here F^D is the efficiency of initiation for processes which proceed under the action of growing $M_nMt \cdot D$ chains. Consequently, the total kinetic effect observed in the presence of an electron donor must, together with other factors, be determined by the ratios k_t/k_t^D and F/F^D. The polymerization of acrylonitrile in a hydrocarbon medium under the action of organomagnesium initiators in the presence of various electron donors is of this type; for biblibography see [27].

Because values of the constants of the events involved are practically nonexistent, only hypothetical conclusions on the contributions of the growth and termination in the transition from M_nMt and $M_nMt \cdot D$ chains are possible. Sometimes in order to do this researchers fall back on a comparison of the molecular mass of polymers which are formed in these and other systems after different intervals of time in the incomplete conversion of the monomer. Such a method enables the essential role of the changes in the magnitude of F to be excluded as a factor which is often used in determining the overall rate of polymerization. Nevertheless, such a simplification does not give grounds for completely simple conclusions since it is necessary to consider the intramolecular complex formation of growing chains of polar monomers (see Section 3.1.4) and, consequently, the coexisting forms AS_ℓ and AS_c. It is very probable that the equilibrium of the general type (3.7) which is displaced in a nonpolar medium in the absence of an electron donor towards AS_c will be displaced, if not completely, to a considerable degree towards AS_ℓ under the action of strong electron donors when the ratio ED–AS is already low. This concept enables the correct interpretation of the facts determined experimentally to be approached. It must be borne in mind that there is a fundamental difference in the dependence of MM of polyacrylonitrile on the concentration of the monomer in processes which are initiated by n-butyllithium (n-BuLi) and lithium dipropylbutylcarbinolate (DPBL). This is in hydrocarbon media with the same duration of experiment and the same stable conditions; higher values of MM are obtained for the alkoxide initiator. These facts have already been cited in the literature [1, 28]; we will limit ourselves therefore to giving the following data relating to polymerization in toluene at −50°C (duration of each experiment 30 sec):

		Concentration of acrylonitrile, moles/liter		
		0.5	1.5	2.0
$\bar{M}_{vis} \cdot 10^{-3}$	n-BuLi	28	51	70
	DPBL	114	315	344

It is necessary to take into account that in both cases the magnitude of F is about 1%, but for n-BuLi this is connected with the fact that it is consumed in side reactions and for DPBL with the relatively low initiation rate. As has already been noted (see Section 3.1), these facts, together with others, enable the structure $M_nLi \cdot (ROLi)_{m-x}$ to be assigned to the active sites which act in the system with alkoxide initiators. Its alkoxide fragment in fact fulfills the function of electron donor. In the same way as for all $M_nLi \cdot D$ chains the form AS_c is incomparably less likely for such active sites than for chains formed by alkyllithium initiators. The interpretation of the above results is based on this concept. In each case which we have examined the growth reaction may be described by the general scheme of Eq. (4.10), in which k^ℓ and k^c are constants corresponding to AS_ℓ and AS_c, respectively.

$$\begin{array}{ccc} AS_\ell & \rightleftharpoons & AS_c \\ M \downarrow k^\ell & & M \downarrow k^c \\ & \text{growth reaction} & \end{array} \qquad (4.10)$$

For both systems the condition $k^{\ell} > k^{c}$ and the difference in the equilibrium concentrations of growing chains seems natural, AS_c being dominant in the n-BuLi system and As_{ℓ} in the DPBL system. On this basis the total effect observed may be connnected with the predominance of one form of the active site participating in Eq. (4.10). It must be borne in mind that in fact active sites which differ not only in the nature of the counterion but also in their conformation are being compared, the latter difference being decisive. This situation is described by the following correlations of the reactivities of the active sites being compared:

$$AS_{\ell} \gg AS_c; \ [AS_{\ell}] \gg [AS_c]; \ [AS_{\ell}] > AS_c;$$

here the symbols without brackets refer to the alkyllithium systems and those in brackets to lithium alkoxide systems respectively. For these ratios the difference in the relative activity which is perfectly possible for the linear forms, namely $AS_{\ell} > (AS_{\ell})$, may be masked by the large scale effects.

The proposed interpretation does not permit the exclusion of the possibility that the differences in the values of MM may depend upon termination reactions. These values are much greater for M_nLi chains than for active sites which are formed by alkoxide initiators. This is shown by the fact that carbonyl groups are absent in polymers which are formed under the action of DPBL but present when n-BuLi is used as an initiator [29].

The differences in the tendency of these growing chains to termination leads to the conclusion that $k_t \gg k_t^D$. This may be regarded as general for typical cases of the deactivation of the anionic active sites containing nitrile or ester substituents. For many polymerization processes of polar monomers in anionic systems it has been demonstrated that deactivation proceeds via a stage in which intermediate complexes of the counterion of the growing chains and the polar group of the same or other macromolecule are formed (see Section 3.1.4). As has already been noted, the increase in the coordination unsaturation of the counterion weakens the tendency to displacement of the equilibrium between the forms of the growing chains towards AS_c which are intermediate deactivation products of the molecular mechanism. The bimolecular termination mechanism, also known for these processes, in principle does not introduce any changes.

The reaction given below is an example of the formation of side products which may affect the activity of the growing chains

$$CH_2{=}C(CH_3)COOCH_3 + BuLi \longrightarrow CH_3OLi + CH_2{=}C(CH_3)(Bu)C{=}O \qquad (4.11)$$

The modifying action of metal alkoxides which affect the behavior of the active sites is well known and we will not deal with them here. The fate of the other reaction product of Eq. (4.11), isopropenylbutylketone (IBK), has become the theme of a special discussion only recently. According to Hatada et al. [30], IBK, which competes successfully with methyl methacrylate (MMA) in the initial stages of polymerization, is responsible for a sharp reduction in the reactivity of the growing chains. From results obtained for toluene at −78°C, only a small fraction of the active sites with IBK end groups undergoes reactivation when MMA is added. The bulk of the deactivated chains are found at the end of polymerization as oligomers containing IBK end groups. The extent of reaction (4.11) which competes with the usual initiation reaction and the degree to which IBK participates in the interaction with the active sites may be judged from the lithium methoxide yield (more than 50%) and the number of groups corresponding to IBK in the polymers and oligomers (about 40%); these values are taken from calculations on the original initiator.

With regard to the question of the nature of side products in these systems, it is interesting to note vet another important fact which was established in [30]. For a long time metallation was included among those reactions which proceeded with the interaction of polar unsaturated monomers with metal alkyls which is accompanied by the splitting off of a corresponding hydrocarbon. i.e., of the butane in the case of butyllithium (see [30]). In order to establish the real cause of such splitting, Hatada et al. compared the features of butane formed when reactive mixtures of MMA–butyllithium–toluene were treated with methanol and methanol deuterated at the hydroxyl group after the mixture had been stored for a short while. The isotopic features of the butane formed corresponded completely with the terminating agent: the use of CH_3OH leads to the formation BuH and the use of CH_3OD, to BuD. The authors therefore regard them as being formed by the unexpended butyllithium and reject the conclusion of other researchers who proposed the metallation reaction in the given system. It is possible that this conclusion, which so far refers to one actual system, is valid for other related systems. It is also pointed out that there is some possibility that passivated but not completely deactivated agents with IBK end groups participate in the modification of growing MMA chains. According to [30], such agents may arise both from the reactions of growing MMA chains with the double bond of IBK and when its ketone group is attacked by the initiator; such an attack leads to products (4.X) and (4.XI), respectively:

$$\begin{array}{c} M_nCH_2C(CH_3)Li \\ | \\ C{=}O \\ | \\ Bu \end{array} \qquad (4.X)$$

$$CH_2{=}C(CH_3)C(Bu)_2OLi \qquad (4.XI)$$

The above are the only data available on the effect of such products on the polymerization process but the participation of such products in the various donor—acceptor interactions with normal active sites cannot be excluded. This is especially true for nonpolar media. Similar conclusions are possible for other processes in the anionic polymerization of polar molecules in which metal-containing compounds of low activity are formed.

4.1.1.4. The Activating Effect of Electron Donors on Organolithium Associates

The majority of data available on the polymerization of nonpolar media show that the weakening of the C—metal bond of growing chains which results from the formation of complexes does not overlap other factors which accompany transformations of Eq. (4.7), and does not give rise to an increase in the reactivity of anionic contact ion pair active sites. There are, however, certain facts which can be more easily made to accord with the notion of an opposite overall result. This is so in cases in which the activity of the associates (M_nLi) and their complexes (M_nLi)·D are compared.

Before giving the data upon which this conclusion is based we note that from the formal point of view it is completely beyond criticism, provided it is accepted that the sequence of particular changes of the initial active site is given in a simplified form for the tetramer form of an organolithium growing chain by

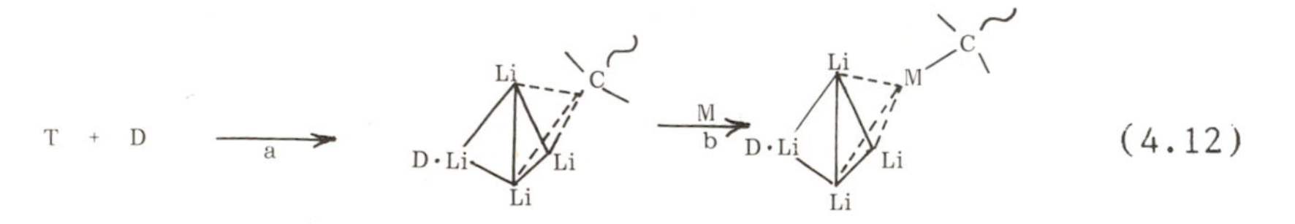

This form, which is the basic form for polybutadienyl and polyisoprenyl M_nLi chains in hydrocarbon solutions at a concentration of $\geq 10^{-2}$ moles/liter, is a tetrahedron with the lithium atoms at the apices. The extremely low reactivity of such active sites (see Section 4.1.1.1) may increase at stage a, since the interaction of electron donors, even with one lithium atom, weakens the whole bonding of the associate and in this way makes possible the greater reactivity of any unscreened C—Li bond at stage b.

There are a number of facts, which we now turn to, which make it possible to assume that the above situation can become real under certain conditions. We have already mentioned in Chapter 3 an ap-

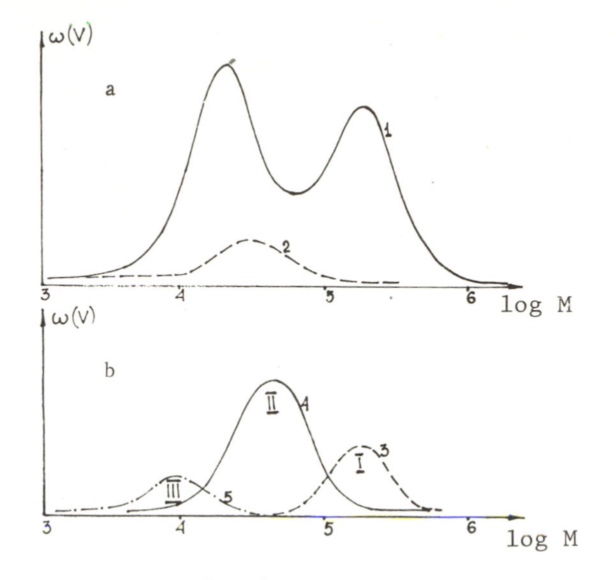

Fig. 4.6. Chromatograms (gpc) of the reaction products formed in the butadiene–TMED–toluene system at −42°C (for conditions see Table 4.3) [32]. a) Experimental chromatogram for the nonfractionated form: 1, changes in total concentration of components (by refractometer); 2, spectrometer curve (for λ_{max} = 269.5 nm) which shows the presence of the polystyrene component. b) Combined gel chromatograms of the fractionation products recorded by refractometer: 3, fractionation product I; 4, fractionation product II; 5, fractionation product III.

proach to the evaluation of the relative reactivity of the coexisting growing chains of different types based on the use of small quantities of strong electron donor during anionic polymerization in nonpolar media. The main conclusion which results from a study of systems which include polysisoprenyllithium chains and bidentate electron donors in the ratio ED–Li = 0.01 is in good agreement with the fact that the reactivity of the M_nLi form is greater than that of $M_nLi \cdot D$. A more detailed examination of the composition of the polymers obtained, the molecular mass, and structural features of individual fractions has led to the conclusion that it is impossible to explain the whole of the experimental data on the assumption that only active sites of the above form took part in the growth reactions. Such a growth mechanism cannot be reconciled with the value of F (27%) calculated for the fractions with the lowest molecular mass (MM = 5000). As was shown in [31], such a high value of F for fractions in whose formation, judging by their structure (about 80% 3, 4 units), complex forms of active site participate to the largest extent, together with a very low ED–Li ratio, shows not only the multiple exchanges of the electron donor between cer-

TABLE 4.3. Fractional Composition of the Products Formed in the Butadiene–Styrene–Oligobutadienyllithium (OBuLi)–Tetramethylethylenediamine (TMED) System for a Total Conversion of 10% [32] (concentration, moles/liter: butadiene, 0.5; styrene, 0.5; molar ratio TMED–OBuLi, 0.01)

t, °C	Concentration OBuLi, moles/liter	Fraction	Fraction by weight, %	$M_p \times 10^{-3}$	M_w/M_n[a]	Styrene content, mole %
−30	1.6	I	14	60	1.7	0
		II	71	20		5.5
		III	15	8		2.6
−42	2.0	I	15	180	5.5	0
		II	73	50		11.5
		III	12	8		2.6

[a]Data for nonfractionated polymers.

tain forms of coexisting active sites. Furthermore, it must be assumed that at least the bulk of the chains which form this fraction cease to take part in growth after the electron donor has split off. Such a phenomenon can be explained by investigating complexes of associates of growing chains with electron donors for active sites which are responsible for the formation of the above low-molar-mass fraction. The general conclusion which can be drawn from this is that in these conditions associates of the form $(M_nLi)_4$ take practically no part in the growth reaction when the complexes of the associated forms of active site make an appreciable contribution to the total result. In defining these forms, mainly complexes of tetramer associates have to be considered, but the possibility that they split giving rise to dimer complexes cannot be excluded.

Results on polymerization in butadiene–styrene systems (1:1) initiated by oligobutadienyllithium in the presence of TMED in a hydrocarbon medium with ED–Li = 0.01 [32] give qualitatively similar results. The results of this study, which are dealt with briefly in Section 3.4.3, illustrate the composition of the reaction products; see Fig. 4.6 and Table 4.3.

As is generally known, in the conditions described but in the absence of electron donors polybutadiene is formed which is practically free from styrene, but when the ratio TMED–Li is of the order of unity a random copolymer comprising equal quantities of each comonomer is formed. The low total styrene content (calculations show that for the whole product it is about 7.5% at −42°C) and the fact that it is concentrated mainly in only one fraction gives evidence of the hybrid nature of fraction II. Obviously, it is a multiblock copolymer which includes blocks of homopolybutadiene (A) and the butadiene–styrene random copolymer (B) which are formed as a result of multiple electron donor exchanges between the growing chains.

From our point of view the formation of two independent fractions I and III is of considerable interest. These are variations of the individual blocks of type A. The considerable difference in the values of MM associated with these fractions lead to the conclusion that they are the product of different active sites. If a simple solution to the problem of the origin of fraction I is taken (the maximum value of MM agrees only with growth on the monomeric forms M_nLi) then the genesis of fraction III must be interpreted taking into account several circumstances. In the first place it is necessary to consider the exchanges between the monomer and associated forms of PBuLi. Incomplete exchange (possible at low temperature and low conversion) may also become apparent when products corresponding to fraction III are formed. However, in an experiment on the homopolymerization of butadiene under the action of OBuLi carried out in the absence of TMED and styrene at −30°C with a conversion of 10%, a polymer was obtained with a value of MM of

$2 \cdot 10^4$ with M_w–M_n = 1.2 [32]. Consequently, the large difference in the same characteristics of fractions I and III is the result of the presence of a subcatalytic quantity of TMED. It is also possible that styrene plays some role. Secondly, in contrast to fraction II, the composition of fraction III is not dependent on the polymerization temperature. The reason for this can be seen in the considerable difference in the concentrations of the organolithium compounds which participate in the equilibria (4.13) and (4.14); for brevity we omit the equilibrium between the associated and monomeric forms of PBuLi:

$$(M_nLi)_m + D \rightleftharpoons (M_nLi)_m \cdot D \qquad (4.13)$$

$$M_nLi + D \rightleftharpoons M_nLi \cdot D \qquad (4.14)$$

The initial ratio TMD–lithium = 0.1 is applicable only to components of equilibrium (4.13). A precise evaluation of the relation between the components of Eq. (4.14) does not seem possible. However, it may be taken as close to unity if it is considered that under the conditions of the experiment the bulk of the TMED must be combined with the great excess of PBuLi. As a result of the state of equilibrium (4.14), the concentration of each component, which probably lies in the range 10^{-7} to 10^{-6} moles/liter, may exhibit temperature sensitivity.

A usual consequence of the reduction in the polymerization temperature is the reduction in the number of exchanges between the associated and monomer forms of PBuLi (both "free" and complexed). This is shown by the MM characteristics of individual fractions and the degree of polydispersity of the nonfractionated polymers (see Table 4.3).

The conclusion that $(M_nLi)_m \cdot D$ associates are responsible for the formation of the low-molecular-weight fraction is in agreement with its increased content of 1,2-polybutadiene groups (70% at −42°C). As has already been noted (see Section 4.1.1.1), the tendency to form such groups distinguishes the PBuLi associates from the corresponding monomer forms.

We now turn to the problem that exchanges of the type (4.3) are rarer in this case than in systems which contain no styrene or electron donors. It will be recalled that opinions regarding this difference are based on the values of M_w/M_n which have already been given for these processes. In the absence of styrene and electron donors these exchanges, which proceed via the dimer forms of associates, are favored by the fact that the coexisting active sites are not complicated by other mutual transformations. On the other hand, in the present case various processes are possible for the monomeric forms, the most important of which are the following (here Bt and St are the butadiene and styrene end units, respectively):

$$M_nBtLi\ (A) \begin{cases} \xrightarrow[]{M_pBtLi\cdot D\ (B^1)} M_nBtLi\cdot D + M_pBtLi \\ \xrightarrow[]{(M_pBtLi)_4\cdot D\ (B^2)} M_nBtLi\cdot D + (M_pBtLi)_4 \\ \xrightarrow[]{M_pStLi\cdot D\ (S^1)} M_nBtLi\cdot D + M_pStLi \\ \xrightarrow[]{(M_pStLi)_2\cdot D\ (S^2)} M_nBtLi\cdot D + (M_pStLi)_2 \end{cases} \qquad (4.15)$$

The experimental conditions make it possible to take the concentration of A as being close to the sum of concentrations ($B^1 + B^2 + S^1 + S^2$). Consequently, the transformations in Eq. (4.15) are capable of competing to an appreciable extent with the dimerization of the monomeric forms (A) which precede their reassociation to tetramers.

Apparently events (4.15) slow down, but do not completely eliminate exchanges between the monomer and tetramer forms of PBuLi. This is shown by the relatively low amounts of fractions I and II in comparison with the hybrid fraction III which comprises more than 70% of the total reaction product.

Data given on the behavior of the various organolithium active sites in conditions which exclude the formation of separated ion pairs support the following reactivity series

$$(M_nLi)_m < (M_nLi)_{m-x} < (M_nLi)_{m-x}\cdot D < M_nLi\cdot D < M_nLi \qquad (4.16)$$

It is possible to observe parallel changes in the structure and the reactivity in other series of related active sites but for a much narrower range of agents. Series (4.16) is unprecedented in the breadth of choice of different types of contact ion pairs which have the same central atom in their counterions. This is why we have devoted so much attention to organolithium compounds. It is possible to find in transition metal compounds a considerable variety of active sites in which the central atom remains constant but which differ with respect to their ligands; however, discussion of the problem is hampered by the lack of relevant information. This is because there is insufficient data on the changes in the reactivity of such sites associated with a change of ligands.

4.1.1.5. The Activating Effect of Electron Acceptors on the Polymerization of Cationic Growing Chains

The great volume of data available on the influence of electron donors on anionic active sites can be supplemented by only a modest amount of data on the action of electron acceptors (EA) on cationic active sites. Both these and other effects are formally grouped under one heading (case 4, Table 4.1). The data of Panayotov and Heublein [33], in which they summarize their earlier work, are fundamental to our present discussion. As has been shown, tetracyanoethylene (TCE), chloranyl, trinitrobenzene (TNB), and other electron acceptors, taken in equivalent quantities with respect to the cation initiators, accelerate the polymerization of various monomers considerably, provided the reactive medium is neither an electron donor nor highly polar. The kinetic effects established do not go as far as the determination of the values of k_p. Nevertheless, the authors, on the basis of the molecular-mass characteristics of the polymers and the efficiency of initiation, conclude that this acceleration results from the changes in propagation-reaction kinetics.

This is well illustrated by the data on styrene. During polymerization under the action of boron trifluoride etherate in dichloroethane (the presence of TCE in the ratio EA:initiator = 1:1), the overall rate constant increases by a factor of about three as compared to that when EA is absent (data for 20°C). Here the total energy of activation falls from 11.1 to 6.9 kcal/mole. The variation of the viscosity of polystyrene with EA has been studied under somewhat different conditions (concentrations of monomer and $BF_3 \cdot OEt_2$ are 1.0 and 0.01 moles/liter, respectively, molar ratio EA:initiator = 1:1, dichloroethane solvent, 0°C) and is given below:

EA	—	TNB	1-Chloro-2,4-dinitrobenzene
[η] of polystyrene	1.8	3.1	3.3

In discussing the mechanisms of these phenomena the authors draw upon UV data which show the formation of EA complexes with the monomer (M·A). The absorption bands corresponding to these complexes disappear when an initiator (C) is introduced into the system. The formation of complexes C·A has been established calorimetrically. The values of the stability constants (K_s) of the latter are higher, as the values obtained for several systems, which include TCE, show (20-25°C):

Electron donor	Styrene	α-Methylstyrene	SBF_6^-	$SbCl_6^-$	$TiCl_5^-$
K_s, liter/mole	0.80	3.18	6.7	26.0	38.0

There are no data on the values of K_s for the counterion BF_3OEt^- but it may be assumed that they are greater than that for styrene.

On the basis of the above, Panayotov and Heublein [33] proposed a scheme for the modification of the active site which, using our notation for convenience, is given below:

$$M + EA \rightleftharpoons M \cdot EA \qquad (4.17)$$

$$M_nY + M\cdot EA \rightleftarrows M_{n+1}^{+}Y\cdot EA^{-} \qquad (4.18)$$

As assumed in [33], stage (4.18) is accompanied by the transition IP_c to IP_s which also leads to activated polymerization under the action of EA. It is difficult to agree with this conclusion and not only on the ground of lack of evidence for such a transformation. The absence of an effect of that magnitude, which usually accompanies the formation of IP_s in the polymerization of unsaturated monomers, is of great significance. In the present case it is incomparably smaller. Therefore it is more likely that there is another reason for the increase in the reactivity of cationic growing chains in the presence of EA. It is sufficient to take into account the stretching of the active bond in the modified active site, i.e., a phenomenon which is qualitatively analogous to the production of a system of M_nMt—D. For anionic sites such a stretching may be masked by other factors, but in the case of cationic agents of this type there are no such inhibiting influences. Such a difference is very remarkable and can be explained by the particular consequences of the donor—acceptor interactions in anionic and cationic systems. Thus in anionic polymerization with a mechanism of the type (4.1), stage (4.1a), raising the electrophilicity of the monomer leads to an increase in its reactivity. The presence of electron donors always weakens this effect. The consequences of this have been discussed in Section 4.1.1.2. It is practically impossible to assume the formation of an intermediate complex of the monomer with the counterion in cationic polymerization.* Consequently changes in the nature of the counterion of cationic active sites cannot be regarded in the same light as those for anionic active sites. Also it is essential that for any $M_nY\cdot M$ complex in a cationic system the reactivity of the monomer which fulfills the role of electron donor should decrease and not increase. Any increase in the electrophilicity of the monomer resulting from its incorporation into the structure of such a complex must only hamper growth.

4.1.1.6. Influence of the Nature of the Reaction Medium

There are a number of cases which differ considerably from one another and which result from a transfer from hydrocarbon to polar media. This is accompanied by a shift in Eq. (4.2) to the right. Sharp differentiation of processes which are due to properties of the reaction medium (which we will call "macroeffects") from interactions such as the action of solvent molecules in forming complexes or transforming IP_c to IP_s is often attained by studying the dependence of polymerization kinetics and microstructure of the products on the presence of small quantities of electron donor or electron acceptor

---

*The events of complex formation with monomers in cationic systems mentioned in Sections 2.1.2 and 5.1.2.4 are fundamentally different.

in media of low or moderate polarity. In the absence of such information the corresponding "microeffects" may escape observation. It must be borne in mind that these terms which are introduced for convenience do not reflect the scale of events. The action of electron donors and acceptors may be very considerable when they are used in catalytic quantities. This question has been touched upon in the previous section and forms the theme of further discussion in several later sections. Here we will attempt to isolate events which may be regarded as "macroeffects" which are typical of polymerization processes. The effect of the polarity of the medium on the course of organic reactions has been examined in detail in a recent monograph by Reichardt [34].

As was pointed out in Section 2.1, the equilibrium between ion pairs and free ions depends essentially upon the solvating properties of the medium. On the other hand, the transformation from IP_c to IP_s is not always accompanied by a clear increase in the value of K_{diss}. It is much simpler to separate the effect of an increase in the polymerization rate, which is produced by such a transfer, from the contribution of the free ions than to give an unequivocal interpretation of the kinetic data which characterize the behavior of ion pairs. This is the more so since the values of k_p, which are related to IP_s, are markedly close to those typical of free ions. Therefore the quantitatively very limited and difficult to control transition from IP_c to IP_s, which occurs when the medium is varied, may be reflected in the values taken for k_p related to IP_c. The results obtained by Szwarc et al., for instance, point to such a conclusion. According to their data the values of k_p for IP_c of sodium polystyryl in such similar solvents as THF and methyl-THF are 80-90 and 11 liters·mole^{-1}·sec^{-1}, respectively, at 25°C [35].

The bulk of information on the reactions which result when pure and mixed solvents are used in anionic polymerization relates to the growing chains of the styrene series. This question has been discussed in detail in the review by Hirohara and Ise [36], which, in spite of being more than ten years old, still retains its significance today. From the detailed material which it contains we will isolate some of the more interesting points.

One of these concerns the interpretation of the dependence of the value of k_p on the nature of the counterion. The changes in these values for polystyryl chains in THF for the series Li > Na > K > Rb > Cs are regarded as consequences of the reduced participation of IP_s in the growth reaction. For the first and last of these the total values of k_p for ion pairs are given as 160 and 22 liters·mole^{-1}·sec^{-1}, respectively. The second point concerns the change in the order of the above series on transfer to media with lower solvating ability. It is reversed in dioxane and tetrahydropyran. The decisive role of this fact follows from the results obtained in different mixed solvents with the same dielectric constant, for example, dimethoxyethane (DME)–benzene and THF–benzene systems, whose composition is ensured by the equality of their values of ε. The values of k_p found are higher in the former. Finally, the third point concerns the solvation of the carbanion component of the growing chains which is usually ignored. The conclusion that this effect plays a considerable role is based on the increase in the rate constants for growth on free ions (k_p^-) when an electric field is applied. Such an occurrence has been established for polystyryl chains with lithium, sodium, and potassium counterions in various media by Hirohara and Ise [36]. A field strength of the order 3 $kV \cdot cm^{-1}$ produces a twofold increase in k_p^- when the k_p values for ion pairs and K_{diss} remain unchanged. Such an effect of the applied field is explained by the authors as due to the desolvation of the free ions when they are displaced to the cathode, i.e., to the relatively low relaxation rate of the solvation shell.

Information which shows that growing diene chains are capable of forming IP_s in anionic systems has appeared only recently and is more appropriately examined in Section 4.1.2.

The elucidation of the nature of the active sites in cationic polymerization of unsaturated monomers in polar media is hampered by the fact that there is practically no K_{diss} data for growing chains. Nevertheless, the use of buffers in the study of such processes sometimes brings some definition into the question of the nature and relative role of active sites in a given system.

One of the first examples is the polymerization of styrene under the action of $HClO_4$, which has in its time served as a basis for the concept of pseudocationic polymerization. Detailed examination of this matter may be found in many sources, the most recent of which is the review by Gandini and Cheradamé [37]. Without dealing with this in detail we note that the use of Bu_4NClO_4 as a buffer (K_{diss} at −97°C in CH_2Cl_2 is about 10^{-3}) by Pepper [38] led to a considerable reduction in polymerization at temperatures close to −100°C. However, this does not affect polymerization in the same system at room temperatures to any appreciable extent. On this basis it was concluded that an equilibrium existed between the three types of active site; the first of these was not ionic in character.

$$M_nOClO_3 \rightleftharpoons M_n^+ClO_4^- \rightleftharpoons M_n^+ + ClO_4^- \qquad (4.19)$$

It must be emphasized that the action of buffer compounds on cationic polymerization cannot always be interpreted on the basis of trivial approaches; some of the results of Higashimura et al. also lead to this conclusion.

Interesting comparisons regarding the dependence of the character of the effects observed on the nature of the medium have been made for the styrene–CF_3SO_3H–$Bu_4NSO_3CF_3$ system [39]. The kinetic

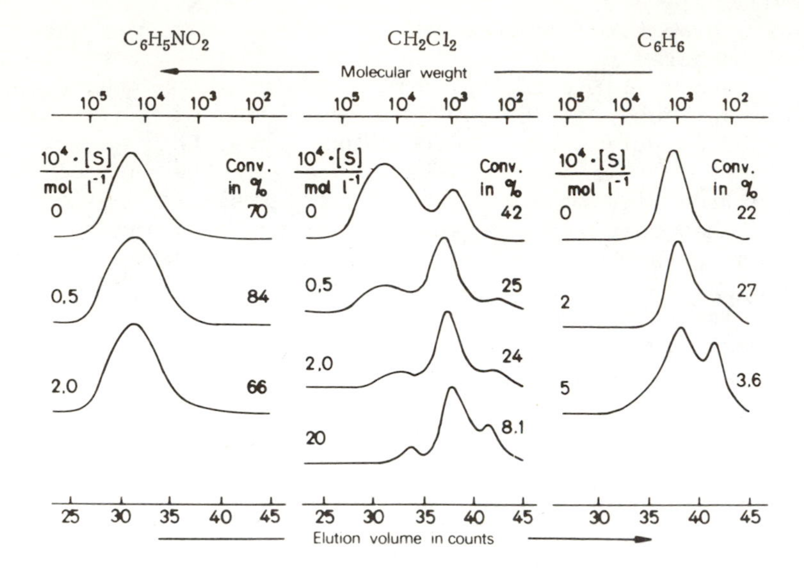

Fig. 4.7. The effect of $Bu_4N^+SO_3CF_3^-$ (S) on the MMD of polystyrene formed under the action of CF_3SO_3H at 0°C [39]. Concentration, moles/liter: monomer, 1.0; initiator, $2.0 \cdot 10^{-4}$ in nitrobenzene and dichloromethane, and $5.0 \cdot 10^{-4}$ in benzene.

characteristics given, namely, the ratio of the total rate constant for polymerization in the absence of a buffer (k) and in the presence of a buffer (k_b), show quite clearly the features of the processes taking place in the various media. By way of illustration we will use values which can be calculated from the graphs from one of the references (concentration in moles/liter: initiator, $(2\text{-}5) \cdot 10^{-4}$; ratio initiator/buffer = 1, temperature 0°C):

Solvent	Nitro-benzene	Dichloro-methane	Benzene
k_b/k	0.5	0.2	0.1

Changes in the MMD only accompany processes in the last two solvents. The introduction of a buffer to nitrobenzene has only an insignificant effect on the mean value of MM (see Fig. 4.7).

The effects established for nitrobenzene are ascribed by the authors to the coexistence of free ions and IP_s whose reactivities differ relatively little. It can be assumed that for polymerization in CH_2Cl_2 the effect of the buffer is more significant. This is because in the presence of buffer only contact ion pairs are effective. When the ratio buffer/initiator = 10 the ratio k_b/k in CH_2Cl_2 falls to about 0.01. Data for other solvents on the effect of increase in the buffer concentration are not quoted. It has so far not been possible to explain the reason for the sensitivity of polymerization in benzene, in which the presence of appreciable quantities of free ions is not possible, to the presence of buffer. One can only assume that in this case the $Bu_4NSO_3CF_3$ compound fulfills the function of a modifying agent.

The results of another work in the same series [40] show that it is necessary to take into account the participation of nonionic active sites (see Eq. 4.19). This deals with the effect of the nature of the medium on the polymerization of p-methoxystyrene (p-MOS) under the action of various cationic initiators. It is interesting that the participation of such ions was deduced from an increase in k_p and a fall in the polarity of the medium when the k_p value was calculated on the basis of data obtained on the concentration of the active sites using UV spectroscopy. For this the optical density at 380 nm for carbocation growing chains is used.

The greatest difference between the values of k_p calculated in this way and those expected has been found for polymerization under the action of methylsulfonic acid when transferring from CH_2Cl_2 to mixed CH_2Cl_2–CCl_4 solvents of different composition. The following data obtained at 30°C are significant in this respect:

Solvent	ε	Optical density at 380 nm	Conversion after 10 sec, %	MM
CH_2Cl_2	10.46	1	55	$5 \cdot 10^4$
CH_2Cl_2–CCl_4	5.49	0.01	20	$6 \cdot 10^3$

We note that the spectroscopic and conversion characteristics were obtained at equimolar concentrations of monomer and initiator, $5 \cdot 10^3$ moles/liter (the kinetics were studied by the diminution in intensity of monomer absorption bands). To evaluate the MM, experiments were carried out at p-MOS and initiator concentrations of 0.1 and $2 \cdot 10^{-3}$ moles/liter, respectively. However, these differences do not affect the main conclusion drawn from the above results, which asserts that there are two types of active site in this system, only one of which is ionic (AS^i) with the typical spectroscopic features of an ionic active site. Such active sites which play a dominant role in a medium of sufficiently high polarity lead to the formation of high-molar-mass polymers. In media of low polarity the growth takes place on nonionic active sites (AS^n). These are not revealed in UV spectra and do not enable polymers of molar mass of the same order to be obtained as do AS^i. In media with values of ε which lie between 10.5 and 5.5 both types of active site are effective. This agrees with the data on the optical density of the band for which λ_{max} = 380 nm and on the values of MM of the corresponding mixed solvents CH_2Cl_2–CCl_4.

The difference in the values of MM may be the result of the increased role played by chain transfer for nonionic active sites. This is deliberately omitted in [40]. However, such a conclusion, which is appplicable to the system under consideration here, cannot be applied to the cationic polymerization of any related monomers where the nature of the active site can be assumed to vary in an analogous fashion when the medium is varied. It is known that the growing chains of styrene series which are obtained using other cationic initiators exhibit features which are completely opposite to those described above when the polarity of the reactive medium is varied (see Section 4.1.2).

The series of studies by Penczek et al., which concentrate on cationic polymerization of oxygen-containing heterocyclic compounds, particularly tetrahydrofuran and oxepane, afford conclusions regarding the role of the polarity of the medium [41]. Using various dioxolanylium derivatives with counterions of the type EF_6^- and $CX_3SO_3^-$ as initiators, the authors obtained detailed information on the kinetics of these processes and the NMR characteristics of the active agents which are effective in these processes. As^i and AS^n compounds are found in cationic systems which contain protic acid derivatives and oxygen-containing heterocyclic compounds. This is reflected in the scheme which has been developed by Penczek and Szymansky [42]. This is essentially the same as the particular case (4.19). However, there is only a formal similarity between the processes of cationic polymerization of styrene and THF when growth takes place with counterions derived from protic acids. Firstly, the reactivities of growing THF chains of the free ion and ion pair are practically identical. This also applied to other monomers of the same type. Secondly, the absolute values of the propagation constants for these and other AS^i, i.e., k_p^+ and $k_p^{\pm}$, are small.

Without dwelling on the reasons for this, we note that such a feature enabled these systems to be characterized qualitatively fairly completely. This is based on a combination of kinetic, electrochemical, and spectroscopic (^1H- and ^{19}F-NMR) measurements.

On the basis of the latter the relative fraction of AS^i is determined. This is used together with the experimental values of k_p to calculate the propagation constant for AS^i (k_p^i) and for the approximate determination of the propagation constants for AS^n, i.e., for active sites of the ether type (k_p^e). The results obtained for $THF{-}CCl_4{-}Ct^+CF_3SO_3^-$ systems (where Ct^+ is a dioxolane cation) give a ratio k_p^i/k_p^e of the order of 10^3 with a mean value of $k_p^i = 4.3 \cdot 10^{-2}$ liter·mole^{-1}·sec^{-1}. Data on the kinetics of the interaction of THF with a model AS^n, for which $EtOS_2CF_3$ compound was used, were taken into account. Therefore, in spite of the fact that in CCl_4 the bulk of the active sites are nonionic (with a THF concentration of 5 moles/liter, 98% of the active sites are of the AS^n type), growth takes place mainly due to AS^i.

The contribution of AS^i increases still further in nitromethane in which, when the concentration of THF is 5 moles/liter, the content of AS_n is not greater than 2%. As regards the value of k_p^i it falls quite slowly and evenly as the permittivity of the medium increases. This has been established for examples of pure and mixed solvents for values of ε from 5.5 to 22. Within this range the extreme values of k_p^i are 4.3 and 2.1 liters·mole^{-1}·sec^{-1}, respectively (data for 25°C). For a more detailed discussion see [43, 44]. It is especially remarkable that this change accompanies the basically opposite path of K_{diss}. The corresponding values differ by about two orders of magnitude (see Section 2.2). In interpreting these facts the authors proceed on the basis of the idea that it is necessary for the monomer to overcome the "solvation barrier" which is greatest in CH_3NO_2 and least in CCl_4.

It follows from the data of other studies on the same series that the kinetics of the polymerization of THF in such systems is characterized by the fact that the value of k_p and the initiation rate constant are practically equal. The nature of the counterion exerts no appreciable effect upon the rate constants [41].

4.1.2. The Role of the Nature of the Y Component in Growing M_n^*Y Chains

In the preceding section special attention was paid to the change in reactivity of ion pairs with modification of the counterion without variation of its central atom. If this restriction is ignored the number of substances and processes which can be considered is extended considerably. This gives some possibility of a decision regarding a connection between the parameters of component Y and the active site M_n^*Y.

4.1.2.1. Influence of the Nature of the Central Atom of the Counterion

The consequence of change in the nature of the central atom of the counterion while all the other conditions are kept constant has been widely studied from the aspect of stereoregulation in ionic polymerization (see Chapter 5). Although the dependence of the reactivity of active sites upon this component is based upon wide and varied material, it is difficult to draw fundamental conclusions from it. There are various reasons for this. Thus in the case of the simplest variation, of which the M_nMt series is an example (here Mt is an alkali metal), attention must be directed mainly towards polymerization in polar media. The necessary kinetic data relating to nonpolar media are very sparse. For nonpolar monomers they are limited almost entirely to processes which are initiated by organolithium compounds. We will mention therefore results which characterize the reactivity of certain members of the M_nMt series with

TABLE 4.4. Kinetic Characteristics of the Interaction between PStMt Compounds and 1,1-Diphenylethylene Monomers [45] (solvent benzene, temperature 24°C)

Monomer	$K_p (K_{diss}/2)^{0.5}$, (liters/mole)$^{0.5}$ $\cdot$min^{-1} for Li	K_p (liters$\cdot$mole$^{-1}\cdot$sec^{-1})	
		K	Cs[a]
DPE	1.27	103.9	56.7
m,m'-DMDPE	1.15	47.8	43.2
p,p'-DMDPE	0.35	18.0	8.5

[a]Average value of those given in [45].

respect to nonpolymerizing nonpolar unsaturated compounds. Examples of this are 1,1-diphenylethylene (DPE) and its derivatives. Data obtained by Busson and van Beylen [45] for model monomer DPE and its m,m' and p,p'-dimethyl derivatives (DMDPE) using polystyryl PStMt chains with different Mt are shown in Table 4.4.

As for the features which are observed in transition from alkali metal derivatives to alkaline earth compounds of the RMtX series (where X is an alkyl or a halogen), these cannot be separated from the presence of the ligand X which also affects the reactivity of the corresponding active sites. A detailed discussion of this question can be found in the recent review by Arest-Yakubovich [46].

Nevertheless, it is possible to come to some conclusions regarding the influence of the metal on the reactivity of the growing chains of this and other series. For this we draw upon values of k_p relating to the polymerization of styrene under the action of lithium and barium initiators, choosing data from the nearest temperature range (20-30°C).

Initiator	Solvent	k_p, liters$\cdot$mole$^{-1}\cdot$sec^{-1}	Ref.
Bis-triphenylmethylbarium (BTMB)	Benzene	0.36	47
n-Butyllithium	Benzene	0.50	17
BTMB	THF	0.019	47
n-Butyllithium	THF	160	36

The values of k_p in benzene which characterize the monomeric forms of growing chains illustrate the differences in the reactivity of single-type active sites. Extrapolating the value of k_p for polystyrylbarium to 20°C yields a value of 0.17 liters$\cdot$mole$^{-1}\cdot$sec^{-1}.

According to the data of [47] the energy of activation for the growth reaction in this system is 13.5 kcal/mole.

On the other hand, the results obtained for the polymerization in THF are not suitable for direct comparison. They indicate the considerable contribution of IP_s in the case of polystyryllithium chains which follows from the much greater value of k_p that might be expected for contact polystyryl ion pairs of the type $M_nLi\cdot mD$. For example, according to data on the polymerization of styrene in benzene under the action of an n-butyllithium complex with tetramethylethylenediamine, the value of k_p at 20°C is 0.08 liter·mole^{-1}·sec^{-1} [17].

The value of k_p for styrene polymerization initiated by bis-triphenylmethylbarium in THF is regarded in [47] as an overall value. A greater number of forms of active site than usual is assumed in these circumstances for alkali metal derivatives. In particular it must be borne in mind that there are two types of contact ion pairs, namely M_n^-,Mt^{2+},M_n^- and M_n^-,Mt^{2+}. However, growing chains of increased activity (IP_s and free ions) are here either completely absent or play a very minor role. Therefore, the value of 0.019 may be taken for the kinetic characteristics of barium IP_c complexes with THF.

From the above it is not difficult to see that the difference in the reactivities of lithium and barium active sites of the same type are, in similar circumstances, the same qualitatively and quite close quantitatively (symbols are explained on page 121):

$$k_M(Li)/k_M(Ba) \sim 3; \; k_M^D(Li)/k_M^D(Ba) \sim 4$$

These values may be taken as indicators of the reactivities of the IP_c of lithium and barium active sites for the polymerization of styrene in nonpolar and polar media.

Changes in the reactivity of the active sites of alkaline earth metals have so far been impossible to illustrate by strictly comparable quantitative data.

Data on the rate of polymerization related to monomer concentration give some idea on the influence of the nature of the counterion on the reactivity of styrene active sites in THF [46]:

Initiator	t, °C	v/[M], sec^{-1}
Benzylpotassium	-30	5.5
Dibenzylstrontium	20	36.0
Dibenzylbarium	20	1.0

In all these cases the order of reaction with respect to the initiator is found to be zero. This may be explained in principle by a mechanism given by Eq. (4.1).

Organomagnesium initiators which are active for polar monomers are inert to dienes and styrene in nonpolar media, even at temperatures which are considerably in excess of room temperature. The same is true for weakly polar media.

In a short discussion of the role of the nature of the counterion in polymerization of polar monomers we will mention first the series of studies by Schultz et al. on the polymerization of MMA by organometallic compounds of the alkali metals. Following the summary of the results obtained in the recent review by Müller [48] new publications from the same school have appeared. A considerable number of these have concentrated on the stereospecificity of the active sites in anionic systems. Questions of a similar nature are considered in Chapter 5. Here we are attempting to focus on the data relating to those processes which can be discussed separately from structural effects which characterize the structure of macromolecules.

Examining some of the kinetic results of the above series of studies we note first the interesting fact of the equalization of the reactivities of growing M_nNa and M_nCs chains in the polymerization of MMA in polar media. In contrast to styrene, the value of k_p for chains with these counterions practically coincides within the limits of error. Individual examples are given below:

Counterion	Solvent	t, °C	k_p, liters·mole^{-1}·sec^{-1}	Ref.
Sodium	THF	−24.5	1820	49
Cesium		−22.1	1660	49
Sodium	DME	−36.6	3100	50
Cesium		−39.0	3000	50

In all these solvents linear Arrhenius curves have been established over a wide temperature range (from room temperature to −100°C in THF and to −65°C in DME). This has led the authors to conclude that these systems have only one type of active site and that it is IP_c in structure.

In order to explain the reasons for the absence of an appreciable difference in the calculated values of k_p the authors draw upon the hypothesis that the final units of the active sites have enol structures. The validity of this concept is borne out by the structural characteristics of the model compounds (see Section 2.2). It is assumed that in such IP_c the counterion lies closer to the enolate fragment of the end unit than to its C_α atom. It is assumed that this can reduce the dependence of the active site reactivity on the nature of the counterion. The authors propose that transition to a solvent with increased solvation (in this case from THF to DME) can modify the active sites while preserving the contact of the counterion with the carbonyl oxygen, at the same time having some effect on the separation of the C_{α}–Mt system. It is proposed that the intermediate structure between the two shown below should be taken for the structure of the active site (here L is the solvent molecule):

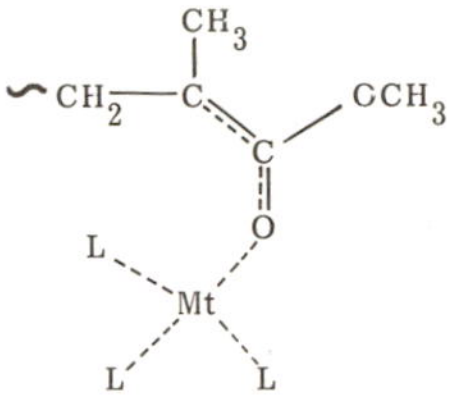

(4.XII)

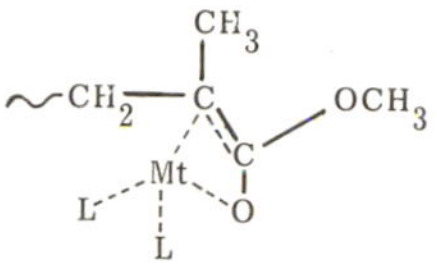

(4.XIII)

Taking these as the characteristics of the structure of the active sites in THF (4.XII) and in DME (4.XIII) it is possible to explain the considerable difference in the values of k_p calculated for polymerization in these solvents. The phenomena observed are regarded as examples of the mutual compensation of the opposing differences in the ionic radii of the counterions ($r_{Na} < r_{Cs}$) and in their ability to be solvated (Na > Cs).

The idea that M_nMt chains which are effective in THF and DME are of the IP_c type is in agreement with the results of another study in the same series. In this a cryptand [222] was used as an additional complex-forming agent in the polymerization of MMA in THF [51]. For growth at −98°C in the case of chains with sodium–cryptate counterions a value of 270 ± 30 liters·mole^{-1}·sec^{-1} was obained. The values of k_p for M_nNa and M_nCs chains at the same temperature were 33 and 21 liters·mole^{-1}·sec^{-1}, respectively [5].

Together with the data of the more recent work of Müller et al. [52] which describe the behavior of M_nLi and M_nK chains in an MMA–THF system, the results quoted above enable one to assert that there is a noticeable connection between the kinetic activity of the growing chains and the interionic distance (a) of the corresponding ion pairs (see Fig. 4.8).

The value of (a) has been calculated from data in crystalline compounds taking the corresponding parameter of the anionic fragment into consideration [52].

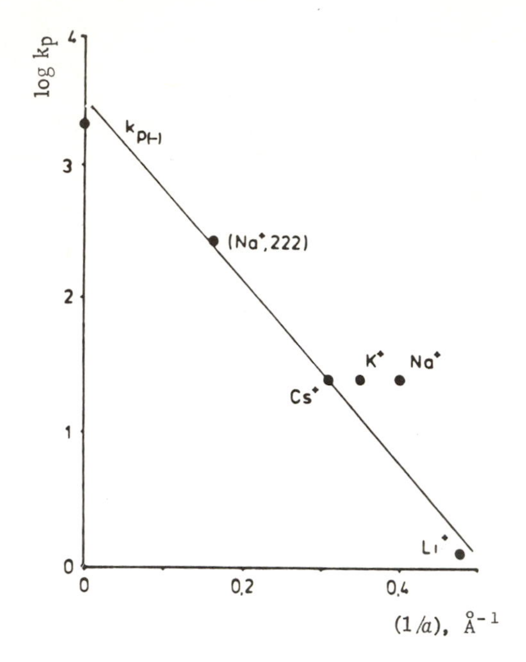

Fig. 4.8. The logarithmic dependence of the growth rate constant of poly(methyl methacrylate) ion pairs on the inverse interionic distance (a) [52]. Solvent, THF; temperature, −98°C.

As we can see, information necessary for an objective description of the dependence of the reactivity of the organometallic active sites in a series of alkali and alkaline-earth-metal derivatives on the nature of the central atom of the counterion is as yet modest. Even less opportunity exists for such evaluation for anionic growing chains based on metallic compounds of other groups in the periodic table. This appears clearly in anionic systems based on organoaluminum compounds. The fact that R_2NAlR_2 compounds are highly active as initiators of low temperature polymerization of acrylates and methacrylates [53, 54], together with the total inability of the common aluminum alkyls to initiate similar processes, is remarkable. This circumstance is all the more interesting since it can be explained by differences in the nature of the atoms connected to aluminum only for the initiation stage. As has been firmly established by Milovskaya et al. [54, 55], in anionic polymerization initiated by alkylamides, growth proceeds on the Al—C bond (the Al—N bonds are completely used up in initiation). Consequently, discussion of the mechanism of these processes from the point of view of the reactivity of the active sites must be based on the features of the Al—C bonds in the growing chains, for example.

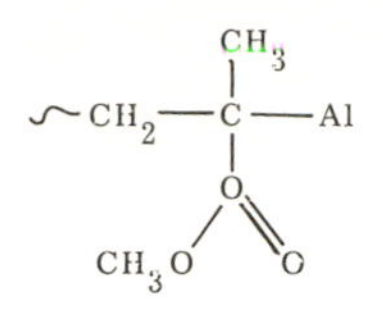

Obviously they differ in some features from similar bonds in the usual aluminum alkyls.

Substances are compared from just this point of view in the quantum-mechanical studies of Milovskaya and Eizner using model compounds

$$(CH_3)_3Al \qquad (4.XIV)$$

$$H_2AlN(CH_3)_2 \qquad (4.XV)$$

$$H_2AlCH(COOCH_3)CH_3 \qquad (4.XVI)$$

Calculations of the optimized structures (4.XIV) and (4.XVI), carried out using the CNDO/2 method, led to the conclusion that the orders and the polarity of the bonds may be used for such characteristics. The polarity of the Al–X bond is regarded as the algebraic semi-sum of the charges on these atoms while the fraction Δq of the total charge of Al (q) was used as the Al component, i.e., for the non-symmetrical compounds $\Delta q = q^{Al}_{H_2AlXH_n} - 2/3q^{Al}_{AlH_3}$ was used and for the symmetrical components, $\Delta q = 1/3q^{Al}_{Al(CH_3)_3}$. These are based on the data of the values of the deaggregation constants in systems

$$(RO)_3Al_2 \rightleftarrows 2(RO)_3Al$$

and the differences in concentration at equilibrium of the monomeric form of the aluminum alkoxide in the conditions selected for the evaluation of the initiating activity of such compounds. From our point of view the passivity of these agents may be connected with their inability to form intermediate complexes with the monomer. According to [54] and [57], this stage is obligatory for the realization of initiation and growth processes which take place under the action of aluminum alkylamides; this is based on the inhibiting action of electron donors in these processes. For example, the ratio THF/Al of equimolar order is sufficient for the complete elimination of polymerization, independent of the stage at which ED is introduced into the system.

There is another reason for the passivity of aluminum alkoxides in these systems which cannot be excluded, namely that the geometry of the $(RO)_3Al{\cdot}M$ complexes is unfavorable for their transformation into active growth sites. Even though such an approach is hypothetical, in this case in particular it is difficult to explain another specific feature of the initiators under consideration without it. We note firstly that among these may be distinguished the commonly used individual compounds (these include for example, $PhNAlEt_2$ [52, 54]) and R_3Al--N systems which contain heterocyclic compounds whose components form particularly active initiators of irreversible interactions. In the case of the system which has been studied in the greatest detail, namely Et_3Al—2,2'-dipiridyl (DP), the following compound is taken as the actual initiator [58]:

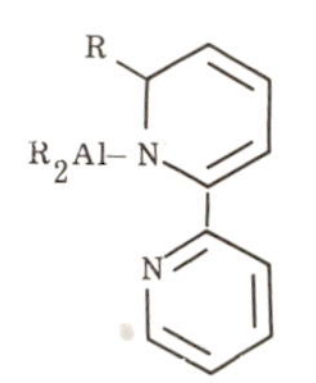

This system is completely comparable in terms of activity to the alkyl derivatives of alkali metals. For the polymerization of MMA in toluene at -50°C the value of k_p = 5 liters·mole^{-1}·sec^{-1} [57] is obtained in this case. However, it is completely inert with respect to acrylonitrile both in the same conditions and also at higher temperatures. Similar behavior is also typical for other initiators of the general type $R_2AlNR'_2$. This is combined with the greater ability of acrylonitrile to form complexes as compared with methacrylate esters. Hence such passivity can only be explained by the fact that for acrylonitrile the interaction of the initiator with monomer is limited by state (4.1a). Obviously the transformation (4.1b) is hindered by the unfavorable mutual orientation of the compounds of the AS·M complexes which are formed.

Together with the high selectivity of the organoaluminum active sites for the polar monomers under consideration (which is not typical for the usual anionic initiators) a formal similarity can be seen between certain derivatives of aluminum and related anionic agents in other respects. For example, aluminum alkoxide is not able to initiate polymerization of polar monomers and this is álso true for magnesium alkoxides [59]. This feature distinguishes these and other compounds from the alkoxides of the alkali metals which are active with respect to acrylate and methacrylate monomers.

The qualitative agreement between kinetic effects when changes take place from C—Al- to N—Al-derivatives in certain anionic systems based on the transition metals is more interesting. The ability of compounds which contain —MtN groups to polymerize unsaturated polar monomers is also shared by the derivatives of various transition metals (see the monograph by Mazurek [60]). However, a comparison of the activity of such initiators with analogous C—Mt compounds has so far been possible only for organochromium systems.

Mazurek et al. [61-64] have shown that very active anionic initiators can be obtained by the interaction of tris-π-allylchromium (All_3Cr) with N-containing heterocyclic compounds and in particular with pyridine and 2,2'-dipyridyl. Examination of the products formed in these circumstances which ensure their maximum kinetic activity enabled the structure of the initiating agents for the pyridyl (4.XVII) and dipyridyl (4.XVIII) systems to be established:

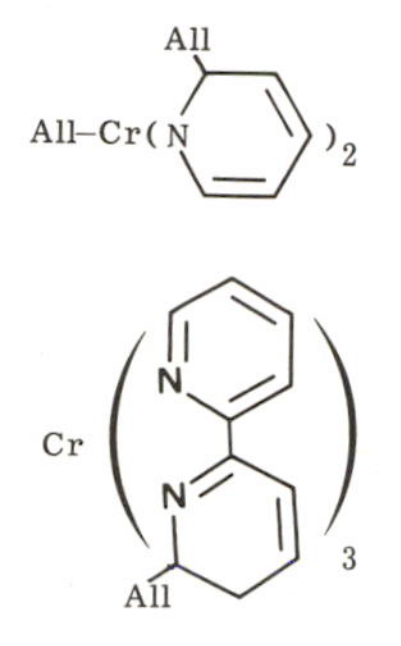

(4.XVII)

(4.XVIII)

The features of these systems which are central to our discussions are as follows. All_3Cr shows only limited initiating activity relative to methylmethacrylate, which is rapidly suppressed in hydrocarbon media, whereas in the presence of the common electron donors (THF, diglyme, etc.), there are not any noticeable changes. On the other hand, compound (4.XVII) with excess pyridine exhibits high activity. The value of k_p when (4.XVII) is used to initiate the polymerization with the molar ratio pyridine–All_3Cr > 100 is 1.7 liters·mole^{-1}·sec^{-1} (toluene, 0°C) [61].

In contrast to alkyl and alkylamide derivatives of aluminum. both All_3Cr and (4.XVII) and (4.XVIII) agents polymerize acrylonitrile efficiently. The first of these compounds affords complete conversion of the monomer in low temperature polymerization in nonpolar (toluene) and polar (dimethylformamide) media. The difference between the two remaining compounds is that the pyridyl derivative (as in the case of MMA) requires a large excess of pyridine, while the active agent (4.XVIII) is formed at a ratio DP/All_3Cr = 2 and produces quantitative polymerization of acrylonitrile in DMF in a fraction of a minute without the introduction of additional amounts of dipyridyl (data for 0°C) [63]. Methyl methacrylate on the other hand does not polymerize at all under the action of (4.XVII) and (4.XIII) compounds in dimethylformamide.

The common origin of the initiating agents in dipyridyl systems based on aluminum and organochromium compounds together with the similarity of their structure makes it natural to attempt to draw some parallels. It is difficult to find a basis for the suggestion that the character of the N–Mt and C–Mt bonds (in the active sites of initiation and propagation respectively) depends upon the nature of the metal. The individual features of the growing chains which we are studying which have aluminum and chromium counterions are more easily explained by the differences in the detailed mechanism of their action. It must be borne in mind that the growth reaction proceeds in the first case via a stage in which AS—M complexes are formed and in the second case when this stage is absent.

4.1.2.2. The Total Effect of the Central Atom and Ligands of the Y Component

The cationic growing chains are distinguished by the great variety of their central atoms. However, in discussing the reactivity of such agents it is difficult to separate the features which are associated only with this atom. For this one would have to examine the counterions with the same number of identical ligands. There are comparatively few examples of this. In particular, it is possible to point to a number of counterions of the general type $[E^5X_6]^-$ (but E^5 is P, As, or Sb and X is either chlorine or fluorine), where the growing chains have more common features than differences. Therefore in the case of cationic active sites it is more appropriate to consider the general nature of the counterion without excluding the associated ligands.

We have already dealt with some events typical of cationic growing chains in connection with dependence of the contributions of the active sites AS^i and AS^n on the polarity of the reactive medium (see page 141). We will mention results which illustrate the features of these processes due to varying the counterion while keeping the reaction medium constant.

An effect which stands out is the great difference in the sensitivity of the growing chains which are formed by protic and aprotic initiators under the same conditions. As has already been noted, in the polymerization of p-methoxystyrene under the action of CH_3SO_3H the change in the permittivity of the solvent roughly from 10.5 to 5.5 reduces the molar mass of the polymer (expressed by M_p) by almost an order of magnitude. When BF_3 etherate is used for initiating the same process with this variation in the permittivity of the medium, the change in M_p in the concentration chosen does not exceed $(2\text{-}3)\cdot 10^5$ [17].

These differences have also been analyzed by Higashimura et al. from the point of view that the growing chains are of ionic and nonionic types. Without entering into a discussion of the features which are typical of pseudocationic processes, we note the possibility that in some cationic polymerization processes (mainly for monomers of the styrene series and protic initiators) there may exist a "nonionic" contribution which is capable of increasing as the permittivity falls, until the ionic component is excluded completely. The necessity of considering these circumstances when interpreting

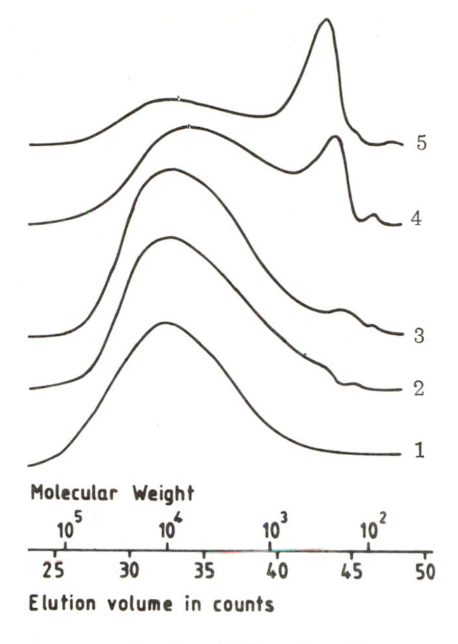

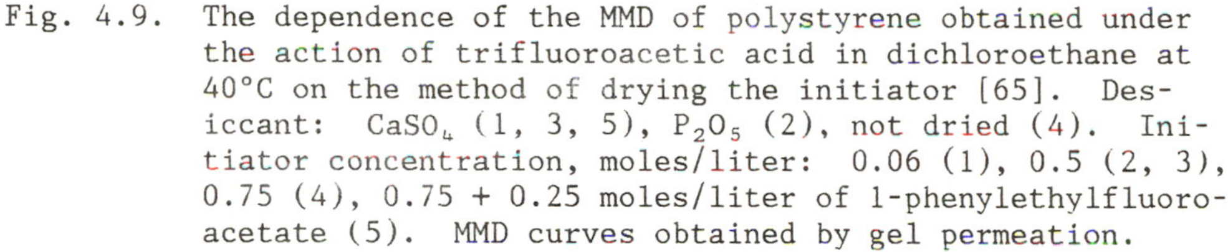
Fig. 4.9. The dependence of the MMD of polystyrene obtained under the action of trifluoroacetic acid in dichloroethane at 40°C on the method of drying the initiator [65]. Desiccant: $CaSO_4$ (1, 3, 5), P_2O_5 (2), not dried (4). Initiator concentration, moles/liter: 0.06 (1), 0.5 (2, 3), 0.75 (4), 0.75 + 0.25 moles/liter of 1-phenylethylfluoroacetate (5). MMD curves obtained by gel permeation.

kinetic data, particularly data relating to moderately polar media, has been examined in detail by Gandini and Cheradamé [37]. This review covers the literature for the period through 1980. We will limit ourselves therefore to the following remarks.

Review [37] contains data of various authors on the polymerization of styrene under the action of trifluoroacetic acid (TFA). The divergence between the kinetic features given in different studies carried out in this field later became the theme of the studies of Obrecht and Plesch [65]. As they show, the kinetics of the polymerization in styrene–TFA–CH_2Cl_2 and the MM parameters of the polymers depend considerably on the method of drying of the original initiator which contained uncontrolled microadditions of water. It can be seen from the data of Fig. 4.9 that this effect is quite remarkable. Obrecht and Plesch [65] emphasize that in many studies on related processes a detailed description of the experimental method is absent and cast doubt upon the absolute reliability of the data characterizing such systems given in the literature.

We will now attempt to discuss some of the questions connected with the variation in the reactivity of Ziegler—Natta active sites with fragment Y, which in these cases is very complex. We emphasize here that, as can be seen from the title of the present chapter, it is not our intention to enter into a special examination of the heterogeneous systems of this type. The great difficulty in interpreting the quantitative features of such processes often forces one to treat the values of the rate constant obtained with caution. The basic reason for this is that values of the concentration of active sites are not sufficiently trustworthy. The various methods used in their evaluation are not universal. This has been recently discussed by Mejzlik and Lesna [66]. We also note that it is very difficult to pinpoint the individual role of the nature of the transition metal as one of the factors determining the reactivity of such active sites. We will therefore concentrate mainly upon the total effect of the Y component in homogeneous initiating systems which are distinguished by increased reactivity.

Until recently the high activity of Ziegler—Natta catalysts in the polymerization of ethylene and other nonpolar monomers was known only for heterogeneous systems. This led to the generally held concept that the heterogeneity of the Ziegler—Natta initiating systems is a necesssary condition for securing maximum selectivity of the growing chains. For some of these systems the propagation rate constants are calculated to be of the order of 10^3 liters·mole^{-1}·sec^{-1} [67, 68]. On the other hand, soluble complexes based on transition metal compounds (in particular $Cp_{n-x}Mt^nR_x$, in which Cp is cyclopentadiene and R is an alkyl group) are distinguished by their moderate reactivity. They were therefore used mainly for the study of the physicochemical parameters of active agents, the mechanisms of certain model reactions, and polymerization kinetics in suitable experimental conditions. The situation in this respect began to change noticeably with the discovery of the cocatalytic action of water on several systems of the Cp-transition metal derivative—aluminum alkyl type. From the detailed studies of the Hamburg group the optimum effects are reached when aluminum alkyl reacts with water in conditions which lead to the formation of cyclic aluminooxanes. For the system $(CH_3)_3Al—H_2O$ with excess of the former, the formation of cyclic compounds of the general type $[Al(CH_3)O]_n$ with values of n from 5 to 15 has been established. The activating influence of aluminooxanes increases with n.

Among systems of this type, those studied by Kaminskii, Sinn, et al. are distinguished by their high catalytic activity. These are systems based on $Cp_2Zr(CH_3)$ or Cp_3ZrH compounds which are formed with aluminooxane complexes soluble in hydrocarbon media. In the polymerization of ethylene it has been found that for these compounds the value of k_p is as high or even higher than those mentioned above for heterogeneous systems. The authors also noted other important features of these complexes, namely the extremely high stability of the active polyethylene chains which preserve their activity for several days after the exhaustion of the monomer, the virtual absence of side reactions which are typical for the majority of such active sites, and the possibility of their being used in a series of practical tasks.

The discovery of soluble Ziegler—Natta complexes which form growing chains whose reactivities exceed those of the most effective heterogeneous systems is especially interesting. According to calculations carried out with the assumption that each zirconium atom forms an active site, the rate at which ethylene molecules are incorporated into the structure of the chain is 10^4 per sec. This value is regarded as the lower limit. According to Sinn et al., not all the zirconium atoms form active sites.

Studies on the kinetics of polymerization at extremely low catalyst concentrations (of the order of 10^{-7} moles/liter calculated for zirconium) make possible a qualitative evaluation of the processes which occur under the action of these very active growing chains. The concentration of the aluminum component in this case is 10^{-3}-10^{-2} moles/liter. The authors emphasize the necessity of such a large excess of aluminooxane in order to achieve the maximum efficiency of the initiating systems and note that the polymerization is of the first order for zirconium and second order for aluminum.* This forms the basis of the suggestion that the active complex is formed when zirconium interacts with the aluminooxane associate [72].

In an attempt to determine the fundamentals of the above phenomena the authors make use of x-ray analysis of certain zirconium aluminum complexes, particularly $Cp_3ZrH \cdot Et_2AlCH{=}CH_2$ and $[Cp_2(Cl)-ZrCH_2CH_2Zr(Cl)Cp_2]2Et_3Al$. The authors stress the particular significance of the interatomic distances and the valence angles. This is shown below for an example of the second of these complexes [74]:

*During a brief discussion on this point with Prof. W. Kaminsky when he visited Leningrad in September 1983, he pointed out that this reaction order applied to the interaction between the Zr and Al components in the absence of monomer (this does not follow from the fragment referred to in [72]). However, the present author does not consider this to be of fundamental importance for the suggestion given below.

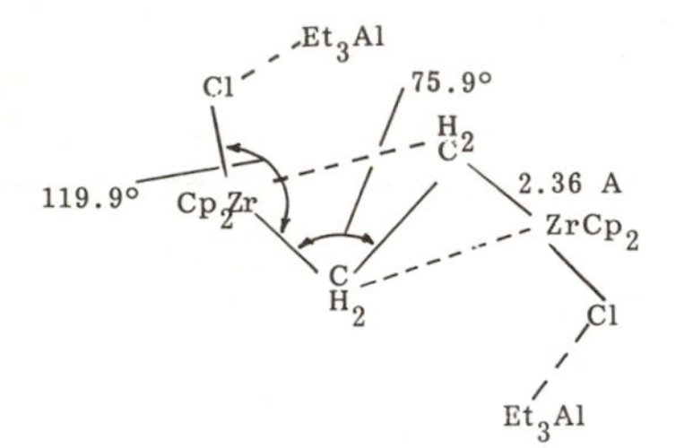

The confirmation of this hypothesis requires information which is as yet unavailable. This would allow us to apply the geometrical features found for crystalline complexes to the above homogeneous systems. It is just possible that the features of the zirconium–aluminum complexes are determined to a considerable extent by the nature of the actual Al component. This cannot be modelled by derivatives which enter into the composition of the above crystalline formation. From this point of view the initiating systems proposed by Sinn et al. possibly ought to be regarded as examples of a special case of a Ziegler–Natta catalyst. The incorrectness of applying the usual views on the functional significance of the non-transition-metal derivative in binary systems based on transition metals to these complexes leads to such a result (see Section 1.3). The particular role of the aluminooxane component may consist in including the O atom in the donor–acceptor interaction with the transition metal. This supplements the usual phenomena associated with the binary system with a particular polydentate effect which facilitates the loosening of the active bond. This assumption is in accord with the promoting influence of aluminooxane when the number of Al–O units increases; see [69-71]. The apparent contradiction between this point of view and the passivating influence of the usual electron donors on the catalytic activity of complexes based on transition metals may be eliminated as follows.

We will begin with the second order reaction which is established in [72]. As these authors propose, such an order may be explained by the diffusion of the organozirconium compounds to the aluminooxane associate, which results in the formation of active sites. It is difficult to agree with such a conclusion; it could stand criticism only if an equilibrium of the type in Eq. (4.20) moves to the left considerably, and there are no grounds whatever for this.

$$m(\underset{\substack{|\\CH_3}}{Al}\text{—}O)_n \rightleftharpoons [(\underset{\substack{|\\CH_3}}{Al}\text{—}O)_n]_m \qquad (4.20)$$

It is much more likely that there exists another equilibrium, namely (4.21) which, when $K_1 \gg K_2$ and the reactivity of AS_1 is much less than that of AS_2, would result in the second order of reaction of polymerization with respect to aluminooxane. For brevity the components of the systems are represented by their central atoms:

$$\mathrm{Zr} \underset{K_1}{\overset{\mathrm{Al}}{\rightleftarrows}} \underset{(\mathrm{AS}_1)}{\mathrm{Al \cdot Zr}} \underset{K_2}{\overset{\mathrm{Al}}{\rightleftarrows}} \underset{(\mathrm{AS}_2)}{\mathrm{Al \cdot Zr \cdot Al}} \qquad (4.21)$$

In assuming such a mechanism for the formation of active sites with increased reactivity (i.e., AS_2), the question arises as to the disposition of the components of the catalytic complex. One possibility is that the original tetrahedral structure of the zirconium component moves close to an octahedral structure with aluminooxane rings at its apices. This would mean that the oxygen atoms of the aluminooxane would fulfill the function of bridge centers. This is shown schematically for a structure with four aluminooxane units. The alkyl groups are omitted:

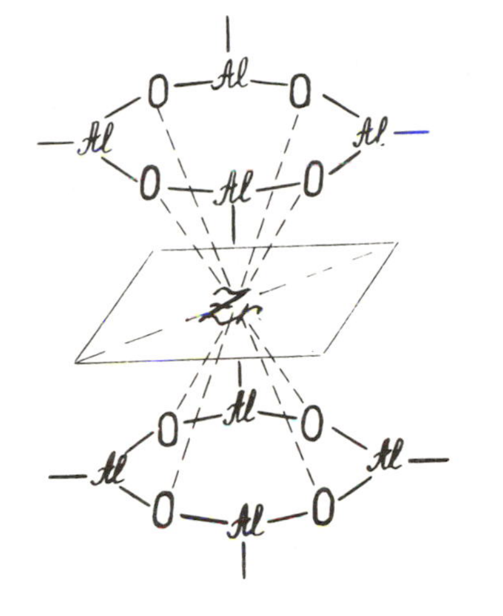

(4.XIX)

This form of representation is chosen more to illustrate the possible nature of the interaction between the components of the complex than to substantiate an octahedral structure. The detailed geometry of the complex seems to us to be less of a problem than the general character of the bonding system. One possible system of bonds is shown in (4.XIX). The "multiple bridging" of the bonds between the transition and non-transition metals, and the nature of the oxygen atoms in the aluminooxane which are different from the usual polydentate oxygen derivatives, may be regarded as the most important elements of the system. Apparently this feature consists of the reduction of the donor capability of the oxygen atoms of the aluminooxane. This may be responsible for the difference in the action of the compounds which contain $\geq$C–O– and $>$Al–O– bonds on the Ziegler–Natta catalysts. The first of these completely or partially passivates such systems as do many other electron donors [75]. The absence of this effect in the case of aluminooxane may be explained either by the difference of Zr–O bonds

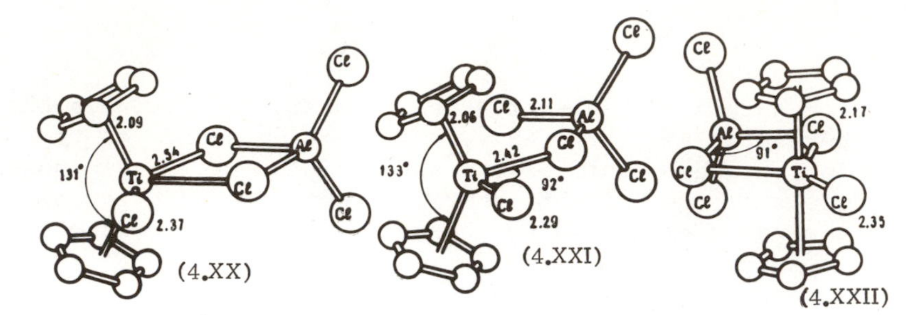

Fig. 4.10. Optimized structure of model titanium–aluminum complexes, calculated using the CNDO method [76].

from the coordination bonds which form the organometallic compounds with the usual ED or by the fact that in general such bonds do not occur.

The second of these assumptions seems the more likely. The particular positions of the oxygen atoms in the aluminooxane

$$\begin{matrix} >\mathrm{Al} & & \\ & \mathrm{O}\cdots\mathrm{Zr}\!< & \\ >\mathrm{Al} & & \end{matrix}$$

may be the reason for this feature of these bonds which closely resemble bridge bonds in character. On the other hand such an assumption does not eliminate the difficulties which arise in attempting a complete explanation of the above experimental facts. The weakening of the influence of the multiple-bridging effect on the active bond must be combined with maintaining the ability of Zr to form intermediate complexes with monomer molecules.

We note that the attempts to explain the dominating factors in an interpretation of the kinetic activity of the Ziegler–Natta catalysts include an attempt to evaluate the relative roles of the separate stages of the two-stage growth mechanism [Eq. (4.1)]. The question of the change in the acceptor ability of the transition metal atom brought about by the formation of the bimetallic complex is not usually afforded special attention from this point of view. It is usually accepted that the tetrahedral cyclopentadienyl derivatives of transition metals form octahedral complexes in Cp_2TiRCl–$RAlCl_2$ systems with one vacant position. The question of the energy advantage of such a reconstruction has been recently investigated theoretially by L'vovskii et al. [76] (using the CNDO method) for systems of the type L_2TiRCl–$AlCl_3$ where L is Cp or Cl and R is Cl, Me, Et, or Pr. In particular, structures of the following types were compared:

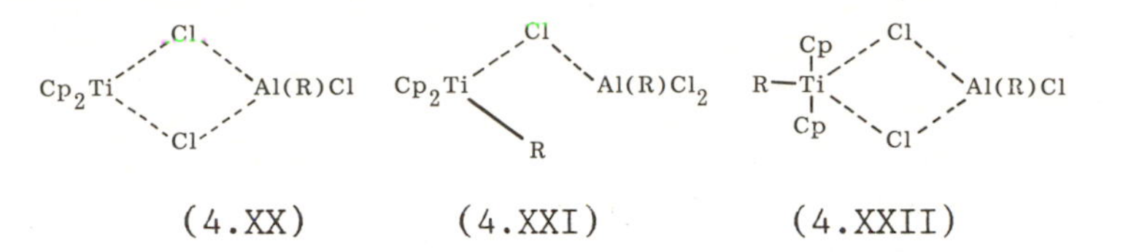

(4.XX) (4.XXI) (4.XXII)

The optimized configurations of these molecules for L = Cp and R = Cl are shown in Fig. 4.10. Calculation shows that the relative stability of these structures clearly depends upon the nature of L. This is shown in the following series:

L = Cp, R = Cl	L = Cl, R = CH_3
(4.XX) > (4.XXI) ≫ (4.XXII);	(4.XX) > (4.XII) > (4.XXI)

From this it seems unlikely that an octahedral complex is formed for cyclopentadienyl derivatives and there is some possibility (although not optimum) of its being formed for alkyltrichloroethane. The authors note the agreement of this conclusion with some known spectroscopic features for related systems.

The reliability of these results can be seen from the fact that the particular variations of CNDO used yield geometrical parameters for such substances which are quite close to those obtained experimentally. This is shown, for example, by the characteristics of the equilibrium geometry of the trimethylaluminum dimer.

Turning to the structural effects at the stage of formation of AS·M complexes, L'vovskii et al. compare variations like the models below:

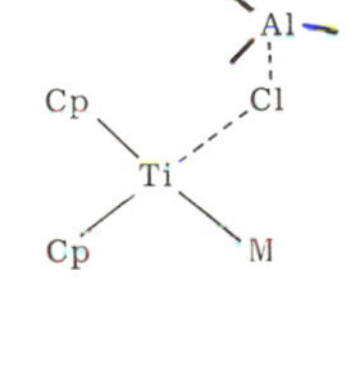

(4.XXIa)

(4.XXIIb)

A detailed analysis of the relative advantage of introducing the monomer into the C—titanium bond of each of these complexes shows the preferred structure to be (4.XXIa). Steric hindrance caused by the H atoms of the cyclopentadienyl groups is reduced as compared with that of structure (4.XXIIb).

The concepts of [76] cannot be transferred to systems containing aluminooxanes as they stand but from our point of view it is reasonable to take them into account.

In attempting to explain the reasons for the high reactivity of the zirconium–aluminooxane system it is necessary to bear in mind the ability of the aluminooxane components to give rise to qualitatively similar effects in other cases. The complexes of analogous titanium derivatives with the same aluminooxane [69, 71] and the $Cp_2TiEtCl$–$AlEtCl_2$–H_2O system which has been studied by Cihlař et al. [77] are of this type. The reactivity of the first of these is rather less than that of zirconium-aluminum complexes (strictly comparable data are not given by the authors). The results of [77] enable the efficiency of the catalytic complexes obtained both in the absence and in the presence of water to be compared. Growth rate constants for the initial stage of the polymerization of ethylene at 0°C are given below:

System	Molar ratio Al/Ti	k_p, liters·mole^{-1}·sec^{-1}
$Cp_2TiEtCl$–$EtAlCl_2$	8	7.2
$Cp_2TiEtCl$–($EtAlCl_2$–H_2O)	15	290

As we see, an aluminooxane component of a completely different type (for which the structure $Cl_2Al[OAl(Cl)]_nCl$ has been established) activates the system considerably. However, quantitatively, the observed effect is considerably less than that described for the zirconium complex. An increase in the ratio Al/Ti may produce a positive but incomparably smaller kinetic effect. This follows from Reichardt and Mayer [78]. According to this, when ethylene is polymerized by a $Cp_2TiEtCl$–$EtAlCl_2$ system which contains no water, with a ratio Al/Ti of about 30, then k_p is 15 liters·mole^{-1}·sec^{-1} (toluene at 10°C).

There is a certain similarity between the Zr–Al and Ti–Al complexes but this is limited by the promoting effect of the aluminooxane component. Processes based on systems containing chlorine initiated by both nonaqueous and aluminooxane systems are distinguished by the occurrence of termination reactions which are manifested by a sharp fall in the rate at an early stage [77, 78].

In papers on the specific influence of the chloroaluminooxane derivatives on the reactivity of the active site, the effects are ascribed to both the high Lewis acidity of this reagent and to the special role of the Al–O–Al group. From our point of view the first of these factors may be decisive in just that very case in which the presence of Al–Cl bonds results in an increase in the acidity of the aluminooxane. Regarding methylaluminooxane complexes, the feature which we have already discussed is much more likely although a full description of the mechanism by which the intermediate complex of the active site with the monomer is formed is difficult. The second order dependence of the polymerization on aluminooxane can be reconciled with the ability of the active site to form complexes with the monomer in the absence of obvious vacant positions [see structure (4.XIX)] and is possible only by assuming that the multi-bridged aluminooxane bonds do not completely exhaust the unsaturation of the central atom.

The increase in the reactivity with increase in the dimensions of the aluminooxane ring is evidence for a structure of the type (4.XIX). It must be borne in mind that the maximum kinetic effect is favored by both the increase in the number of $\begin{matrix} >Al \\ >Al \end{matrix}\!\!>O\cdots Zr$ bonds which are distinguished by their low stability and their lengthening. In its turn, the second of these effects facilitates the approach of the central atom of the active site to the monomer at the stage of π-complex formation. The geometry of this is obviously particularly favorable to the insertion of the monomer molecule into the structure of the growing chain.

The concepts discussed here represent, in a somewhat extrapolated form, suggestions of the Hamburg school which are not distinguished for their precision. In particular the possible role of the Al—O—Al bonds and the oligomeric state of the aluminooxane [72], which has so far not been used to arrive at a precise interpretation of the action of the active sites, were noted.

Thus there are grounds for regarding aluminooxanes as specific activators of transition metal compounds whether they contain halogens or not. Regarding the role of the nature of the central atom in these active sites, it can be said that, of the number of systems studied to date, the zirconium complexes differ markedly with respect to the reactivity of the growing chains which they form.

4.1.3. The Role of the Nature of the M_n^* Component in Growing M_n^*Y Chains

The dependence of the reactivity of a given type of active site on the nature of the M_n^* fragment is one of the important elements of the mechanism of ionic copolymerization although it sometimes appears somewhat masked. An example of such masking is the determination of a group of phenomena such as the "activity inversion" of monomers in M_1—M_2 systems. This term, which appeared about twenty-five years ago as a result of certain errors, appears from time to time in the modern literature and this makes it necessary to deal with it briefly here.

The efficiency of homopolymerization with fixed components and conditions may be evaluated with sufficient objectivity by the overall rate at which polymer is formed. If the problem is to compare the efficiency of processes, then stating parallels between systems containing different monomers is quite correct. Features based on the overall rate of polymerization are such that they may be even more useful than the value of the growth rate constants, especially in the solution of practical problems. However, these concepts lose their meaning when the same results are discussed from the point of view of the reactivity of real coreagents which are active in the processes compared. None of the kinetic characteristics which relate to homopolymerization, including the values of the absolute rate constants of elementary processes, can be used as a basis for rigorous conclusions for M_1–M_2 systems. Cross propagation in which one of the reactants is constant may provide such a basis for these conclusions. We emphasize that we have in mind an evaluation of the reactivities of active sites which lies outside the framework of single M_1–M_2 systems, i.e., consists of more than the determination of the copolymerization constants.

Making use of the symbols used for describing copolymerization, the values which characterize the reactivity of active sites and monomers may be written as the series (4.22) and (4.23), respectively:

$$k_{11},\ k_{21},\ k_{31},\ k_{41},\ \ldots \tag{4.22}$$

$$k_{11},\ k_{12},\ k_{13},\ k_{14},\ \ldots \tag{4.23}$$

The use of the term "activity inversion" essentially asserts the appearance of such a total kinetic effect in the M_1–M_2 system which could not be foreseen on the basis of data on the values of k_{11} and k_{22} only, or the total rate constants in the corresponding homopolymerization processes. It is clear that such effects are the consequence of making the system more complex and deviate from normal behavior. There is no possibility of making a judgment about these deviations on the basis of the data on individual processes of the polymerization of monomers M_1 and M_2.

An appreciable volume of data on the cross-propagation constants has been established only for those values corresponding to the series (4.23). On the other hand, for series (4.22) the data are only fragmentary and clearly insufficient even for the most modest generalizations. One of the important restrictions is connected with the necessity of using the values characterizing the growing chains of the corresponding monomers (i.e., M_1, M_2, etc.) of a single definite type (see Chapter 3) for the formulation of series (4.22). Any belief that this condition holds can strictly be based only in comparatively rare cases and an attempt to use them to choose parameters for comparison (the simplest of these is the ratio k_{11}/r_2)

is premature. Such a problem could form the subject of a special experimental research program.

4.1.4. Concluding Remarks

Many of the important questions connected with the reactivity of active sites are inseparable from the relative activity of monomers. Nevertheless, before proceeding to Section 4.2, it is worthwhile to discuss certain individual features of the active sites which differ in the form and in the nature of their original components.

Apart from the classification of active sites according to the processes to which they give rise, this question may be reduced to a choice of criteria necessary for the establishment of the connection between the nature of the fragment M_n^* and the relative activity of M_n^*Y in the growth reaction. Attempts which have been undertaken earlier in this field have usually been limited to grouping ionic initiators according to their selectivity in the polymerization of the various unsaturated monomers. The interpretation of the differences in behavior of the initiating agents and of growing chains which give rise to polymerization processes which differ from free radical processes was based on the qualitative features of the ionic character of the active bond. This parameter has not been used as an index of reactivity (IR) until now.

The question of the IR which we touched upon at the beginning of the present chapter has so far not been widely discussed when applied to ionic active sites. There are two basic reasons which cause this. The first is the almost complete lack of data on the physicochemical characteristics of real ionic reagents, particularly data relating to the usual conditions of polymerization. The factors which are important for this purpose are first and foremost the geometry of the M_n^*Y reagents and their electron density distribution. The second, which is no less important, is confidence in the correct use of known constant parameters as criteria of the dynamic behavior of these reagents.

The possibility of obtaining the necessary values experimentally in the near future cannot be counted on and so theoretical studies of well-chosen model compounds make even more sense. The use of the order of the active bond as an IR seems hopeful, but this needs to be tested on a large number of substances. If this were done sufficiently widely it would allow one to see whether the very modest results obtained so far have any significance beyond the limits of the particular examples studied. The determination of an IR along these lines is particularly useful because it yields the possibility of obtaining large amounts of information relatively quickly.

The other side of the problem is the necessity of an accurate and unequivocal correlation between completely defined active sites and the kinetic effects which they produce. This is of primary significance for processes in which growth takes place on different active sites which each give an appreciable contribution to the total effect. This question may even become an end in itself in the general problem of the reactivity of these sites. There are many ways of approaching this problem. There are a number of studies in which attention has been concentrated on the discussion of the reliability of kinetic data. There is also the difficulty of which of the available data are to be preferred even when the reactivity is omitted from the problem. Reminders of this are the discussions of the experimental values obtained for the polymerization of styrene under the action of free ions [79-81], the physical significance of the kinetic data for various cationic systems [82], and the reasons for the scale of the differences in the reactivity of free ions and ion pairs in hydrocarbon and heterochain anionic and cationic active sites ([83, 84, 85], respectively). The differences in conclusions which are often found are the result of the methods employed or differences in interpretation of the effects.

We will now deal with the question of the differences in the values of k_p for active sites of the type IP_c and free ions. Writing these as $k^{\pm}$ and k^{*}, respectively, we introduce the symbol Q for the ratio $k^{*}/k_p^{\pm}$. The average value of Q for ionic polymerization of various monomers are shown by the following data:

Processes	Anionic polymerization					Cationic polymerization			
Monomers	Styrene	α-Methyl-styrene	Buta-diene	Iso-prene	Ethyl-ene oxide	Styrene	Butyl vinyl ether	THF	Dioxo-lane
Q	10^2-10^3	10^2	10^4	10^4	10	10^4-10^5	10^4-10^5	1	1
Ref.	36	86,87	88	89	83	37	37	85	85

The features found for THF and dioxolane are obviously the result of the specific nature of their active sites which may exist in the oxonium form.

The discussion of whether the growing chains which play the dominant role in cationic polymerization are carbenium or oxonium in nature has in recent years become clearer. This has resulted from the appearance of a considerable amount of new data on the physicochemical features of compounds and model systems. This has been examined in great detail by Penczek et al. [42] who, on the basis of their own research and published data, came to the conclusion that carbenium active sites possibly played some part in growth in the case of cyclic acetals but not in the case of epoxides. Evidence favoring this is provided by ^{1}H- and ^{13}C-NMR spectra which are used to evaluate equilibrium constants of the type

$$>C^{+} + O \rightleftarrows >C-\overset{+}{O}- \quad (4.24)$$

and the degree of electrolytic dissociation of the carbenium and oxonium ions when polymerizing the above monomers. In particular the constants of the following equations were evaluated using ^{13}C-NMR:

$$CH_3OCH_2^+ + OEt_2 \rightleftarrows CH_3OCH_2\overset{+}{O}Et_2 \quad (4.25)$$

$$CH_3OH_2^+ + CH_3OCH_2OCH_3 \rightleftarrows CH_3OH_2-\overset{+}{O}(CH_3)-CH_2-O-CH_3 \quad (4.26)$$

Some of the values in [42] are given below:

Reaction	t, °C	K, liters/mole
(4.25)	−70	$2.6 \cdot 10^6$
	−30	$6.8 \cdot 10^4$
(4.26)	−70	$3 \cdot 10^3$
	−30	$4 \cdot 10^2$

Reactions (4.25) and (4.26) may be taken as precise models for the growth process up to the stage which precedes the transformation of the M_nY chain to M_{n+1}. In the majority of research which concentrates purely on the oxonium mechanism, the act of ring opening is in general not defined. Hence the detail which is the most important for the correct interpretation of the growth mechanism remains essentially undiscussed.

The scheme for oxonium growth is usually formulated [42, 85] in the form of the equation

$$\sim\overset{+}{O}\bigcirc \;\overset{O\bigcirc}{\rightleftarrows}\; \sim O\text{———}\overset{+}{O}\bigcirc \quad (4.27)$$

It is possible to add the unimolecular stage (4.28a)

$$\sim\overset{+}{O}\bigcirc \;\underset{a}{\rightleftarrows}\; \overset{+}{C}\!< \;\underset{b}{\overset{O\bigcirc}{\longrightarrow}}\; >C-\overset{+}{O}\bigcirc \quad (4.28)$$

A single-stage reaction (4.27) in a system of cyclic active sites (which exist as ion pairs) is so improbable that even with extremely low values of the equilibrium constants for reaction (4.28a), this mechanism must be given preference over reaction (4.27). Contrary conclusions cannot be regarded as being beyond dispute. Such conclusions are based on the absence of indications of the existence in cationic polymerization of heterocyclic monomers of carbenium growing chains which appear in the polymerization of unsaturated monomers. Such a feature is easily explained by the suppression of side reactions which are typical of C^+ centers. This is due to conditions which are extremely favorable to interactions of the type (4.28b) with the monomer or with units of the macromolecule by a mono- or bimolecular mechanism.

From this point of view the scheme proposed by Penczek et al. [37] for the mechanism of reoxonization when active sites react with monomer, which is shown below for THF,

THF → CH_2—CH_2—CH_2—O ⋯ C^+ ⋯ O(THF) (4.29)

may be regarded as a possible way of describing reaction (4.28).

However, an interpretation closer to scheme (4.27) is not excluded if it is based on the distribution of the electron density calculated for the oxonium active sites of cyclic monomers (Section 2.3). On the basis that the bulk of the charge is concentrated on the C_α atoms of the active sites, it is possible to propose that the growing chain interacts with the monomer via a pathway, the first stage of which approximates to scheme (4.29):

δ+ O δ+ $\xrightarrow[a]{M}$ ~O ⋯ O $\xrightarrow{b}$ ~O ⋯ δ+ O (4.30)

Such a reaction path distinguishes these active sites from the usual carbenium-ion sites and simplifies the growth mechanism in the polymerization of heterocyclic monomers. Even without carrying out special calculations (which have so far not been done) it is not possible to be in any doubt that the orientation of the active site and the monomer corresponding to stage (4.30a) is energetically very favorable.

Similarly it is possible to approach the mechanism of the splitting off of "foreign" monomers from the growing oxyalkylene chain. This mechanism is similar to the formation of intramolecular complexes of anionic active sites. Of the example given earlier (see Section 3.4) we mention again the splitting off of macrocycles from growing ε-caprolactone chains. In constructing an analogous scheme for the cationic polymerization of oxygen-containing heterocycles there can be no question of the formation of complexes with the direct participation of the counterion in a donor—acceptor interaction. The function of the reagent which attacks the oxygen atoms of the growing chains in such processes must be fulfilled by the end group. Formulation of the mechanism for these processes leads to alternative conclusions which reflect the transformations (4.28) and (4.29) which have been discussed above. Taking, for simplicity, the formulation of the first of these it is possible to conceive a picture of the structure of the corresponding AS_c as a structure of the types (4.XXIII) and (4.XXIV) in which the O atoms belong to other arbitrary units of the chain (counterions omitted).

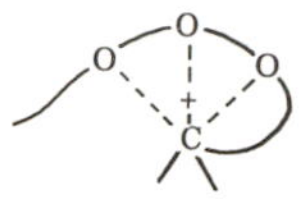

(4.XXIII)

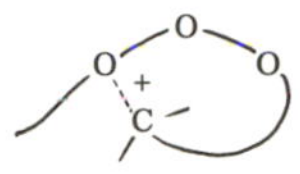

(4.XXIV)

Although these structures cannot strictly be regaded as complexes, for the sake of brevity we will refer to them, analogous to anionic reagents, by the symbols As_c—PC (4.XXIII) and AS_c—SC (4.XXIV). Using these it is possible to propose scheme (4.31) which reflects the interaction between the various forms of cationic heterochain active sites. Here M_x is any possible product which might split off, including the monomer itself:

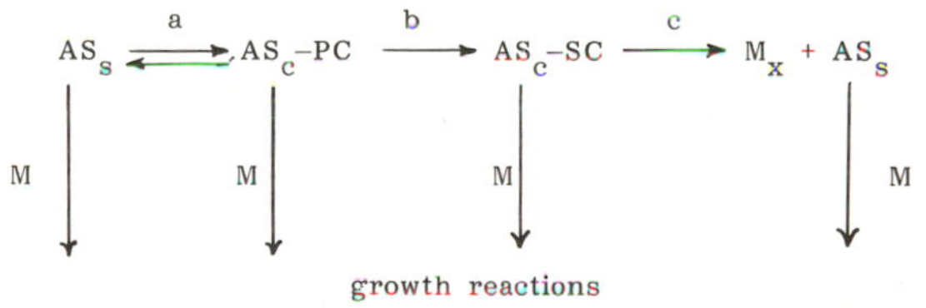

$$AS_s \overset{a}{\rightleftharpoons} AS_c\text{–PC} \overset{b}{\longrightarrow} AS_c\text{–SC} \overset{c}{\longrightarrow} M_x + AS_s \qquad (4.31)$$

(each of AS_s, AS_c–PC, AS_c–SC and AS_s → M ↓ growth reactions)

The hypothetical nature of this scheme makes further comment necessary. First, for brevity the reversibility of the reactions which is characteristic for such processes is not shown [see Eq. (4.27)]. Secondly, in scheme (4.31) the different symbols usually used for the active sites in the initial and final states are not used here, since the corresponding reagents, although similar, are not identical. Thirdly, and finally, this scheme emphasizes the irreversibility of stages (4.31b) and (4.31c) which from our point of view may be closer to a limiting case than to the normal situation. Special attention needs to be paid to the probability of various rings splitting off as a result of the periodic formation of an AS_c—SC structure. The form of proposed scheme (4.31) shows this concept quite clearly.

We return now to the condition Q = 1 which is found for several processes in the cationic polymerization of cyclic oxides (see page 164). The possibility of a determining role for the unimolecular event of ring opening in two-stage growth reactions of the type (4.28) or (4.30) must be regarded as a possible reason for this phenomenon. It may not be very significant whether the active sites are ion pairs or free ions. Regarding attack of the monomer by the growing chain, here the superiority of free ions over ion pairs is beyond question. For this stage the condition $Q \gg 1$ must hold.

The dependence of the modification of anionic active sites (provided the ion pairs are in an undissociated state) on the nature of the components is an interesting phenomenon but one for which so far an exhaustive explanation has not been found. Without considering crown ethers and cryptands it is possible to assert that there exist bidentate complex-forming compounds which are active separators of ion pairs (DME) and those which are not distinguished by such an ability (TMED). To some extent these differences also depend upon the initial growing chains. As an illustration we put forward the dependence of the reactivity of PBL in THF on the presence of small quantities of DME [19]. With a ratio DME/PBL = 10 and a temperature of −70°C a fivefold increase in the total value of k_p is observed while the value of K_{diss} did not change to such a degree that this effect could be ascribed to the increase in the concentration of free carbanions. This has led Vinogradova et al. [19] to conclude that a certain amount of IP_s appears in the system. This is also supported by data from the same study on polymerization in hexane.

As has already been noted (see page 122), the values which can be given for k_p for complexes of the monomeric form PBL with TMED and DME in hydrocarbon media are 0.20 and 0.38 liter·mole^{-1}·sec^{-1}. respectively. Only the second of these continues to increase when further excess of electron donor is introduced, reaching a value close to 0.8 liter·mole^{-1}·sec^{-1} with a ratio DME/PBL = 200 (the concentration of the PBL being of the order of 10^{-3} moles/liter at room temperature). Polymerization in TMED medium does not cause the value of k_p to deviate from 0.20 liter·mole^{-1}·sec^{-1}. For these conditions there is no doubt that IP_s are formed under the action of DME while such an effect is absent for TMED.

Regarding the polymerization in the butadiene–PBL–THF system, the reason for the activating action of moderate amounts of DME can be seen from the following scheme:

$$M_nLi\cdot mTHF + DME \underset{}{\overset{K_1}{\rightleftharpoons}} THF + M_nLi\begin{matrix}\swarrow (m-1)THF \\ \nwarrow DME\end{matrix} \quad \overset{K_2}{\leftrightharpoons}$$

$$M_nLi \;\vdots\vdots\; DME \;\vdots\vdots\; Li\cdot(m-1)THF \qquad (4.32)$$

The dependence of the value of k_p on the concentration of DME accords with the relatively large value of K_1 when $K_1 \gg K_2$.

The dependence of k_p on the D/Li ratio for scheme (4.32) can be contrasted with the action of TMED in the polymerization of butadiene and styrene induced by the corresponding M_nLi chains. In both cases the maximum and constant value of k_p is reached when the D/Li ratio is close to unity. The low value of this constant enables the active sites in the system to be assigned as IP_c with confidence.

The action of the other nitrogen-containing polydentate electron donor, tetramethyltetraazacyclotetradecane (TMTAC), is distinguished by an interesting feature. According to Helary and Fontanille [90], in the polymerization of styrene the value of k_p increases linearly with an increase in the ratio TMTAC/Li to unity, but remains constant if this ratio increases further. However, the value of k_p (it is found to be equal to 750 liters·mole^{-1}·sec^{-1} in cyclohexane at 20°C), which considerably exceeds the values of k_p known for IP_c, is much smaller than those of IP_s established for the anionic polymerization of styrene in other systems. Thus according to Szwarc et al. [91] the value of k_p for sodium polystyryl ion pairs separated by dimethoxyethane is 14,000 liters·mole^{-1}·sec^{-1} (at 0°C). This circumstance led Helary and Fontanille [90] to assume the existence in the TMTAC system of a special type of ion pair with a stretched active bond which corresponds to an intermediate state between IP_c and IP_s. Although correct in principle, it is difficult to regard this as a final conclusion since the values compared above relate to different counterions and reactive media which possibly introduce features of their own. Furthermore, only the value of k_p established for systems containing TMTAC can be regarded as strictly reliable. The kinetic characteristics of IP_s in DME, as was shown in [91], are no more than provisional and subject to large errors.

Nevertheless it must be admitted that the variety of forms of ion pairs which have been studied in detail in [92] and [93] is still not exhaustive. In particular for such complex-forming compounds as TMTAC, whose backbones are less flexible than those of their oxygen analogs (crown ethers), the possibility of the separation of ion pairs, in the usual sense in which this term is understood, is not obvious.

4.2. REACTIVITY OF THE MONOMERS

If the factors which are most important for characterizing the relative reactivity of monomers are separated from those features of copolymerization which have been frequently discussed in the literature [94-97], then the first and most general conclusion seems to be that the concept of reactivity itself is very complex in ionic as compared to radical processes.

The two-stage growth reaction which is quite usual for ionic polymerization [Eq. (4.1)] often forces one to be satisfied with data on overall effects. Only occasionally can they be broken down, and stage (4.1a) either excluded or eliminated from the selectivity of the active sites for comonomers.

The features of ionic systems associated with their multicenteredness are extremely important for the evaluation of the reactivity of monomers (see Section 3.4).

These circumstances lead to the conclusion that only the use of experimental values which may be regarded as strictly comparable is valid when evaluating the reactivity of monomers in ionic systems. We emphasize that this concept concerns only the question of the suitability of experimental data for establishing the connection between the physicochemical features of the monomer and its relative activity in growth.

In ionic polymerization it is possible to treat as ideal those conditions which ensure that all the systems studied are single-centered and that no appreciable role is played by intermediate complex-forming processes. Naturally, the term single-centered is to be understood here as meaning that the two varieties of growing chain, m_1 and m_2, belong to the same type of active site. Among the conditions must also be included the absence of noticeable differences in reactivity (and selectivity) of the growing chains of the general type $m_x m_x$ and $m_y m_x$. This can sometimes be arranged (see, for example, [98]). Such a choice severely limits the range of substances which can be studied. This however does not make such an approach inappropriate.

Proceeding to actual systems we will discuss the polymerization of unsaturated and heterocyclic monomers separately.

4.2.1. Unsaturated Monomers

Taking into account the fact that the reactivity of unsaturated monomers has been discussed many times we will concentrate on those aspects of the problem which require further examination.

4.2.1.1. The Principles of Selection and the Indices of Reactivity

The dependence of the growing chains (for example, $M_n^- \cdot Y^+$ or $M_n^+ \cdot Y^-$) on the nature of the Y fragment and the state of the active bond can often be identified in copolymerization as changes in selectivity of the active site for different comonomers. In its turn these changes reflect the differences in the selection principles which result from changes in the central atom of the active site, its ligands, and the form in which it exists.

From the point of view of the reactivity of the monomers, these same effects may be regarded as the result of changes in the relative roles of the various features of the compounds being polymerized. This leads to the necessity of choosing dissimilar parameters as the reactivity indices (RI) of the same monomers (M_1 and M_2) when going from one type of reactive M_n^*Y group to another. This question does not always arise in a clearly identifiable form. For example, in the case of a pair of monomers such as styrene and methacrylate, which is often used as an indicator of the mechanism, the simplest selection principle operates. It may be defined as selection by electrophilicity or nucleophilicity of the atom of the $\rangle C{=}C\langle$ system being attacked. These same parameters which are the most natural indices of the reactivity of unsaturated monomers preserve their value in many real ionic polymerization processes (reactions in styrene–MMA–M_n^*Y systems are not actually of this type). When copolymerization processes are compared which share a common mechanism but in which the initiating agents are of different types or exist in different forms, it is found that these indices can be used in a variety of ways. One example is the copolymerization of styrene (M_1) with butadiene (M_2) when the initiating compounds are of the RLi and RLi·D type in a hydrocarbon medium (toluene at 20°C).

Initiator	r_1	r_2	Ref.
n-Butyllithium (BuLi)	0.004	12.9	99
BuLi·TMED	0.6	1.0	100
BuLi, 2,3-DMB	1.2	1.7	100

The above values given for BuLi are preferred to any previously quoted data (see [99] also for bibliography).

These results, which go beyond purely quantitative differences, illustrate the inapplicability of the same criteria when trying to find the reactivity ratios of monomers both in the absence of and in the presence of ED. When interpreting the nature of phenomena which result from the strong dependence of the values of r_1 and r_2 on the degree of coordination saturation of the Li atoms, it is simplest to fall back on the mechanism of Eq. (4.1). M_nLi chains may be assigned a decisive role in selecting monomers at the stage of formation of $M_nLi{\cdot}M$ complexes (4.1a). On the other hand, for $M_nLi{\cdot}D$ chains, whose ability to form complexes with monomers is sharply reduced or is not manifested at all, the selection of a single-stage growth reaction may become decisive. Without some reservations, this interpretation is difficult to accord with the scale of the corresponding effects. In particular, the differences in the values of r_1 and r_2 in copolymerization in the absence of ED cannot be assigned to only one factor, i.e., to stage (4.1a). For this it would have to be assumed that butadiene predominates strongly over styrene in complex formation with the active sites, but there are no serious grounds for such an assumption. Therefore, accepting that complex formation at the stage of the initial selection of monomers is of overwhelming significance, it must be assumed that, for a given system, butadiene has a considerable advantage over styrene with regard to insertion in the chain. Of course such a forced conclusion would have to be made to agree with the almost complete removal of the differences between the same monomers in copolymerization with active sites of the type $M_nLi\cdot D$.

Certain arguments directed towards a detailed interpretation of these phenomena are possible but without special theoretical calculations they remain debatable. We direct our attention first to the importance of the mutual orientation of the components of each of the four possible $M_nLi\cdot M$ complexes and secondly to the changes of the favored point of attack by the monomer on the butadiene (Bt) end units as they change from M_nBtLi to $M_nBtLi\cdot D$ active sites, i.e., from the C_α to the C_γ atom of the Bt unit. Competition between growth events may be considerably affected by these and other events.

We will limit ourselves therefore to noting that it is not enough to ascribe the RI of butadiene and styrene to only one complex-forming activity with the lithium atom in the absence of ED and to the electrophilicity of the growing chain in the case of $M_nLi\cdot D$ active sites.

Thus the features which distinguish ionic copolymerizations from the analogous reactions in free-radical systems cannot be explained simply even in comparatively straightforward cases. Data on the variations of r_1 and r_2 within the limits of some general mechanisms, with changes in the type of active site, have not as yet revealed the true significance of these effects. Their interpretation more often than not is based only on the electronic features of the comonomers. In isolated studies of a limited number of polymerizations this sometimes appears sufficient. However, for the parallel determination of the features of active sites (which determine their selectivity) and the ability of comonomers to compete it is necessary to draw considerably upon spatial characteristics of the reagents and the stereochemistry of the growth reaction.

In attempting broader generalizations it is also necessary to differentiate processes which are free from kinetic transformation events and those which are accompanied by the deactivation of the growing chains. This is envisaged in the following exposition.

TABLE 4.5. Constants for Cross Growth in PVB–Mt–M_2–Diethyl Ether Systems at 25°C [101]

M_2	k_{12}, liter·mole^{-1}·sec^{-1}	
	PVB–Li	PVB–Na
Butadiene	4.6×10^{-2}	-
Isoprene	4.7×10^{-3}	4.4×10^{-2}
2,3-DMB	4.4×10^{-4}	2.7×10^{-3}
1,3-Pentadiene	2.1×10^{-3}	1.0×10^{-2}
2-Methyl-1,3-Pentadiene	9.5×10^{-5}	9.7×10^{-4}
3-Methyl-1,3-Pentadiene	2.0×10^{-4}	2.0×10^{-3}
2,4-Hexadiene	1.2×10^{-6}	1.1×10^{-5}
1,3-Cyclohexadiene	2.2×10^{-3}	1.2×10^{-2}

4.2.1.2. Nonterminating Copolymerization Processes

The most objective indices of monomer reactivity with respect to some standard active site are the rate constants of the corresponding reactions [see series (4.23)]. The first series of such values can be found in the well-known studies of Szwarc on the addition of sodium polystyryl to a number of unsaturated monomers [92]. Analogous characteristics have been recently obtained by Al-Jarrah and Young [1] in which polyvinylbiphenylsodium (PVB-Na) and polyvinylbiphenyllithium (PVB-Li) were used as standard chains. The rate constants (k_{12}) for the addition of these chains to diene monomers (M_2) are given in Table 4.5.

Each standard chain has the same monomer reactivity series, which, as noted in [101], may agree with the changes in the electronic and spatial characteristics on transition from butadiene to its homologs or analogs. The relative role of the second of these factors is the more important. Data favoring this conclusion can be found in Table 4.5. The difference in the values of k_{12} for 2,4-hexadiene and 1,3-cyclohexadiene is significant. Furthermore, changes in the electronic structure of butadiene when a substituent is introduced are not so significant that they can be assigned a decisive role; this follows from the results given below which have been obtained for the monomer series (method CNDO/2) [25]. For brevity only the π-components of charge on the C atoms are shown (in e units):

$$\overset{-0.02}{CH_2}=\overset{+0.02}{CH}-CH=CH_2 \qquad \overset{-0.05}{CH_2}=\underset{+0.04}{\overset{CH_3}{C}}-\overset{+0.02}{CH}=\underset{-0.02}{CH_2}$$

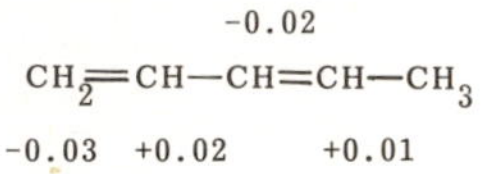

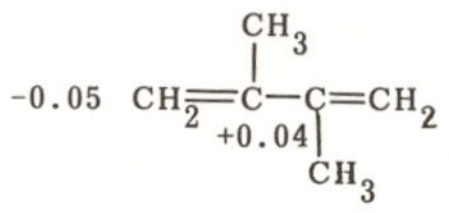

$$\overset{-0.03}{CH_2}=\overset{+0.02}{\underset{-0.02}{CH}}-CH=\overset{+0.01}{CH}-CH_3 \qquad \overset{-0.05}{CH_2}=\overset{CH_3}{\overset{|}{C}}-\underset{CH_3}{\underset{|}{\overset{+0.04}{C}}}=CH_2$$

It will be recalled that the differences between the charge characteristics of the individual compounds can be considered to be sufficiently objective theoretical values. The absolute values depend considerably on the quantum-chemical method used and the choice of parameters (see Section 3.3).

Quantitative evaluation of the differences in the reactivity of butadiene and isoprene with PVB-Li based on the values of k_{12} (Table 4.5) are in good agreement with similar values taken from the kinetic data for the polystyryllithium–diene system [92, 101]. In both cases the constant for butadiene is about ten times greater than that for isoprene. At the same time the difference between the growth constants in the homopolymerization of butadiene and isoprene in polar media initiated by polydienyllithium chains is much smaller; the ratio is 1:1.5 (see page 122). Such a result is difficult to foresee. On the basis of the spatial factor playing a greater role than the electronic factor, a rather greater superiority of butadiene over isoprene ought to have been expected when they react with the standard active site. This is because in homopolymerization, making the structure of the monomer more complex leads to a more complex end unit. On the other hand, the changes in the distribution of the electron density of the monomer induced by the introduction of the substituent and the active site act in just the reverse manner. In the present case this redistribution exerts a negative influence on the reactivity of the monomer and a positive influence on the activity of the growing chain; this produces at least partial compensation of the above effects.

From our point of view these aspects of the interaction of these monomers with the standard growing chains and the growing chains proper may be explained by drawing upon the conformational features of polybutadienyl- and polyisoprenyllithium. These can be found from the results of theoretical calculations from certain model compounds. As has already been noted in Section 2.3, it has been established for examples of alkenyllithium derivatives that the "cissoid" conformation is superior to the extended one. Obviously in real systems which include polydienyllithium chains equilibrium occurs between these and other shapes, similar to the equilibrium considered in Section 3.1.4 which for the specific case may be briefly explained by the scheme (4.33), in which R is H or CH_3:

$$\underset{A}{\sim CH_2-CR=CH-CH_2 \sim Li} \rightleftharpoons \underset{B}{\sim CH_2-CR=CH-CH_2 \sim Li \text{ (cyclic, CR=CH} \dashrightarrow \text{Li)}} \qquad (4.33)$$

Taking the linear form as the most reactive, among the factors which determine the behavior of the butadiene (R = H) and isoprene ($R = CH_3$) chains must be included the difference in the relative fractions of structures A and B. There are no experimental data which would enable this difference to be evaluated. Theoretical calculations of the optimum geometry have been so far carried out only for structures with model end units for these chains, i.e., $CH_3CH{=}CHCH_2Mt$ and $CH_3C(CH_3){=}CHCH_2Mt$, where Mt is lithium or sodium (CNDO/2 method; see [102]). In all cases the cissoid conformation seems to be the optimum. Of course from our point of view such a result might be regarded as disappointing. However, the models studied have been greatly simplified in order to answer the question. In this case (in contrast to other actual problems; see Section 3.3) at least the energy characteristics are necessary for the two-unit models. Only by starting with these is it possible to count on obtaining results necessary for a discussion of the effects of π-complex formation corresponding to scheme (4.33). The cissoid form of the above structures does not in general fall within the concept of a π-complex. Therefore our suggestion that there is a connection between the rate constants compared above and the conformation of the growing chains remains possible to some extent.

However, such a conclusion must be judged against the reservations which still remain about the reliability of the kinetic data compared. The data given in Table 4.5 do not give rise to doubts on these grounds, since it is the ratio of the values of k_{12} evaluated by exactly the same method which is important and not their absolute values.

In the homopolymerization of butadiene and isoprene the rate constants are evaluated very approximately using different approaches and so the situation cannot be compared with that of the first paragraph. Nevertheless we consider it possible to also draw upon those data which have up to now been the only data available in the literature on homopolymerization.

The data of [45] (see Section 4.12.1) on the kinetic features of the growth of polystyryl chains PStMt (where Mt is K or Cs) with compounds of the 1,1-diphenylethylene series (DPE) also deserve attention. The interaction between the above reagents is limited to the addition of one molecule of model monomer, which, together with the great difference in the magnitudes of λ_{max} for PStMt and the final reaction products (see Section 2.1) favors reliable results. The rate constants for DPE and its derivatives with respect to polystyryl ion pairs (with the counterions of K and Cs) and free polystyryl anions are given in Table 4.6. In contrast to the conclusions drawn on the basis of the data of Table 4.5 it is obvious here that the decisive influence is the effect of substituents on the electron density of the vinyl monomers.

TABLE 4.6. The Interaction between PStMt Chains and DPE Derivatives at 24°C [45]

Substituent in DPE	k, liter·mole^{-1}·sec^{-1}		
	PStK–C_6H_6 [a]	PStCs–C_6H_6 [a]	PStK–THF [b]
H	103.9	56.7	2130
m,m'-$(Me)_2$	47.8	23.2	745
p,p'-$(Me)_2$	18.0	8.6	340
p,p'-$(t\text{-}Bu)_2$	12.5	7.5	310
p,p'-$(MeO)_2$	3.2	0.75	53

[a] Average values given.
[b] Values from R. Busson and M. van Beylen [45]; under these conditions the contribution from the free ions is about 90%.

The reasons for the differences in the relative reactivity of potassium and cesium ion pairs are discussed by Busson and van Beylen [45]. They regard the higher reactivity of the former as due to the formation of intermediate $M_nMt \cdot M$ complexes (see Eq. 4.1) which cause the redistribution of the electron density in the monomer so as to favor reaction. Here it is accepted that in the case of cesium growing chains the growth reaction is single-staged as a consequence of the inability of the Cs counterion to function as an acceptor with respect to DPE and its derivatives. On the basis of such concepts the higher reactivity of PVB-Na compared with PVB-Li (see Table 4.5) can be explained by the fact that the series of experiments were carried out in ether, i.e., under the action of $M_nMt \cdot M$ chains. Obviously in these conditions the superiority of Li atoms over Na atoms as acceptors disappears.

The choice between the determining role of the electronic and spatial factors in copolymerization of monomers of a similar nature is distinguished by a special peculiarity in the alkenyl ethers. As far back as the beginning of the seventies Higashimura et al. discovered a quite unexpected effect in cationic systems which contain corresponding vinyl and propenyl derivatives as comonomers [103-105]. This is in contrast to cationic polymerization of monomers of the styrene series, where the reactivity of the β-substituted derivatives is considerably lower than those of styrene; in the unsaturated ethers a contrary phenomenon is observed. 1,2-Dimethoxyethylene is distinguished by a particularly high reactivity:

$$CH_2{=}CHOCH_3 < trans\text{-}CH_3CH{=}CHOCH_3 \ll cis\text{-}CH_3CH{=}CHOCH_3 < CH_3OCH{=}CHOCH_3$$

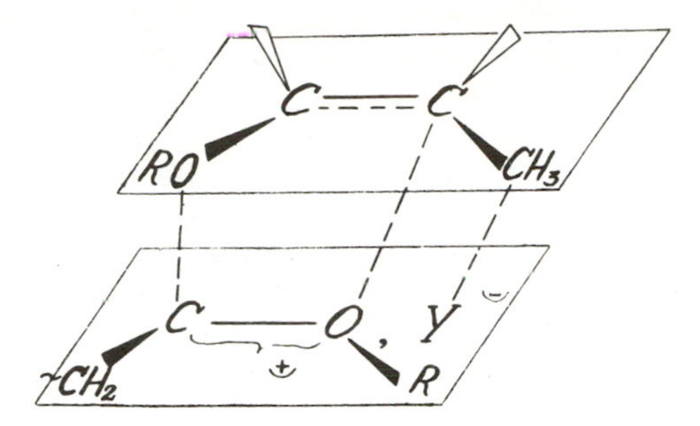

Fig. 4.11. Hypothetical mutual orientation of the monomer and the active center which precedes the act of growth in the cationic polymerization of cis-alkenyl ethers.

Such a sequence is ascribed to the participation of alkoxy monomer groups in the formation of intermediate complexes. For these the following schematic structures are proposed:

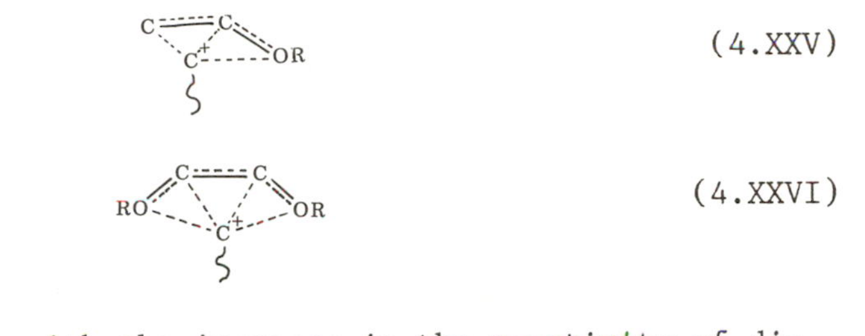

This suggestion agrees with the increase in the reactivity of dialkoxy derivatives, but not with the superiority of the cis-monoalkoxy compounds over the trans forms. The explanation of this superiority is not possible without additional concepts. These might possibly include the special character of the cationic active sites of vinyl ethers and the role of counterions. The former is distinguished by the even distribution of the positive π-charge between the C and O atoms of the end units (see [106]) and the latter is included in neither of the formulas (4.XXV) or (4.XXVI).

There is still one more circumstance which cannot be ignored. That is the distribution of the electron density in the monomers. Consideration of these points can lead to a more precise view of the geometry of the intermediate products and growth reaction paths which are responsible for the differences in the behavior of the cis and trans forms of monoalkenyl ethers. In particular at the intermediate stage the interaction between the counterion and the substituents on the C_β atom of the alkenyl group of the monomer is not excluded. Acccording to calculations carried out for propenylmethyl ether (using the CNDO/2 method of [103]) this substituent carries the positive π-charge. The possibility of including this substituent in the donor—acceptor reaction with the counterion is real only for the cis form of the monomer (see Fig. 4.11).

The difference in the reactivity of the cis and trans forms of these monomers manifests itself especially strongly only in the copolymerization in nonpolar media. In a polar medium the situation changes considerably in a quantitative and sometimes qualitative manner. Some examples from [103] are given below (initiator $BF_3 \cdot OEt_2$ at −78°C):

	r_{cis}		r_{trans}	
	toluene	nitro-ethane	toluene	nitro-ethane
Ethylpropenyl ether	2.76	1.14	0.27	0.88
Isopropyl-propenyl ether	4.70	0.67	0.14	1.09

The reason for this is seen in the reduced formation of intermediate complexes of active sites with monomer in polar media. Similar effects and also the influence of the nature of the counterion can be found in many other cationic copolymerization processes. A summary and analysis of the majority of results obtained up to 1975 are given by Kelen et al. [107]. Taking into account the critical remarks which this contains, we will refrain from comments on the connection between the nature of the initiator, the reaction medium, and the selectivity of the active sites for one of a pair of monomers in these conditions. We note that the sharp changes that occur in anionic polymerization when the sites are changed have not yet been noted in cationic copolymerization. This may be caused by the much lower ability of Y^- counterions to form complexes as compared with Y^+ counterions. The most typical feature of the first of these is coordination unsaturation and the tendency not to exchange ligands. Therefore, for cationic active sites the formation of complexes of this type, which is typical for anionic systems (in particular for $M_nMt \cdot M$), cannot be regarded as natural. It would require either an increase in the coordination number of the counterion or the squeezing out of one of the ligands. For the above processes the probability of either of these events is very small.

Digressing from differences in the scale of the effects observed in anionic and cationic systems when the conditions in which the copolymerization of the standard monomer pair is carried out are varied, it is possible to arrive at qualitatively similar conclusions regarding the mechanism of the processes. Generally it amounts to the realization of the parallel contributions of the reversible preselection of the monomers and their irreversible entry into the growing chain when the relative roles of these contributions are varied. For the relative reactivities of the monomers this means the simultaneous appearance of their various properties which usually reflect opposing tendencies and which correlate with the different RI. With these remarks we would hope to emphasize the necessity of regarding the phenomena observed as a result of the more or less balanced effects of different tendencies. Copolymerization with free ions should possibly be regarded as one of the few exceptions. As yet data necessary for a full discussion of this problem are unavailable.

4.2.1.3. Copolymerization in Processes with Termination

The problem of the reactivity of monomers acquires a particular character when applied to copolymerization where termination reactions cannot be neglected. The coincidence of the tendency of growing chains m_1 and m_2 to termination may be accidental in such processes. Therefore the possibility of each of these elementary processes will be determined apart from the values of r_1 and r_2 by the "lifetime" of each active site, i.e., by the constants of reactions

$$\sim m_1^* \xrightarrow{k_{t,1}} \text{deactivation} \qquad (4.34)$$

$$\sim m_2^* \xrightarrow{k_{t,2}} \text{deactivation} \qquad (4.35)$$

The influence of these reactions on the values of r_1 and r_2 must become greater as the difference between the constants $k_{t,1}$ and $k_{t,2}$ becomes larger. However, for the breakdown of the steady state in copolymerization which is often expressed by the equation

$$k_{12}\ [m_1^*]\ [M_2] = k_{21}\ [m^*_2]\ [M_1]$$

the existence of reactions (4.34) and (4.35) is sufficient; they can change the physical sense of r_1 and r_2. On the other hand, the absolute rate constants for homo- and cross propagation (both of which may be determined for such processes) may be taken as objective indices of the relative reactivities of the comonomers but not as elements in the usual equations of the copolymer composition.

The peculiarity of the situation, which is the result of including the termination reactions in the competition of the growth reactions, may lead to the dependence on the nature of the initiator even within the limits of an identical general mechanism. Data relating to the anionic copolymerization of polar monomers enable one to see just how considerable effects of this type may be.

The influence of the growing chains on the composition of the copolymer may not be sufficiently clear to allow unqualified conclusions to be drawn. Some effects characteristic of processes which are accompanied by reactions (4.34) and/or (4.35) are difficult to interpret simply. We will examine from this point of view the copolymerization of acrylonitrile (AN) with acrylates initiated by n-butyllithium in nonpolar solvents [108].

At temperatures between −80°C and about −70°C copolymers which contain mainly AN are formed, but at −30°C the selectivity of the growing chains practically disappears. The interpretation of this phenomenon assumes that the difference in the heats of complex formation when the comonomers interact with the growing chains is the deciding factor [stage (4.1a)] [109]. The explanation for the temperature dependence of the composition of the copolymers may be approached from another viewpoint. This involves taking into account an effect which is quite common in anionic polymerization of polar molecules, namely the decrease in the ratio k_p/k_t with rise in temperature. Before dealing with this it is compelling to point out yet another feature which is typical of these systems. This is the formation of nonhomogeneous copolymers which result from the copolymerization of AN with methyl-, ethyl-, and butylacrylates in nonpolar media. Therefore it is not completely correct to apply simple methods when using the usual experimental results to calculate copolymerization constants. The values obtained in this way must be regarded as no more than formal; hence from now on they will be referred to as "observed" copolymerization constants and designated r^o.

For data characterizing the anionic polymerization of NA(M_2) and acrylates it is possible to propose that the decrease in k_p/k_t as the temperature increases is more significant for M_1. This must be reflected in the r^o values as a reduction in r_1^o/r_2^o as the copolymerization temperature increases. This corresponds to the values calculated in the usual way (M_2 is methacrylate and toluene is the solvent, see [109]):

	−70°C	−60°C	−50°C
r_1^o	2.40 ± 0.10	1.98 ± 0.21	1.95 ± 0.13
r_2^o	0.20 ± 0.08	0.65 ± 0.13	1.35 ± 0.14

The following physical meaning may be given the above values:

$$r_1^o = f(\frac{k_{11}/k_{t,1}}{k_{12}}); \; r_2^o = f(\frac{k_{22}/k_{t,2}}{k_{21}}). \qquad (4.36)$$

The values of k_t are taken as the total rate constants of termination.

The dependence of these values on the inverse temperature may serve as a qualitative test of such an assumption. A relation of the type (4.36) must cause the nonlinearity of Arrhenius curves.

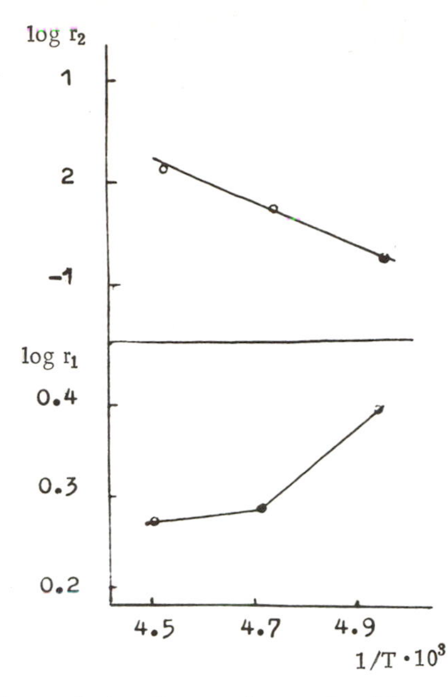

Fig. 4.12. Arrhenius plots for polymerization constants obtained in the acrylonitrile (M_1)–methacrylate–toluene system [109]. Initiator, n-butyllithium.

In the case of this system this condition is found for values of r_2^0 (see Fig. 4.12).

The facts and concepts agree qualitatively with the composition of the fractions of the nonhomogeneous copolymers under discussion. We thus draw upon data obtained from the copolymerization of acrylonitrile with methyl acrylate (MA) at −50°C with a monomer ratio of 1:1 and a conversion of 10% (duration of experiment = 15 sec). Average values are given for molar masses, evaluated using diffusion, sedimentation, and viscometric methods [108]:

	Fractions		
	dissolved in ethanol	dissolved in acetone	not dissolved in acetone
Molar mass	2500	7000	18000
MA content, mole %	75	40	11

The above feature of the copolymerization of acrylonitrile with acrylates also appears in a hydrocarbon medium with lithium alkoxides. However, in dimethylformamide, copolymerization of the same monomers with lithium alkoxides is not temperature sensitive and is not accompanied by the formation of nonuniform copolymers [110]. This is to be regarded as a consequence of the dependence of the final result upon the ratios k_p/k_t and k_p^D/k_t^D. The first of these characterizes copolymerization in toluene and the second in dimethylformamide. The consequences of the differences in these ratios, which may vary over quite wide ranges but generally obey the relations $k_p > k_p^D$ and $k_t^D < k_t$, are briefly discussed in Section 4.1.1.3.

We select for more detailed discussion those monomer pairs for which information is available for the different conditions and different initiators. As the examples already discussed show, such a variation makes uncertain the character of the reactivity indices of the constant components of the system, i.e., of the monomers themselves.

4.2.2. Heterocyclic Monomers

While considering the two-stage growth mechanism in the polymerization of unsaturated monomers, we have not paid special attention to the reversibility of the events [Eq. (4.1a)]. For the discussion of those questions touched upon in Section 4.2.1 it had no significance because the intermediate AS·M complexes, in the case of unsaturated monomers, do not difer from the initial active sites in the nature of their end units. Equation (4.1) may be regarded as applicable to anionic copolymerization of heterocyclic monomers which are, for this class of compounds, of limited interest. Cationic systems which cover an extremely large range of copolymerizing heterocyclic monomers and the intermediate stages of the growth reaction have a different character. In particular the reversibility affects the nature of the end units

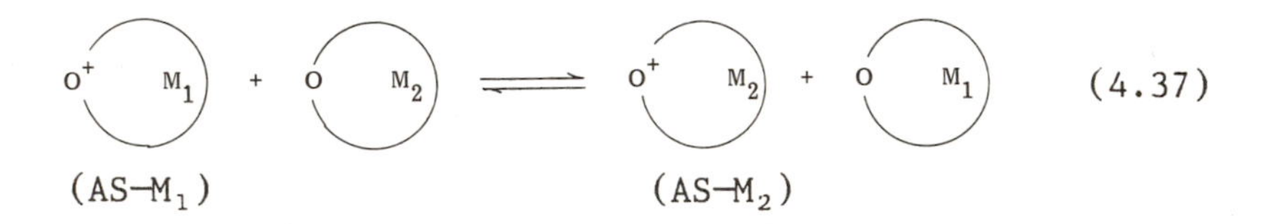

The difference between reactivity and selectivity of $AS{-}M_1$ and $AS{-}M_2$ is not particularly great for unsubstituted homogeneous rings, but cannot be neglected in monomer pairs of different types. These include primarily all copolymers with different numbers of oxygen atoms (for example THF/dioxolane) or with different substituents on the C_α atoms.

Another circumstance which relates to all heterocyclic pairs M_1/M_2 is the difference in the ability of $AS{-}M_1$ and $AS{-}M_2$ to spontaneous [(Eq. (4.28)] and induced [(Eq. (4.29)] severance of the terminal rings, i.e., to an irreversible stage which coincides with Eq. (4.1b).

However, the most important feature of copolymerization including the equilibria (4.37) which distinguishes them from the mechanism of type (4.1) is the parallel and not sequential course of the initial selection of monomers [Eq. (4.38a)] and the transformation of the cyclic end groups to penultimate groups of the growing chains [Eq. (4.38b)]. An example of one of the active sites is shown below in abbreviated form

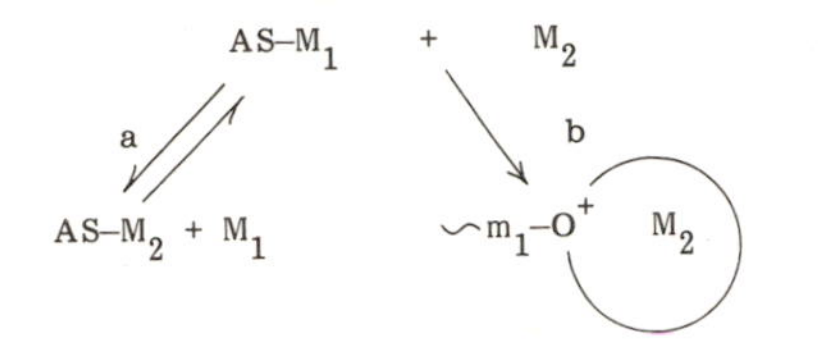

(4.38)

As a consequence of this feature the role of Eq. (4.38a) differs from that of Eq. (4.1a). The latter influences the selection of monomers but does not affect the physical meaning of the values of r_1 and r_2. Competition between these parallel processes [see scheme (4.38)] is connected with the basicity of reagents M_1 and M_2 and with the stability of the terminal rings in $AS{-}M_1$ and $AS{-}M_2$. However, this is so vague that a sufficiently rigorous analysis of the connection between these factors is scarcely possible.

Until recently such competition could only be assumed on the basis of an analogy with acyclic oxygen-containing compounds. Information has now begun to appear which describes some systems of this type directly. Thus Penczek et al. [111] investigated the vulnerability of the exo- and endo-CH_2 groups of a model trialkyloxonium active site consisting of salts of the ethyloxepane ion (counterion EX_6^- (where E is P, Sb, or Sn and X is F or Cl) when it interacts with THF. Analysis of the ^1H-NMR spectra of this system has led to evaluation of the rate constants of the competing events shown in Scheme (4.39). This scheme coincides with that of the general type (4.38):

$$\text{Et–}\overset{+}{O}\text{(7)} + \text{THF} \begin{cases} \xrightarrow{k_{exo}} \text{Et–}\overset{+}{O}\text{(5)} + \text{O(7)} \\ \xrightarrow{k_{endo}} \text{Et(CH}_2)_6\overset{+}{O}\text{(5)} \end{cases} \qquad (4.39)$$

At 35°C the following values were obtained ($\times 10^4$, liters·mole^{-1}·sec^{-1}.

Solvent	k_{exo}	k_{endo}
Dichloroethane	1.2	5
Nitromethane	3	6

The authors note the agreement of the values of k_{endo} and k_p which have been obtained by the study of the polymerization of oxepane under similar conditions ($5 \cdot 10^{-4}$ liters·mole^{-1}·sec^{-1} in nitromethane at 35°C).

Equation (4.37) considerably complicates the evaluation of the relative role of basicity as a factor responsible for the behavior of heterocyclic monomers in cationic copolymerization. This question has apparently never been examined in detail. To carry out such a task would require the overall equilibrium (4.37) to be broken down into its intermediate stages [for example, of the type (4.28) or (4.30)] for which there is no quantitative experimental data.

In the case of monomers which are sharply distinguished in this respect it is comparatively easy to approach these factors. One of the monomer pairs which correspond to this condition, namely ethylene oxide (EO) and trioxane (TO), was recently examined in detail by Collins et al. [112]. We give here some of their results which illustrate the features of such processes. In the copolymerization initiated by boron trifluoride dibutyletherate in a melt of TO (70-90°C) and with a large excess of the monomer, the initial stage of the process proceeds practically without the participation of TO. It takes part in the reaction only after almost all the EO has disappeared, i.e., at the moment when initial concentration of EO has fallen to about 0.0025 moles/liter. At the stage preceding the polymerization of TO the formation of "foreign" monomers of 1,3-dioxane (DO) and 1,3,5-trioxepane (TOP) is found. We have dealt with reaction mechanisms of this type in Section 3.1.4.

The concentrations of DO and TOP increase noticeably only at the very start of the reaction. Some of the data of [112] are given below (the initial concentration of EO is 0.5 moles/liter, that of the initiator is of the order of 10^{-3} moles/liter, temperature, 70°C):

Duration of experiment, sec	Concentration in solution, %	
	DO	TOP
20	0.01	0.10
40	0.033	0.18
50	0.43	1.9

When TO begins to take part in the polymerization, i.e., in the case of the above experiment after about 60 seconds (this is easily determined by the sharp jump in temperature and by the loss of homogeneity of the system), the concentrations of DO and TOP begin to fall, gradually approaching some equilibrium value. The above phenomenon in fact corresponds to the successive polymerization of EO and TO and not to their copolymerization. Each of these stages is complicated by additional reactions. The first of these is the formation of DO and TOP and the second is the depolymerization of

TO to formaldehyde (FA) and the copolymerization of the three heterocycles (TO, DO, and TOP) in which FA plays some role. The effects connected with the depolymerization of TO and competition between the growth of TO and FA have been discussed earlier by Jaacks [113]. It is sufficient here to draw attention to the correlation between the known phenomenology and the physicochemical features of the initial and "foreign" monomers.

The initial selection of monomer is carried out, according to Collins et al., under the action of a Brönsted acid which arises in the system as a result of the presence of small amounts of water (according to Collins et al. this is of the same order as the concentration of the initiator). The higher basicity of EO as compared with TO ensures that the initiator is used practically exclusively on the formation of the initial active sites designated in [112] by $H\text{–}\overset{+}{O}\!\lhd$, X^- where the counterion is assigned a series of structures of the form $[BF_3OH]^-$ to $[BF(OH)_3]^-$ depending upon the actual water content of the system. For the same reason these initial active sites react only with EO, forming active sites upon which growth takes place. Naturally such a succession may be regarded as the only possibility until the fall in concentration of EO to the level at which, in great excess of TO, the ability of the reactants to compete manifests itself. In this sense the considerable difference in ring strain may be considered to have no influence.

In a detailed discussion of the mechanisms in this system, Collins et al. [112] include "ring expansion" as shown below as responsible for the formation of new monomers DO and TOP. For brevity we will show this only for the formation of DO (counterions omitted):

$$\sim\overset{+}{O}\!\lhd \xrightarrow{\text{HCHO}} \sim\overset{+}{O}\langle\text{ring}\rangle O \qquad (4.40)$$

$$\sim OCH_2\text{–}\overset{+}{O}\langle\text{ring}\rangle O \longrightarrow \sim\overset{+}{O}CH_2 + O\langle\text{ring}\rangle O \qquad (4.41)$$

An equilibrium concentration of HCHO has been known in such systems since the work of Kern and Jaacks [114], and the formation of dioxane active sites as given by Eq. (4.41) may be connected with the entry into the reaction of FA or TO. Nevertheless, the reality of the sequential events (4.40, 41) is not obvious. The authors justify this mechanism by proposing the existence of as yet undiscovered large rings among the reaction products. However, the splitting off

TABLE 4.7. Theoretical Charcteristics of Some Cyclic Oxides and Their Corresponding Active Sites [116, 117]

Monomers	P_{O-C_α}	$P_{O-C_\alpha'}$[a]	S[b]	q_O, $\bar{e}$	Active site	P_{O-C_α}	$P_{O-C_\alpha'}$[a]	S
 O	0.915	-	0.094	-0.212	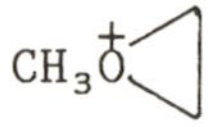 $CH_3\overset{+}{O}$	0.805	-	0.29
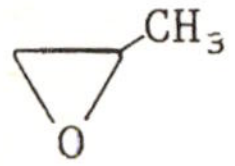 O, CH_3	0.920	0.886	0.15	-0.223	 $CH_3\overset{+}{O}$, CH_3	0.804	0.743	0.38
 O	0.900	-	0.06	-0.28	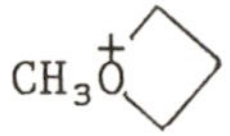 $CH_3\overset{+}{O}$	0.740	-	0.33
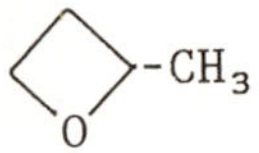 O, $-CH_3$	0.910	0.873	0.11	-0.292	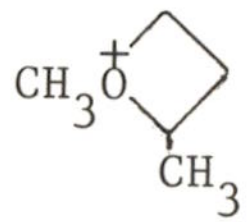 $CH_3\overset{+}{O}$, CH_3	0.774	0.707	0.37
 O	0.985	-	-0.11	-0.234	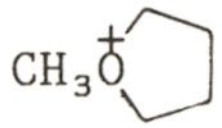 $CH_3\overset{+}{O}$	0.854	-	0.15

[a] C_α' designates the C atom of the C_α substituent.

[b] Ring strain.

of rings which are larger than TOP is a fact which has been established by researchers working in the field of cationic polymerization of oxygen-containing heterocyclic compounds ([113, 115], for example). This does not exclude the mechanism of splitting of DO and TOP which is the more probable and which is the result of intramolecular events similar to those considered in Chapter 3. Thus, it is obvious that for an EO/TO system ring strain as an independent parameter is assigned to a secondary role. It is necessary to take this factor into consideration in the copolymerization of monomers which are not so sharply differentiated as EO and TO. Also it is not possible to ignore yet another "subparameter" which is comparatively rarely considered; this is the possible release of ring strain when ◯O is transformed into the active site $Ct{-}\overset{+}{O}$◯.

Results of recent theoretical studies by Geller et al. [116, 117] concentrate on the different oxides (3-, 4-, and 5-membered), their substituents, and the active sites of the general types $CH_3{-}\overset{+}{O}$◯. The results of calculations on about 15 monomers and corresponding active sites show a marked increase in ring strain (S) when they are transformed into active sites. The degree of change found is shown for several examples in Table 4.7. This also contains values of the orders of the C—O bonds in the initial and final substances and also the charges on the O atoms of the monomers. The latter values may be taken as theoretical features of the basicity of the oxides studied.

In discussing these results the authors draw a series of conclusions about the correlation of the features obtained with experimental data. It is noted that the lack of correspondence which is sometimes observed between the theoretical and experimental parameters may be explained by steric hindrance caused by substituents in the monomer during copolymerization. This effect may exceed the inductive effect of the substituent which increases the basicity of the monomer and reduces the order of its C—O bond. In carrying out such calculations the authors are guided by earlier results; for bibliography see [117].

The characteristics given in Table 4.7 are a modest selection from the data given in [116, 117]. We have used some of these data in Chapter 3 in discussing the electronic structure of oxonium active sites. We note that the charge characteristics of a large number of other active sites which have been studied confirm the concepts which we have presented in Chapter 3.

The theoretical characteristics of many three-membered oxides have been calculated using the CNDO/2 method in [118]. Here is stressed the dependence of the electronic structure of the monomer on the nature of the substituent. The question of the active sites corresponding to these monomers is not touched upon.

Instead of attempting to summarize the whole material of the present chapter we have preferred to review its main section which deals with the reactivity of active sites (see Section 1.4). A more complete summary of the questions connected with the mechanism of ionic polymerization is examined after the discussion of stereospecificity which forms the theme of the following chapter.

REFERENCES

1. B. L. Erusalimskii, "Structure and reactivity of anionic active centers," J. Polym. Sci., Polym. Symp., 62, 29-50 (1978).
2. B. J. Wakefield, The Chemistry of Organolithium Compounds, Pergamon Press, New York (1974).
3. V. N. Zgonnik, E. Yu. Melenevskaya, and B. L. Erusalimskii, "The study of active centers in anionic polymerization using spectroscopic and quantum-chemical methods," Usp. Khim., 47, 1479-1503 (1978).
4. A. Streitwieser, Jr., J. A. Williams, S. Alexandratos, and J. M. McKelvey, "Ab initio SCF-MO calculations of methyllithium and related systems. Absence of the covalent character in the C—Li bond," J. Am. Chem. Soc., 98, 4778-4784 (1976).
5. J. D. Dill, P. v. R. Schleyer, J. S. Binkley, and J. A. Pople, "Molecular-orbital theory of the electronic structure of molecules. 34. Structure and energies of small compounds containing lithium or beryllium. Ionic, multicenter, and coordinate bonding," J. Am. Chem. Soc., 99, 6169-6173 (1977).
6. T. Clark, J. Chandrasekhar, and P. v. R. Schlery, "^{7}Li-^{13}C NMR coupling constants and the nature of the carbon—lithium bond: INDO MO calculations," J. Chem. Soc., Chem. Commun., 671-673 (1980).
7. G. D. Graham, D. S. Maryninch, and W. N. Lipscomb, "Effects of basic set and configuration interaction on the electronic structure of CH_3Li with comments on the nature of the C—Li bond," J. Am. Chem. Soc., 102, 4572-4578 (1980).
8. G. D. Graham, S. Richtmeister, and D. A. Dixon, "Electronic structure of the alkyllithium clusters $(CH_3Li)_n$ (n = 1-6) and $(C_2H_5Li)_n$ (n = 1-2)," J. Am. Chem. Soc., 102, 5759-5766 (1980).
9. M. Morton and L. J. Fetters, "Homogeneous anionic polymerization. V. Association phenomena in organolithium polymerization," J. Polym. Sci., A2, 3311-3326 (1964).
10. R. Ohlinger, "Kinetischer Untersuchungen der mit Lithium-organischen Washstumskatalysatoren initiierten Copolymerisation von Butadien und Styrol mit dem Ziel der Darstellung von statistischen Copolymeren mit bestimmter Butadien-Styrol-Zusammensetzung," Dissertation, Hamburg (1974).

11. V. V. Shamanin, E. Yu. Melenevskaya, and V. N. Zgonnik, "The influence of the concentration of growing chains on the polymerization rate and the microstructure of the polymer formed in the polybutadienyllithium–butadiene–aliphatic hydrocarbon system," Acta Polym., 33, 175-181 (1982).
12. A. Hernandez, J. Semel, H.-Ch. Broeker, H. G. Zachmann, and H. Sinn, "The determination of the degree of association of polyisoprenyllithium in heptane," Makromol. Chem. Rapid Commun., 1, 75-77 (1980).
13. S. Bywater, "The preparation and properties of star-branched polymers," Adv. Polym. Sci., 30, 89-116 (1979).
14. K. Matsuzaki, Y. Shinohara, and T. Kendai, "Nuclear magnetic resonance studies on polymer carbanions. 1. Living polystyrene and its model compounds," Makromol. Chem., 181, 1923-1934 (1980).
15. S. Dumas, B. Marti, J. Sledz, and F. Shué, "The influence of N,N,N',N'-tetramethylethylenediamine on the anionic polymerization of isoprene in cyclohexane," J. Polym. Sci., B16, 81-86 (1978).
16. W. Gebert, J. Hinz, and H. Sinn, "Umlagerungen bei der durch Lithiumbutyl initiierten Polyreaktion der Diene Isopren und Butadien," Makromol. Chem., 144, 97-115 (1971).
17. E. Yu. Melenevskaya, V. N. Zgonnik, V. M. Denisov, E. R. Dolinskaya, and K. K. Kalnin'sh, "The nature of the active centers in the copolymerization of styrene with butadiene initiated by an n-butyllithium–tetramethylethylenediamine complex," Vysokomol. Soedin., A21, 2008-2016 (1979).
18. L. V. Vinogradova, N. I. Nikolaev, and V. N. Zgonnik, "The nature and reactivity of the active centers in the system butadiene–n-butyllithium–tetramethylethylenediamine–hydrocarbon medium," Vysokomol. Soedin., A18, 1756-1761 (1976).
19. L. V. Vinogradova, N. I. Nikolaev, V. N. Zgonnik, and B. L. Erusalimskii, "Forms of existence and relative activity of polybutadienyllithium in the polymerization of butadiene," Eur. Polym. J., 19, 617-620 (1983).
20. A. A. Davidyan, N. I. Nikolaev, V. N. Zgonnik, and K. K. Kalnin'sh, "The reactivity and physicochemical features of the active centers in the system isoprene–butyllithium–tetramethylethylenediamine–hexane," Vysokomol. Soedin., B17, 586-590 (1975).
21. A. A. Davidyan, N. I. Nikolaev, V. N. Zgonnik, and V. I. Petrova, "The reactivity and physicochemical features of the active centers in the system isoprene–oligoisoprenyllithium–dimethoxyethane–hexane," Vysokomol. Soedin., A18, 2004-2010 (1976).
22. M. Morton, L. Fetters, and E. Bostick, "Mechanisms of homogeneous anionic polymerization by alkyllithium initiators," J. Polym. Sci., C1, 311-323 (1963).
23. S. Bywater, "Anionic polymerization," Adv. Polym. Sci., 4, 66-110 (1965).

24. A. A. Davidyan, "The reactivity and physicochemical features of the active centers in the system isoprene–alkenyllithium–electron donor," Dissertation, Inst. Vyskomol. Soedin. Akad. Nauk SSSR, Leningrad (1977).
25. Yu. E. Eizner and B. L. Erusalimskii, The Electronic Aspect of Polymerization Reactions [in Russian], Nauka, Leningrad (1976).
26. H. F. Ebel, "Structure and reactivity of carbanions and carbanioid compounds," Fortschr. Chem. Forsch., 12, 387-439 (1969).
27. I. G. Krasnosel'skaya (Krasnoselskaya) and B. L. Erusalimskii (Erussalimsky), "Sequences of complex formation in the polymerization processes induced by organomagnesium compounds," Eur. Polym. J., 13, 775-781 (1977).
28. B. L. Erusalimskii, "Overall and individual effects in systems including organolithium compounds," Makromol. Chem., 182, 911-915 (1981).
29. B. L. Novoselova and B. L. Erusalimskii, "Mechanisms der durch Lithiuminitiatoren angeregten Polymerisation von Acrylnitril," Faserforsch. Textiltech., 26, 293-300 (1975).
30. K. Hatada, T. Kitayama, K. Fumikawa, K. Ohta, and H. Yuki, "Studies on the anionic polymerization of methyl methacrylate initiated with butyllithium in toluene using perdeuterated monomer," in: ACS Symposium Series 166, J. E. McGrath (editor), 327-341 (1981).
31. A. Davidyan (Davidjan), N. I. Nikolaev, V. N. Zgonnik (Sgonnik), B. G. Belen'kii (Belenkii), V. V. Nesterov, V. D. Krasikov (Krasikow), and B. L. Erusalimskii (Erussalimsky), "Subkatalystische Effekte im System Isopren–Oligoisoprenyllithium–N,N,N',N'-Tetramethylethylendiamin. 2. Umsatzabhangigkeiten der Molekulargewichtsverteilung und Mikrostrucktur der Polymere," Makromol. Chem., 179, 2155-2160 (1978).
32. B. L. Erusalimskii, A. A. Davidyan, N. I. Nikolaev, V. N. Zgonnik, V. G. Belen'kii, V. D. Krasikov, V. V. Nesterov, and M. L. Kononenko, "Polymerization in the butadiene–styrene system under the action of organolithium active centers with subcatalytic quantities of tetramethylethylenediamine," Vysokomol. Soedin., A25, 2121-2125 (1983).
33. I. M. Panayotov and G. Heublein, "Cationic polymerization in the presence of π-electron acceptors," J. Macromol. Sci., Chem., A11, 2065-2086 (1977).
34. C. Reichardt, "Solvent effects in organic chemistry," in: Monographs in Modern Chemistry, H. F. Ebel (editor), Vol. 3, Verlag Chemie, Weinheim (1979), pp. 1-355.
35. T. Shinohara, J. Smid, and M. Szwarc, "Effect of solvation of ion pairs," J. Am. Chem. Soc., 90, 2175-2177 (1968).
36. H. Hirohara and N. Ise, "On the growing active centers and their reactivities in "living" anionic polymerization of styrene and its derivatives," J. Polym. Sci. D., Macromol. Rev., 6, 295-336 (1972).

37. A. Gandini and H. Cheradamé, "Cationic polymerization. Initiation with alkenyl monomers," Adv. Polym. Sci., 34/35, 1-289 (1979).
38. J. P. Lorimer and D. C. Pepper, "A stopped-flow study of the "free-ion" polymerization of styrene by $HClO_4$ in CH_2Cl_2 at low temperature," Int. Symp. on Cationic Polymerization, Rouen (1973), prepr. C23.
39. M. Sawamoto, T. Masuda, and T. Higashimura," Cationic polymerization of styrene by protic acids and their derivatives. 2. Two propagating species in the polymerization by CF_3SO_3H," Makromol. Chem., 177, 2995-3007 (1976).
40. M. Sawamoto and H. Higashimura, "Stopped-flow study of the cationic polymerization of p-methoxystyrene. Evidence for the multiplicity of the propagation species," Macromolecules, 11, 502-504 (1978).
41. S. Penczek, P. Kubisa, and K. Matyjaszewski, "Cationic ring-opening polymerization," Adv. Polym. Sci., 37, 1-149 (1980).
42. S. Penczek and R. Szymansky, "The carbenium—onium ion equilibrium in cationic polymerization," Polym. J., 12, 617-628 (1980).
43. K. Matyjaszewski, S. Slomkowski, and S. Penczek, "Kinetics and mechanism of the cationic polymerization of tetrahydrofuran in solution. THF—CH_2Cl_2 and THF—CH_2Cl_2—CH_3NO_2 systems," J. Polym. Sci., Chem. Ed., 17, 2413-2422 (1979).
44. K. Matyjaszewski, T. Diem, and S. Penczek, "Rate constants of propagation of THF on macroesters and macroions," Makromol. Chem., 180, 1827-1829 (1979).
45. R. Busson and M. van Beylen, "The Hammett relation in anionic polymerization. Reaction of polystyryl alkali salts with disubstituted 1,1-diphenylethylenes," Macromolecules, 10, 3120-3136 (1977).
46. A. A. Arest-Yakubovich, "Alkaline earth metals as initiators of the anionic polymerization of unsaturated monomers," Usp. Khim., 1141-1167 (1981).
47. B. I. Nakhmanovich, V. A. Korolev, and A. A. Arest-Yakubovich, "The kinetics of the polymerization of butadiene and styrene under the action of bis-triphenylmethylbarium in THF," Vysokomol. Soedin., A18, 1480-1485 (1976).
48. A. H. E. Müller, "The present view of the anionic polymerization of methyl methacrylate and related esters in polar solvents," ACS Symposium Series 166, J. E. McGrath (editor), 441-461 (1981).
49. R. Craft, A. H. E. Müller, V. Warzelhan, H. Höcker, and G. V. Schulz, "On the structure of propagating species in the anionic polymerization of methyl methacrylate. Kinetic investigation in tetrahydrofuran using monofunctional initiators," Macromolecules, 11, 1093-1096 (1978).
50. R. Craft, A. H. E. Müller, H. Höcker, and G. V. Schulz, "Kinetics of anionic polymerization of methyl methacrylate in 1,2-dimethoxyethane," Makromol. Chem. Rapid Commun., 1, 363-368 (1980).

51. C. Johann and A. H. E. Müller, "Kinetics of anionic polymerization of methyl methacrylate using cryptated sodium counterions in tetrahydrofuran," Makromol. Chem. Rapid Commun., 2, 687-691 (1981).
52. H. Jeuk and A. H. E. Müller, "Kinetics of the anionic polymerization of methyl methacrylate in tetrahydrofuran using lithium and potassium as counterions," Makromol. Chem. Rapid Commun., 3, 121-125 (1982).
53. S. Murahashi, H. Yuki, H. Hatada, and T. Okata, "Polymerization of methyl methacrylate by diethylaluminumdiphenylamide. II. Initiation and stereoregulation in the polymerization reaction," Chem. High Polym., 24, 309-317 (1967).
54. E. V. Milovskaya, M. N. Makarychev-Mikhailov, and E. P. Skvortsevich, "Organoaluminum compounds as initiators in the anionic polymerization of methacrylates," Vysokomol. Soedin., A17, 1217-1222 (1975).
55. E. P. Skvortsevich, E. L. Kopp, and E. B. Milovskaya, "AlR_3–2,2-dipyridyl systems as initiators in anionic polymerization," Vysokomol. Soedin., A19, 1736-1743 (1977).
56. E. B. Milovskaya and Yu. E. Eizner, "Electronic structure, conformation, and reactivity of the active centers of anionic polymerization with an aluminum counterion," Eur. Polym. J., 15, 889-893 (1979).
57. E. P. Skvortsevich, E. L. Kopp, V. V. Mazurek, and E. B. Milovskaya, "The kinetics of the polymerization of methyl methacrylate under the action of triethylaluminum–2,2'-dipyridyl system," Vysokomol. Soedin., A21, 1554-1561 (1979).
58. E. L. Kopp, E. P. Skvortsevich, V. M. Denisov, A. I. Kol'tsov, and E. B. Milovskaya, "The catalytic activity of AlR_3–2,2-dipyridyl systems," Izv. Akad. Nauk SSSR, Ser. Khim., No. 9, 2055-2058 (1977).
59. I. G. Krasnosel'skaya, B. L. Erusalimskii, and G. N. Novinskaya, "The effect of magnesium alkoxides on polymerization in polar monomer–organomagnesium initiator systems," Vysokomol. Soedin., A16, 1730-1735 (1974).
60. V. V. Mazurek, Polymerization under the Action of Transition Metals [in Russian], Nauka, Leningrad (1974).
61. N. A. Shirokov and V. V. Mazurek, "The tris-π-allylchromium–pyridine system as an initiator of the polymerization of methylmethacrylate," Vysokomol. Soedin., A18, 1687-1690 (1976).
62. L. A. Fedorova, V. V. Mazurek, N. A. Shirokov, and L. D. Turkova, "The nature and specificity of the active centers in the tris-π-allylchromium–pyridine–acrylonitrile system," React. Kinet. Catal. Lett., 15, 361-365 (1980).
63. L. A. Fedorova, V. V. Mazurek, N. A. Shirokov, and L. D. Turkova, "Systems based on tris-π-allylchromium as initiators in the polymerization of acrylonitrile," Vysokomol. Soedin., A23, 1749-1754 (1981).
64. L. A. Fedorova, V. V. Mazurek, N. A. Shirokov, and L. D. Turkova, "The tris-π-allylchromium–2,2-dipyridyl system as an initiator of acrylonitrile polymerization," React. Kinet. Catal. Lett., 23, 343-347 (1983).

65. W. Obrecht and P. H. Plesch, "The polymerization of styrene by trifluoroacetic acid," Makromol. Chem., 182, 1459-1473 (1981).
66. K. Mejzlik and M. Lesna, "A comparison of the methods used to determine the active centers in the polymerization of Ziegler—Natta olefins," V Int. Mikrosymp. Fortschr. in Ionenpolymerization, Prague (1982), prepr. 50.
67. V. A. Zakharov, T. D. Bukatov, and Yu. I. Ermakov, "A mechanism of the catalytic polymerization of olefins based on data on the number of active centers and rate constants of the individual stages," Usp. Khim., 49, 2213-2240 (1980).
68. H. Franz, H. Meyer, and K.-H. Reichert, "An attempt to determine the concentration of active sites in supported Ziegler—Natta catalysts," Polymer, 22, 226-230 (1981).
69. J. Herwig, "Olefinpolymerisation mit löslischen insbesondere halogenfreien Ziegler Katalysatoren unter Verwendung von oligomerem Methylalumoxan als Aluminiumalkylkomponente," Dissertation, Hamburg (1979).
70. J. Pein, "Untersuchungen von Systemen aus Cyclopentadienylzircon(IV)- Verbindungen mit n-Propylaluminium-Verbindungen," Dissertation, Hamburg (1980).
71. A. Andresen, "UV-Spektroskopische Untersuchungen an homogen Ziegler—Natta Katalysatoren mit Methylalumoxan als Katalysatorkomponente," Dissertation, Hamburg (1980).
72. H. Sinn, W. Kaminsky, H.-J. Vollmer, and R. Woldt, "'Lebende Polymere' bei Ziegler Katalysatoren extremer Productivität," Angew. Chem., 92, 396, 401-402 (1980).
73. H. Sinn and W. Kaminsky, "Ziegler—Natta catalysis," Adv. Organomet. Chem., 16, 99-149 (1980).
74. W. Kaminsky, H. Sinn, and H.-J. Vollmer, "Extrem verzerrte Bindungswinkel bei organozirkonium Verbindungen, die gegen Ethylen aktiv sind," Angew. Chem., 88, 688-689 (1976).
75. J. Boor, Jr., Zielger—Natta Catalysts and Polymerizations, Academic Press, New York (1979).
76. V. E. Lvovsky, E. A. Fushman, and F. S. Dyachkovsky, "A study of the structure and reactivity of the complexes of cyclopentadienyltitanium derivatives with alkylaluminum halides," J. Mol. Catal., 10, 43-56 (1981).
77. J. Cihlař, J. Mejzlik, and O. Hamřik, "The influence of water on ethylene polymerization catalyzed by titanocene systems," Makromol. Chem., 179, 2333-2358 (1978).
78. K. H. Reichert and K. R. Meyer, "Zur Kinetik der Niederdruckpolymerisation von Äthylen mit löslischen Ziegler-Katalysatoren," Makromol. Chem., 169, 163-176 (1973).
79. K. J. Toelle, J. Smid, and M. Szwarc, "The absolute rate constants of propagation of the free living polystyrene ions and the dissociation constant of the $\sim S^-, Na^+$ ion pair," J. Polym. Sci., B3, 1037-1041 (1965).
80. H. Hostalka and G. V. Szwarc," Some remarks on the comments by Toelle, Smid, and Szwarc," J. Polym. Sci., B3, 1043-1044 (1965).

81. B. J. Schmitt and G. V. Schulz, "Über zwei formen des Initiators Na-Naphthalin und die Bestimmung der 'lebenden' Kettenenden in der anionischen Polymerisation," Makromol. Chem., 121, 184-204 (1969).
82. P. H. Plesch, "Propagation rate constants in cationic polymerization," Adv. Polym. Sci., 8, 137-154 (1971).
83. K. S. Kazanskii, A. A. Solov'yanov, and S. G. Entelis, "The nature of the active centers and the mechanism of the anionic polymerization of epoxides," in: Advances in Ionic Polymerization [in Russian], Warsaw (1975), pp. 77-87.
84. P. Sigwalt, "The mechanism and kinetics of anionic polymerization of episulfides," IUPAC Int. Symp. on Macromol. Chem., Budapest (1969), pp. 251-280.
85. S. Penczek, P. Kubisa, and K. Matyjaszewski, "Cationic ring-opening polymerization," Adv. Polym. Sci., 37, 1-149 (1980).
86. A. A. Korotkov and A. F. Podolskii, The Catalytic Polymerization of Vinyl Monomers [in Russian], Nauka, Leningrad (1973).
87. F. S. Dainton, G. A. Harpell, and K. J. Ivin, "The kinetics of anionic polymerization of α-methylstyrene in tetrahydrofuran and dioxane," Eur. Polym. Sci., 5, 395-403 (1969).
88. L. V. Vinogradova, V. N. Zgonnik (Sgonnik), N. I. Nikolaev, and E. P. Vetchinova, "The polymerization of butadiene by polybutadienyllithium in the presence of tetrahydrofuran," Eur. Polym. J., 16, 799-801 (1980).
89. S. Bywater and W. J. Worsfold, "Anionic polymerization of isoprene. Ion and ion pairs contribution to the polymerization in THF," Can. J. Chem., 45, 1821-1824 (1967).
90. G. Helary and M. Fontanille, "The activation of styrene by crown tertiary amines in cyclohexane," Polym. Bull., 3, 159-165 (1981).
91. T. Shimomura, J. Smid, and M. Szwarc, "Reactivities of contact and solvent-separated ion pairs. Anionic polymerization of styrene in dimethoxyethane," J. Am. Chem. Soc., 89, 5743-5749 (1969).
92. M. Szwarc, Carbanions, Living Polymers and Electron Transfer Processes, Interscience, New York (1968).
93. M. Szwarc (editor), Ions and Ion Pairs in Organic Reactions, Interscience, New York (1972).
94. B. L. Erusalimskii, Ionic Polymerization of Polar Monomers [in Russian], Nauka, Leningrad (1970).
95. B. L. Erusalimskii, "Über einige Besonderheiten der anionischen Polymerisation polarer Monomerer," Plaste Kautsch, 15, 788-792 (1968).
96. G. E. Ham (editor), Copolymerization, Interscience, New York (1964).
97. G. E. Ham, "Ionic copolymerization," J. Macromol. Sci., Chem., A11, 227-230 (1970).
98. P. Kubisa and S. Penczek, "Penultimate unit influence in the cationic copolymerization of tetrahydrofuran with oxetanes," J. Macromol. Sci., Chem., A7, 1509-1524 (1973).

99. R. Ohlinger and F. Bandermann, "Kinetics of the propagation reaction of butadiene–styrene copolymerization with organolithium compounds," Makromol. Chem., 181, 1935-1947 1980).
100. V. N. Zgonnik, N. I. Nikolaev, E. Yu. Shadrina, and L. V. Nikonova, "Copolymerization of butadiene with styrene on butyllithium complexes with tetramethylethylenediamine and 2,3-dimethoxybutane," Vysokomol. Soedin., B15, 684-686 (1973).
101. M. M. F. Al-Jarrah and R. N. Young, "Anionic copolymerization of vinylbiphenyl: kinetics of a system having spectroscopically distinguishable ion pairs," 26th Int. Symp. on Macromolecules, Mainz (1979), Vol. 1, pp. 373-376.
102. S. R. Rafikov, Z. M. Sabirova, O. A. Ponomarev, G. S. Lomskii, Yu. B. Monakov, and K. S. Minsker, "The connection of the stereospecific effects with the nature of the counterion in the anionic polymerization of dienes," Dokl. Akad. Nauk SSSR, 259, 1139-1143 (1981).
103. T. Higashimura, J. Masamoto, S. Okamura, and T. Yonezawa, "Cationic polymerization of 1,2-dialkoxyethylenes," Polym. J., 2, 154-160 (1972).
104. T. Higashimura, K. Kawamura, and T. Masusada, "Cationic polymerization of α,β-disubstituted olefins. Part 17. Effect of polymerization conditions on the reativity of alkenyl ethers relative to vinyl ethers," J. Polym. Sci., Polym. Chem. Ed., 11, 713-722 (1973).
105. T. Higashimara and K. Yamamoto, "Cationic polymerization of α,β-disubstituted ethylenes. Investigation of the propagation reaction," Makromol. Chem., 175, 1139-1156 (1974).
106. Yu. E. Eizner and B. L. Erusalimskii, "The electron structure of the active centers of a linear oxonium ionic type," Vysokomol. Soedin., A12, 1614-1620 (1970).
107. T. Kelen, P. Tudos, B. Turcsány, and J. P. Kennedy, "An analysis of the linear methods for determining copolymerization reactivity ratios. IV. A comprehensive and critical reexamination of carbocationic copolymerization data," J. Polym. Sci., Polym. Chem. Ed., 15, 3047-3074 (1977).
108. I. Artamonova, S. Klenin, A. Troitskaya, and B. Erusalimskii (Erussalimsky), "Zum mechanismus der anionischen Copolymerisation polarer ungesättigter Monomere," Makromol. Chem., 175, 2329-2338 (1974).
109. I. L. Artamonova, V. V. Mazurek, and B. L. Erusalimskii, "The influence of temperature on the composition of copolymers formed in the acrylonitrile–methylacrylate–butyllithium system," Vysokomol. Soedin., B19, 179-181 (1977).
110. I. L. Artamonova, A. V. Novoselova, S. I. Vinogradova, B. L. Erusalimskii, H.-J. Adler, and W. Berger, "Copolymerisation von Acrylnitril mit Acrylaten mittels Lithiumalkoxiden," Faserforsch. Textiltech., 28, 511-514 (1977).
111. K. Brzezinska, K. Matyjaszewski, and S. Penczek, "Macroion pairs and macroions in the kinetics of the polymerization of oxepane," Makromol. Chem., 179, 2387-2395 (1978).

112. G. L. Collins, R. K. Greene, F. M. Berardinelli, and W. H. Ray, "Fundamental considerations on the mechanism of copolymerization of trioxane with ethylene oxide initiated with boron trifluoride dibutyl etherate," J. Polym. Sci., Polym. Chem. Ed., 19, 1597-1607 (1981).
113. V. Jaacks, "Anomalien bei der kationischen Copolymerisation von Trioxan, 32. Mitt. über Polyoxymethylene," Makromol. Chem., 101, 33-57 (1967).
114. W. Kern and V. Jaacks, "Some kinetic effects in polymerization of 1,3,5-trioxane," J. Polym. Sci., 48, 399-404 (1970).
115. M. Okada, S. Kozawa, and Y. Yamashita, "Kinetic studies on the polymerization of 1,3-oxepane initiated with triethyloxonium tetrafluoroborate," Makromol. Chem., 127, 271-281 (1969).
116. N. M. Geller, Yu. E. Eizner, and V. A. Kropachev, "Change in electron structure of cyclic oxides during their interaction with electron acceptors. Quantum chemical investigation," Acta Polym., 32, 144-149 (1981).
117. N. M. Geller, Yu. E. Eizner, and V. A. Kropachev, "The effect of substituents on the electronic structure of the complexes of cyclic oxides with an electron acceptor. Quantum-chemical investigation," Acta Polym., 34, 584-588 (1983).
118. E. G. Furman and A. P. Meleshevich, "A study of the influence of the nature of the substituent on the electronic state of the epoxide ring using the CNDO/2 method," Teor. Eksp. Khim., 13, 328-333 (1977).

Chapter 5

The Problem of Stereospecificity

The creation of macromolecules which are distinguished by their high degree of structural regularity became part of the everyday practice of synthetic polymer chemistry due to the wide use of ionic initiating systems. Their variations allow the selective regulation of the structure of the majority of monomers which are capable of ionic polymerization. On the other hand, ionic processes which determine the strict homogeneity of the structure of macromolecules constitute only a tiny fraction of the total number of reactions of this type which have been studied. Essentially, in known cases of the formation of macromolecules in ionic systems, stereospecific polymerization is a comparatively rare exception. The mechanism of these and other processes may be regarded as having two aspects. These are the explanation of the factors which inhibit the stereoregulation and the evaluation of the nature of the factors which eliminate them. Regarding the first of these approaches we will note that the formation of active polymers still does not constitute evidence for the absence of stereospecific active sites in any given system. The same result can be brought about by the coexistence of certain active sites, not all of which are capable of forming stereoregular polymer chains. Of course we are concerned primarily with systems in which different active sites are subject to frequent mutual transformations [see Eq. (3.1)]. If such events are absent or their role is insignificant, structurally inhomogeneous polymers are formed. The fractionation of these can lead to the discovery of stereoregular macromolecules in reaction mixtures. It will be recalled that in the stereochemistry of the individual events of the growth reaction discussions cannot be conclusive without reliable data on the homogeneity of the polymer chains. The absence of clear effects of stereoregularity can be the result of the nonstereospecificity of active sites or the masking of the stereospecificity of one of the active sites of a multicentered system by one of the others. The large number of possible variations in such cases is represented below by some examples which illustrate the formation of macromolecules of mixed iso- and syndiotactic structure (AS_{ns} is a nonstereospecific growing chain and AS_i and AS_s are active sites with iso- and syndiotactic selectivity respectively):

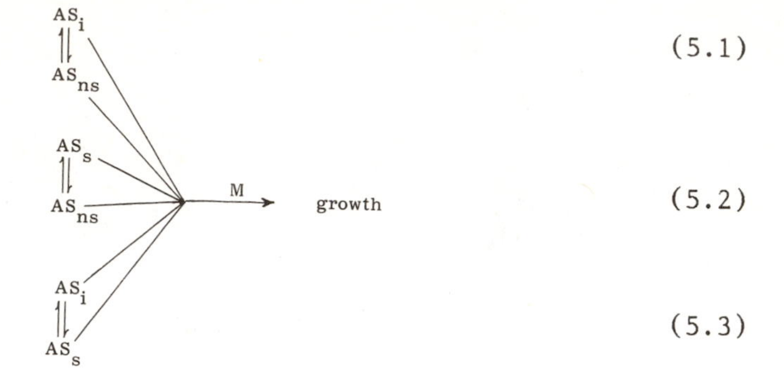

As far as the nonstereospecific growing chains are concerned, their action is illustrated by the scheme

$$AS_{ns} \xrightarrow{M} \begin{cases} \xrightarrow{k_i} \\ \xrightarrow{k_s} \end{cases} \text{growth} \qquad (5.4)$$

In describing the structural effects which are inherent in ionic polymerization, we considered it necessary not to limit ourselves only to those systems which are distinguished by increased stereospecificity but to include the causes of the inhibition of spatial directivity in the formation of macromolecules. From the general point of view it is no less important than the explanation of those factors which ensure the creation of stereoregular polymer chains upon which attention is usually focused.

Further discussion is an attempt to approach the mechanism of stereoregulation through an evaluation of the relative roles of the directing action of the individual active sites and the "degree of masking" of their behavior by the effects of multicenteredness. It must be borne in mind that the relatively high, but to some extent disrupted, spatial homogeneity of the structure of a chain may, within limits, be caused by one or more of the equilibria (5.1) to (5.3) and also by the corresponding ratios k_i/k_s [see Eq. (5.4)]. The explanation of the true situation may provide a basis for the strict interpretation of the structural effects observed and lead to a more precise understanding of the mechanisms of the extremely stereospecific active sites which are found in certain real systems.

With all the variety of parameters, the variation of which is reflected in the nature of the structure of polymer chains which are formed by nonradical polymerization processes, it is possible to distinguish one common feature in these growing chains which is common to practically all known monomers. This is the minimal spatial selectivity of free ions and separated ion pairs (IP_s). Only some of the growing chains of the general type M_n^*Y are distinguished by a high degree of stereospecificity and these are not usually true ion pairs but polar compounds.

It is easy to associate this with the obligatory participation of the Y fragment in the events of directed stereoregulation, at least in homogeneous systems. Quite simply, in this respect it consists of reactions with free ions. In the first place, scheme (5.4) is perfectly natural and in the second place, because of the considerable differences in the values of $k_p^{\pm}$ and k_p^* (see Chapter 4), it is sometimes possible to determine the complete kinetics of the system. Reactions with IP_s are often accompanied by considerable contributions of other active sites. Therefore it is not possible to isolate structural features of macromolecules formed by separated ion pairs in any process which takes place with the participation of such active sites. Those systems in which crown ethers or cryptands figure as agents which separate ion pairs can be assigned with the greatest confidence to systems which include IP_s and which act as single-centered systems. The ability of such compounds to ensure the complete conversion $IP_c \longrightarrow IP_s$ in nonpolar media excludes complications associated with the formation of free ions.

Nevertheless, to liken IP_s with highly screened counterions to the more common active sites of the same general type is somewhat risky. There are examples which indicate that the structure of the macromolecule depends upon the nature of the separating component in the IP_s (see Section 5.1.2.1). Nevertheless, for the differences in the structure of the polymer chains which result from just these factors, the concept of degree of atacticity is more applicable than that of regularity. The corresponding structural fluctuations are the comparatively small advantages of one of the possible types of monomer sequences over the other.

These features of the various active sites appear in the polymerization of all monomers. However, the detailed description of the stereochemistry of the growth reaction requires that the features of each coreagent be considered. This is easy to do by a separate consideration of the polymerization of compounds of the vinyl and diene series.

5.1. THE POLYMERIZATION OF VINYL AND VINYLIDENE MONOMERS

5.1.1. Nonpolar Monomers

5.1.1.1. Propylene

In the polymerization of propylene and its homologs the high degree of regularity of the macromolecules is attained only under the action of some of the active sites formed in heterogeneous

Ziegler–Natta systems. Apart from rare exceptions, the building of stereoregular chains is not the only way in which macromolecular structures are formed. Usually they face competition from the parallel reactions of nonstereospecific growth on other types of active site.

Those phenomena which are found as a result of the absence of mutual transformations of active sites which coexist in heterogeneous conditions were earlier usually regarded as showing the dependence of the total stereospecificity of the initiating systems on the nature of their initial components. An evaluation based on multicenteredness is more correct when taking the kinetic features of the active sites into consideration. This remark can be understood on the basis of some old-established facts, in particular the data of Natta et al. on the isotactic fraction in polypropylene (PP_i), obtained under the action of various α-$TiCl_3$–R_nMt systems [1].

R_nMt	Et_2Be	Et_3Al	Et_2Zn
PP_i, %	94-96	80-90	30-40

These provided the basis for the conclusion that the greatest stereospecificity is exhibited by complexes based on organic compounds of those metals which are distinguished by very small ionic radii. The same results may be approached from the point of view of multicenteredness which is exhibited particularly strongly in the latter example. The dependence of the relative contribution of each coexisting active site is not only on the ratio AS_i–AS_{ns} but also on their propagation rate constants [these are represented by k_p^i and k_p^{ns}; the latter in fact is the ratio of the constants shown in scheme (5.4)].

Such an interpretation (which so far has been impossible either to confirm or reject) may mean that the cause of the differences observed is the superposition of the action of the AS_i by even a small fraction of AS_{ns} provided that the reactivity of the growing chains obeys the relation $k_p^i < k_p^{ns}$.

The often noted reduced reactivity of the varieties of growing chains which act in Ziegler–Natta systems and which are distinguished by selectivity in the formation of stereoregular macromolecules show that such a condition is perfectly possible. Apparently such an inverse relationship between the reactivity and the stereospecificity of the active sites is especially characteristic for multicentered systems in which growth takes place at an interface.

An analogous interpretation of the stereospecific effects in systems which are differentiated by their initial transition metal compounds may seem less fundamental. Thus the example of fundamental differences in the structure of polypropylene formed under the

TABLE 5.1. The Isotactic Fraction Content in Polypropylene Formed in Heterogeneous Ziegler–Natta Systems

System	Modifier	t, °C	PP_i, %	Ref.
$TiCl_4$–Bu_2Mg–$(i\text{-}Bu)_3Al$	-	40	50	3
	Ethylacetate	60	74	
$TiCl_4$–$MgCl_2$–Et_3Al	Ethylbenzoate	60	92-84	4
	p-Methoxyethyl-benzoate	60	95-96	
$TiCl_4$–$MgCl_2$–Et_3Al	-	41	57	5
	Ethylbenzoate	41	94[a]	

[a]Data from S. Kvisie et al. [5]; the increased ratio Al/Ti in the same system reduced PP_i considerably.

action of titanium–aluminum complexes based on α- and β-$TiCl_3$ is difficult to examine separately from the structure of the crystal lattice of the titanium halides under discussion. Nevertheless, it is necessary to take the multicenteredness of the above systems into account. Particular data characterizing the influence of the crystal modification of titanium chloride on the polymer structure in propylene–Et_3Al–$TiCl_3$ systems point first and foremost to the quantitative difference of the relative contributions of the stereospecific active sites in growth reactions [1]:

Variation of $TiCl_3$	α, γ, δ	β
PP_i, %	80-93	40-50

Consequently there are AS_i in each of these systems.

The same is true for bimetallic complexes based upon other transition metal halides, vanadium, for instance. The majority of these ensure the formation of at least some isotactic material in polypropylene. The results which are characteristic of many such systems show that in the polymerization of propylene both high overall stereospecificity and the complete absence of stereospecific effects are relatively rare. The data summarized in [2] are supplemented by isolated examples from more recent works (see Table 5.1).

As may be seen in the literature, the variation of the components which form insoluble catalytic complexes clearly affect (for the rare exceptions see below) only the relative participation of AS_i and AS_s in parallel growth reactions. It is almost always impossible to judge the differences in the action of the very different modifications of each of these. With the present level of understanding, it is easier to assume that the individuality of the AS_i which are formed, for example, in $TiCl_3$–Et_2Be and $TiCl_3$–Et_3Al systems appears mainly if not completely in their relative reactivity and consequently in their ability to compete with the coexisting nonstereospecific active sites. A more detailed discussion of the question would be at best speculative. We note only the impossibility of a simple explanation of the reasons for the difference in the behavior of such systems as those being compared. They may be differentiated either in concentration or kinetic superiority of the active sites AS_i in one system over another. Of course such an alternative does not exclude completely the possibility of some difference between the same AS_i in the purely stereospecific sense. Variations in the nature of the components of the catalytic complexes which are distinguished by a high stereospecificity may to some extent touch upon geometric features of intermediate structures during growth. However, apparently, it affects the rate at which transformations take place to a much greater extent. From this point of view differences in the values of k_p^i for the different AS_i are not in themselves essential for the microstructure of the polymer but as parameters which determine the relative role of the reactions

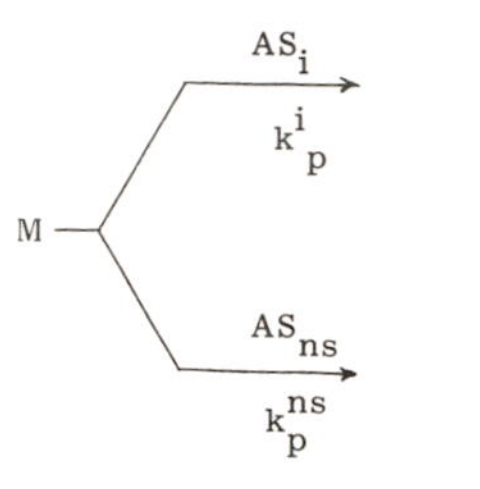

growth (5.5)

in each of the multicentered systems.

An exposition of all probable concepts cannot be regarded as free from criticism. The structural features of the majority of the propylene polymers formed in Ziegler—Natta systems are based upon the so-called isotacticity indices, taken as the weight fraction which is insoluble in boiling heptane. Direct structural analysis is carried out in certain cases and this is not sufficient for unconditional generalizations. Consequently, there may be structural differences in such fractions in various propylene polymers. The discovery of quite large differences in the microstructure of the polypropylene fractions may serve as a basis for explaining the connection between the characteristics of the real stereospecific complexes and the degree of stereoregularity of the molecules which they form.

There is also the question of the degree of atacticity of propylene polymers which form the major product in most of the processes under discussion. There has not been a great deal of research done along these lines, but in certain systems a difference has been noted in the microstructure of atactic polymers (discussed later).

TABLE 5.2 The Polymerization of Propylene under the Action of $TiCl_4-MgCl_2-Et_3Al-PhCOOEt$ Systems [8]

Fraction	Solubility	Mass of fraction, %	Pentads, %						
			mmmm	mmmr	mmrr	mrmm	rrrr	rrrm	mrrm
I	In boiling hexane[a]	19.7	28	12	16	11	15	8	4
II	Insoluble in boiling hexane, soluble in boiling heptane	8.6	68	10	10	3	3	2	4
III	Insoluble in boiling heptane, soluble in boiling octane	47.8	89	4	4	1	-	-	2
IV	Insoluble in boiling octane	23.9	90	5	4	-	-	-	2

[a]Pentads rmmr and rmrm are also found in this fraction (up to 3-4%).

Such phenomena can naturally be ascribed to the individual character of AS_{ns} in different systems which affects the ratio k_p^i/k_p^s; see Eq. (5.4). We also note the possibility of the coexistence of several AS_{ns} in some systems. That such a possibility exists is shown in particular by the discovery in certain binary mixtures which form bimetallic complexes of transition metal derivatives in various valence states (for example Ti^{2+}, Ti^{3+}, and Ti^{4+}) [6, 7]. Therefore, the symbol As_{ns} is to be understood as a certain selection of the various nonstereospecific active sites.

Of the precise structural features of polypropylene the most careful attention must be given to data obtained by ^{13}C-NMR spectroscopy, which is now examined from the general point of view in [2]. From these it follows that the isotactic dyad content in fractions which are insoluble in boiling heptane is about 98%. Apparently when triads are used for evaluation the isotacticity is 94-95%.

More recently, Doi et al. [8] have evaluated contents of the different pentads in polypropylene formed in different systems. They compared fractions isolated by successive extraction in boiling hexane, heptane, and octane. The microstructure of the fractions which are insoluble in the two latter hydrocarbons proved to be practically identical (87-90% isotactic pentads). They were differentiated only according to weight fraction. The results obtained by detailed study of one of the polymers are shown in Table 5.2.

The structural characteristics given are interesting in several ways. In the first place the considerable difference in the isotacticity of fractions I and II point to the presence of at least two types of AS_{ns}. In the second place, the difference in the solubility of fractions III and IV with the same structural characteristics show that they fractionate according to molar mass. The fraction with the highest molar mass (fraction IV) leads to the curious result of the coexistence of two types of AS_i with similar stereospecificity but with different reactivity. We emphasize that the discovery of the weight ratio of the fractions being compared has led to such a conclusion. With their inverse ratio one would have to take into consideration the possibility of slow initiation as a reason for the coexistence of such fractions. This fact is far from trivial, although possibly it appears in a latent form in a large number of real systems.

We also note that the isotacticity of fractions III and IV expressed as isotactic triads (with the total fraction of all pentads which contain the sequence mm) is 98-99%; this has been established by Doi in another work [9] for polypropylene obtained from δ-$TiCl_3$–$AlEt_2Br$ or $AlEt_2I$ systems with the appearance of a fraction of type III equal to 97-98%. The use of Et_3Al or Et_2AlCl as organoaluminum content has led to an isotactic triad content of 94% with the appearance of corresponding fractions of 65 and 91%.

These differences are not so great that an additional role can be assigned to an inorganic carrier which participates in the stereoregulation of the above complex system. This is also the opinion of Doi et al. [8, 9]. Apparently the action of the magnesium component (in this case $MgCl_2$, but other magnesium derivatives are also used and also other metallic salts, for example, $MnCl_2$) amounts mainly to an increase in the kinetic activity of the system which is the result of the increase in the number of atoms of the transition metal participating in the formation of active sites. Such a possibility stems from the ability of the corresponding salts to cocrystallize. This is shown particularly by the $TiCl_3$–$MgCl_2$ systems, for example.

The question of the structure and mechanism of active sites in such systems has been discussed by Giannini [10]; the following structure has been proposed for the titanium–magnesium components (used later in combination with R_3Al):

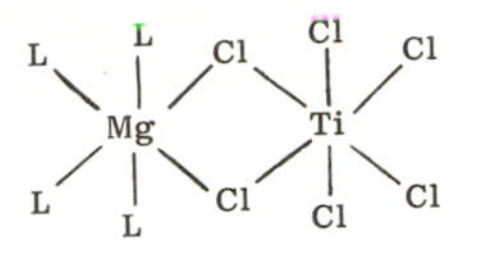

Here L is a Lewis base, for example an ester. These compounds are often used as components for such catalytic systems. A more general examination of such systems (the so-called "third generation catalysts") is given by Galli et al. in [11].

Such a detailed structural analysis as given above from the data of [8] has been carried out on a comparatively small number of propylene polymers formed with Ziegler–Natta catalysts. However, it is difficult to assume that in these processes growth is complicated by isomerization. Such events have been established only for the cationic polymerization of olefins. They attracted attention about twenty years ago in connection with the discovery of the complete or limited occurrence of reactions of the type (5.6) for certain unsaturated monomers:

$$n\ CH_2{=}C\langle^{R}_{H} \longrightarrow (-CH_2-\overset{R}{\underset{H}{C}}-)_n \qquad (5.6)$$

The particularly apt name of "phantom monomer" coined by Kennedy [12] for macromolecules which are formed exclusively or principally by a path different from that of Eq. (5.6) is applied to olefins of the general type $CH_2CH{=}(CH_2)_nCHRR'$ (5.I) and to certain of their halogen derivatives. It will be recalled that the polymerization of 3-methyl-1-butene at −130°C under the action of $AlCl_3$ leads to the formation of a polymer which is constructed entirely of $CH_2CH_2C(CH_3)_2$ units [12].

The small number of individual isomerization processes which are the result of the presence of tertiary C atoms in such monomers and which easily attract cationic sites enable the correct structural features of the macromolecules formed from monomers (5.I) to be determined. As was shown by Cesca et al. for the polymerization of propylene [13] and 1-butene [14, 15], the courses are much more varied for typical cationic active sites of normal olefins. A combination of ^1H- and ^{13}C-NMR methods using selectively labelled monomers (in particular propylene: $^{13}CH_3CH{=}CH_2$, $CH_3{}^{13}CH{=}CH_2$, and $CH_3CH{=}^{13}CH_2$) yielded extremely rich information on the structure of the corresponding macromolecules. The structural features of the groups, other than the initial monomer unit, which sometimes appear (see below) are unprecedented in terms of detail and these lend interest to the brief examination of the results given in the above references.

The basic data for propylene has been obtained during the formation of low-molar-mass products (the mean value of M is about 600) in a medium of n-pentane and dichloromethane with initiators of the aluminum series at −78°C. In contrast to monomers (5.I) the structures of the polypropylene chain in the cationic systems cannot effectively be controlled. Thus the structural units found, usually in the order of tens, consist of fractions which are sensitive to variations in the nature of solvent and initiator to quite a limited extent. The greatest number of components has been found for the following structures:

Type	$>CH-CH_2-CH_3$ (5.II)	$>CH-CH_2-CH_2-CH_3$ (5.III) $>C(CH_2)_2CH_3$ (5.IV)	$-CH_2-C(CH_3)_2-CH_2-$ (5.V)
%	20-34	17-26	6-18

For one of these systems the differences in the structural effects which are caused by changes in temperature are compared. This proves to be quite significant for the fragments given below (initiator $AlBr_3 \cdot HBr$, solvent $n\text{-}C_5H_{12}$):

Structure		(5.II)	(5.III) + (5.IV)	(5.V)	$-C(CH_3)_3$
Content,	−78°C	35	23	2	10
%	70°C	9	13	22	0.5

On the other hand the content of the fragments $-\!\!>CCH_2CH_3-CH_2(CH_2)_2CH_3$ and $>CHCH(CH_3)_2$ do not suffer serious changes even over such a wide range of temperature.

Even greater difference in the structural units is found in studies of the products of the oligomerization of 1-butene. It can be seen that, in parallel with the methide shifts in the end units (Scheme 5.7), first the growth reactions take place on each of the forms (A-D), and second, chain transfer takes place on the polymer [Scheme (5.8)]:

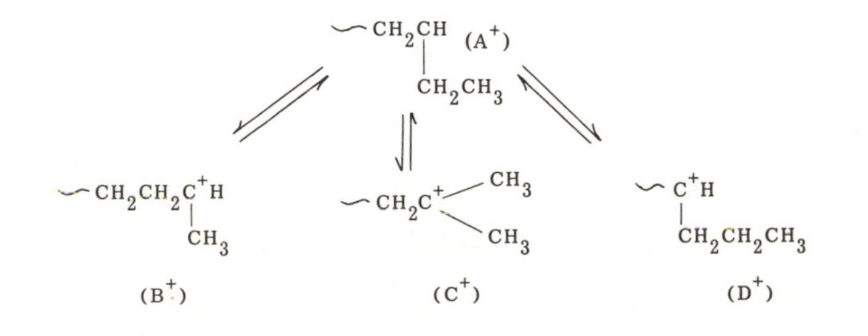

(5.7)

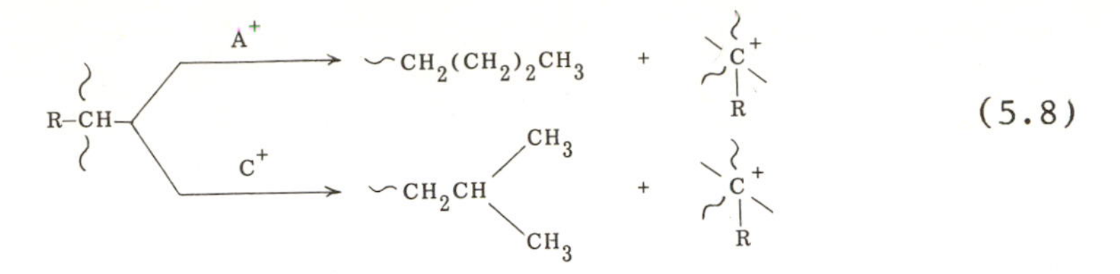

(5.8)

The latter scheme illustrates only a number of individual cases which are possible in principle.

Furthermore, the overall result is complicated by isomerizations which become possible after the transformation of any of the anomalous end groups into penultimate groups, for example:

$$D^+ + C_3H_6 \longrightarrow \sim\underset{\substack{| \\ C_3H_7}}{CH}CH_2\underset{\substack{| \\ CH_2CH_3}}{\overset{+}{C}H} \longrightarrow \sim\underset{\substack{| \\ C_3H_7}}{CH}-\overset{+}{C}H-C_3H_7 \longrightarrow \sim\overset{+}{C}\begin{smallmatrix} C_4H_9 \\ C_3H_7 \end{smallmatrix} \qquad (5.9)$$

Monomers with their third and fourth carbon atoms labelled were used for the precise determination of the origin of the structures found. This enabled the possibility that α,ω units are formed to be ruled out (this is also true for polypropylene).

It follows from the quantitative features of the products obtained in n-pentane under the action of $AlCl_2$–i-C_3H_7Cl systems that the content of nonisomerized structural groups (A) remains practically unchanged over the temperature range −20°C to 70°C at about 10%. The greatest change observed with increase in temperature has been found for the $>CHCH(CH_3)_2$ fragment (an increase from 9 to 23%). None of the structural groups found in polybutene formed under the above conditions are dominant.

We note yet another circumstance connected with the above data. A quantitative evaluation of those structural units which arise from intramolecular isomerization could, in principle, provide a basis for the relative reactivity of certain end units (B^+, C^+, etc.) in the growth reaction. The necessary information is either the contribution to conversion of such fragments or their dependence on the monomer concentration chosen; the latter may vary over wide limits. With the absence of such data the determination of a definite role of a given isomerized form in the structure of the macromolecule seems to be an obscure function of the competing events of isomerization and growth. This is expressed for each of the coexisting active sites by the ratio $k_{is}/k_p[M]$. Here k_{is} and k_p are the rate constants of the isomerization and growth events related to a specific AS form. It will be recalled that this question was covered as long ago as the earlier studies of Kennedy et al. on the polymerization of type (5.I) monomers. For our purposes it is the great variety so distinctive of the final products which is of interest. This variety is limited by those cases which are possible in real systems, i.e., for

$$i > k_{is}/k_p[M] > 0.$$

It must be borne in mind that when $k_{is} \gg k_p[M]$ the initial form of the active site becomes unobservable. Consequently the differences in the contributions of the structural units discovered may generally be determined by the lack of agreement of the relative roles of the factors which characterize the extent to which these forms of active site are formed and the degree to which they participate in growth.

From the formal point of view this situation has some similarity with the anionic polymerization of dienes. For one example it has been shown of these that it is possible to evaluate the difference in the relative activity of the isomeric forms of the active sites using structural effects inherent in these processes (see Section 5.2).

The question as to the precise definition of the structural variations of the active sites which figure in Scheme (5.1) and that of the stereochemistry of the growth events has been the theme of many studies. A large portion of these have been summarized by Boor [16]. No new elements have been introduced into the concepts presented by various authors over ten years ago. Even the IUPAC Symposium on Macromolecules (Florence, 1980) concentrated to a considerable extent upon the stereospecific polymerization of α-olefins and in this respect it went no further than the addition of more detail or the making of minor adjustments to ideas which had been presented earlier. However, the connection between the stereospecificity of the active sites and the lattice structure was developed further. Corradini et al. [17] has carried out a semiempirical calculation for the different modifications of $TiCl_3$, ascribing the increased coordination energy to those active sites which protrude above the neighboring layers of the lattice. Here it is accepted that the mutual repulsion of the substituents in the monomer and in the end unit plays an insignificant role when the active sites are strongly screened. Under these conditions a certain type of asymmetric orientation of the end unit is fixed which is regarded by these authors as a fundamental factor determining the formation of isotactic sequences.

Within the framework of Corradini's concepts the difference in the total stereospecificity of the Ziegler–Natta catalysts, which include nonidentical modifications of $TiCl_3$, can be explained by changes in the number of active sites which protrude above the crystalline lattice.

Concepts on the low accessibility (i.e., the large degree of screening) of sites which differ in their increased stereospecificity agree with data characterizing the connection between the yield of the polymer and its isotacticity in the polymerization of propylene with various catalysts based on δ-$TiCl_3$.

It will be recalled that δ-$TiCl_3$ usually consists of a mixture of cubic (α) and hexagonal (γ) forms of titanium trichloride which is activated by grinding at high temperature. This leads to an increase in surface inhomogeneity of the crystal lattice. At the same symposium Zanetti et al. [18] reported on a similar treatment applied to γ-$TiCl_3$ free from other components which led to the appearance of features typical for the α-form (data from X-ray analysis). From this it follows that the δ-form of titanium trichloride is not distinguished by such a high degree of individuality as its other modifications. This has to be taken into account when comparing the corresponding catalytic complexes.

The results given by Locatelli et al. [19] provide further grounds for the determination of the mechanism of insertion of the monomer into the active bond. A combination of techniques using labelled atoms and ^{13}C-NMR enabled earlier concepts on the introduction of the monomer into the structure of the growing chains via the C—transition-metal σ-bond to be confirmed. The features of the low-molar-mass syndiotactic and isotactic polymers led to the same conclusion. In the first case the labelled monomers $^{13}CH_2{=}CHCH_3$, $CH_2{=}CH^{13}CH_3$, and the system VCl_4—$(CH_3)_2AlCl$—anisole were used and in the second case, the unlabelled propylene and a catalytic complex based on the $TiCl_3(^{13}CH_3)_3AlI$ system.

We now return to the question which we touched upon earlier. This concerns the great variety of structures of atactic propylene polymers (see page 203), which have been recently investigated by Doi [20]. He compared the structural forms of polymers obtained with six different systems based on titantium—aluminum complexes. The common features of these polymers was the low content of both iso- and syndiotactic sequences. The fractions of these and the other pentads were contained within the interval 16-42% and 11-18%, respectively. Of the other pentad sequences (in all about ten variations were examined) the greatest relative content was found for the structures mmmr, mmrr, and rrrm (about 10-15%) and the least for rmmr and mrrm (mainly 3-5%). As these values show, isotactic sequences are especially sensitive to changes in the nature of the initiating complex.

For a more complete description of the features of the catalysts used it is necessary to note the considerable differences in the relative content of the atactic fractions. We give two examples from the original results of [20]. Below are the indices relating to the fractions dissolved in boiling heptane:

System	Weight of fraction, %	Pentads, %			
		mmmm	mmmr	rrrr	rrrm
$TiCl_4$–$MnCl_2$–PhCOOEt–Et_3Al	54.9	42	11	12	5
$TiCl_3$–Et_3Al	45.0	27	11	14	9

The remaining part of these polymers may be taken as the considerably enriched isotactic pentads. The conclusion which can be drawn from this on the parallel actions of AS_i and AS_{ns} is defined by Doi on the following basis. The high stereospecificity of the former is ascribed to coordination unsaturation of the Mt atoms which enter the structure of the catalytic complex (i.e., connected to the central Ti atom via Cl) and are situated on the surface of the lattice. The absence of a definite direction of the growth reaction in the case of the latter type of active site Doi explains by the presence of a vacancy on the "surface" Mt atoms. This creates the conditions for reversible migration of the Al component. These exchanges are accompanied by mutual transformations of the catalytic complexes. In one such transformation the central Ti atoms are connected with six and in the other with five ligands. The individuality of their action is explained by the symbols As_i^6 and AS_s^5 and the structures given in Fig. 5.1.

It is not difficult to see that the mechanism proposed for the atactic macromolecules is formally of the type given by scheme (5.3).

In [20] are also given the results of a statistical analysis of the structure of polypropylene which was carried out on the assumption of an enantiomorphic distribution of the isotactic sequences (enantiomorphic-site propagation statistics) and that the distribution of the syndiotactic pentads obeys Bernoulli statistics. Apart from one small exception (the pentad mmrr) an excellent correspondence is found between the experimental data and the values calculated in this way. The results given by Doi [20] lead to the conclusion that atactic polypropylene has a stereoblock structure but with microblocks of very small length. This follows from the absence in the IR spectra of the atactic polypropylene absorption bands which are characteristic of the crystalline form in iso- and syndiotactic sequences. This circumstance in its turn indicates a fairly high frequency of exchanges of the type shown in Fig. 5.1.

In concluding the present section we note that a concise exposition of modern views on the mechanism of the polymerization of olefins under the action of Ziegler–Natta catalysts has been recently given by Reichert [21].

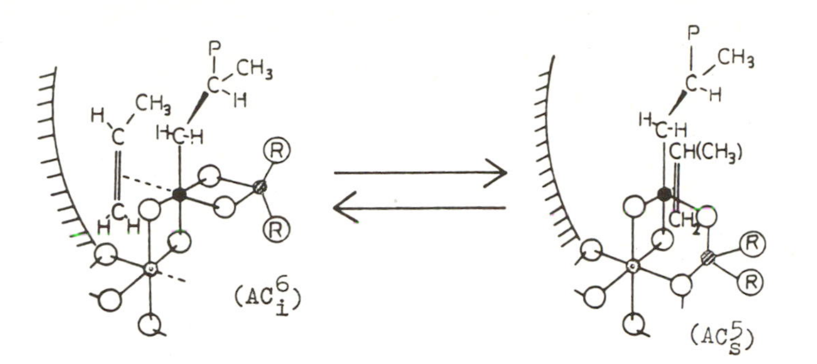

Fig. 5.1. The orientation of propylene while interacting with iso- (AC_i^6) and syndiotactic (AC_s^5) active sites during polymerization under the action of a three-component Ziegler–Natta system (from Doi [20]). •) Titanium; ⊘)aluminum; ⊙) magnesium or manganese; o) chlorine.

5.1.1.2. Styrene Monomers

A quite peculiar situation has arisen in the problem of the control of the chain structure in the polymerization of styrene in ionic systems. Studies of this aspect of anionic processes which was of interest in the fifties due to the stereospecificity of alfin catalysts almost ceased completely when these catalysts were replaced by the Ziegler–Natta catalysts. However, research into these processes did not proceed very far because isotactic polystyrene, contrary to the initial assumptions, did not prove to be a good practical polymer. Consequently, information on the effect of the nature of the Ziegler–Natta complexes on the structure of polypropylene greatly overshadows the modest amount of earlier data on polystyrene formed in similar systems. It is only possible to assume that the ability of styrene to form stereoregular polymers must depend less on the actual type of initiator than does that of propylene. The probability of phenyl groups participating in the donor–acceptor interactions with the Mt components which facilitate stereoregulation points to this conclusion. Such interactions are well known for anionic styrene active sites.

The mechanisms of chain formation in the polymerization of styrene have recently again become the subject of thorough research. This is dealt with in detail in [22]. The fact is that several of these aspects are open to question and the appearance of new work in this field gives point to a further examination of the existing data.

Matsuzaki et al. [23-27] have obtained a considerable quantity of data on the microstructure of styrene polymers formed by anionic polymerization. These results, which reflect the dependence of the

TABLE 5.3. Content of Racemic Diads (r) in Polystyrene Formed in Anionic Systems [23, 24]

Expt.	Initiator	Solvent	t, °C	r, %
1	n-Butyllithium[a]	Toluene	–20	70
2	Sodium naphthalenide	Toluene	30	66
3	" "	Pentane	30	71
4	" "	THF	30	67
5	" "	THF	–78	71
6	" "	Dimethoxyethane (DME)	30	68
7	Potassium naphthalenide	Toluene	30	46
8	" "	Toluene	–78	38
9	" "	Heptane	30	41
10	" "	THF	30	67
11	" "	THF	–78	67
12	" "	DME	30	68
13	Cesium naphthalenide	Toluene	30	45
14	" "	Toluene	0	37
15	" "	THF	30	58

[a]Under the same conditions but with wet solvent (molar ratio H_2O/Li about 4) polymers are formed which contain about 30% of the isotactic fraction [25]; the possible causes for this are discussed in [22].

structural effects on the nature of the initiator, solvent, and temperature in the case of styrene itself, are given in Table 5.3. As can be seen, the indices are within the range typical for atactic polymers but it is clear that there is some tendency towards changes in the microstructure of polystyrene when the conditions are varied.

These authors regard it as secondary that while potassium and cesium initiators give rise to the greatest content of isotactic sequences (expts. 8 and 14), they are insoluble in hydrocarbon media. The insolubility of the sodium initiator in these circumstances does not give rise to a similar increase in meso-dyad polymers (expt. 3). Matsuzaki et al. regard the ionicity of the C–metal bond of the growing chains as the decisive factor in the formation of isotactic sequences. The conclusion that the negative charge on the phenyl group on the end unit increases in growing chains with the potassium and cesium counterions is confirmed by Matsuzaki et al. [28] on model compounds using the quantum-chemical method CNDO/2 and ^{1}H- and ^{13}C-NMR spectroscopy. Some of the total charge characteristics on the C atoms of the phenyl group (q_{Ph}) in $RCH_2CH(Ph)Mt$ compounds are given below:

Counterion	q_{Ph} (in fractions of electron)	
	calculated values*	NMR data†
Li	−0.15	−0.29
Na	−0.19	-
K	-	−0.46
Cs	-	−0.40

*R = H.
†R = CH_3.

The strengthening of the interaction of the phenyl group of the model active site with the counterion, which is caused by the increase in the negative charge, is regarded as a stereoregulating factor.

The ^{13}C-NMR spectra of EtCH(Mt)Ph compounds depend on the nature of the metal and also show a difference in the chemical shifts of the two ortho (Δ_o) and the two meta (Δ_m) C atoms. This difference is greatest for cesium [28]:

Mt	Δ_o	Δ_m
Li	7.4	1.4
K	8.8	1.8
Cs	9.2	2.1

Such a dependence corresponds qualitatively with the assumptions on the connection between the strengthening of the interaction of the counterion with the phenyl group of the end unit and changes in the microstructure of the polymer. The influence of the nature of the metal on the restriction of rotation of the phenyl group in the compounds is especially striking. In the lithium derivative the difference between o, o' and m, m' proton signals is quite distinct at −40°C but disappears completely at room temperature. In potassium and cesium derivatives such an effect is not observed even at 80°C and there are no signs of a rotation of the phenyl group in the ^{1}H-NMR spectra.

Apart from the charge characteristics, of all the results of the quantum-chemical calculations, the differences in the geometry of the Li and Na derivatives of the model styrene active sites are of the most interest (see Fig. 5.2). At first sight the closeness of approach of the Ph groups to the sodium atoms is at variance with the acceptor properties of the different alkali metals. It is sufficient, however, to look only at the above values of q_{Ph} to see that lithium is a much better acceptor than sodium. As for the difference in the geometry of these compounds, this is determined by the dependence of the efficiency with which the π orbitals of the aromatic nucleus overlap with the Mt orbitals of the cation.

The results of Matsuzaki et al. (op. cit.) which the authors themselves discuss mainly in connection with the formation of polymers which they call "isotactically rich" ought to be regarded from another point of view. The question arises that in spite of indications that the counterion has a considerable influence on the structure of model active sites, even those which are most enriched with polystyrene meso dyads (62-63%) are essentially atactic. This leads to the question of the cause of the small-scale changes in the microstructure of the polymers when the influence of the counterion on the geometry and charge characteristics of the $RCH_2CH(Mt)Ph$ compounds increases considerably.

The foregoing makes the following assumptions possible. First, the tendency of the counterion to closely approach the Ph group of the end unit in real active sites may be weakened as a result of the inclusion of other units of the growing chains in similar interactions. Both the theoretical and experimental data on various M_nMt chains (see Chapter 2) and certain low-molecular-weight organometallic compounds of the aromatic series (see below) favor the existence of such polyligand effects. Secondly, this tendency which is characteristic of model conditions may lose its significance (considerably if not completely) at the stage of introduction of the next monomer molecule into the active bond. These circumstances can reduce the role of the regulating action of the counterion when the structure of the chain is laid down, in particular during the formation of $(CH_2CHX)_n$ macromolecules with nonpolar substituents X. Similar effects which weaken the influence of the counterion on the stereochemistry of the growth reaction is sometimes suppressed on transition to heterogeneous polymerization or to monomers with polar substituents. This follows, in particular, from the fact that isotactic polystyrene can only be synthesized under the action of certain insoluble ionic initiators and the stereospecific polymerization of a number of polar monomers in homogeneous ionic systems (see Section 5.1.2).

In connection with the possibility that polyligand effects may appear in "living" styrene chains, it is appropriate to recall the results obtained by Dils and van Beylen [29] for the model compounds $RC(Ph)_2Li$ (5.VI) and $RCH(Ph)CH_2C(Ph)_2Li$ (5.VII), where $R = C_5H_{11}$. A comparison of kinetics when these reagents are added to styrene showed that the first is fifteen times more reactive than the second (nonpolar medium). The authors ascribe this difference to the passivating action of the phenyl group of the penultimate unit which is explained by the structure below (counterion omitted):

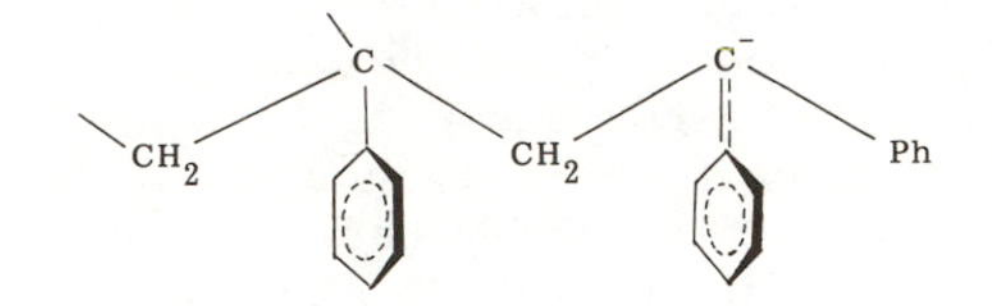

(5.VIII)

This assumption is based on the UV characteristics of the compounds (5.VI) and (5.VII) (λ = 410 and 420 nm, respectively; data for 22°C in cyclohexane). It is possible, however, that the individual features of these compounds have another origin, namely the interaction of the lithium atom with the phenyl groups of both units in reagent (5.VII):

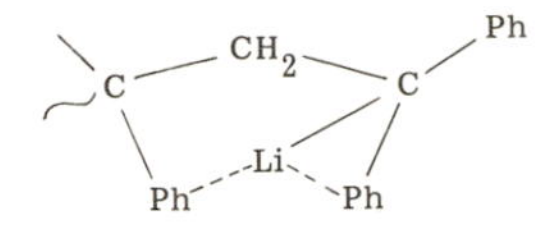

Whether such an interpretation is to be preferred or not, the facts established in [29] indirectly favor the influence of the phenyl groups of the penultimate unit on the growth reaction. The data of Walckiers and van Beylen [30] obtained from the study of the kinetics of the interaction of styrene with 1-phenyl-n-hexyllithium (PHL) and polystyryllithium (PStL) in different media are distinguished by their great definition. The average values of these constants for 24°C are given below:

Reagent	Solvent	Rate constant for addition of styrene
PHL	Cyclohexane	1.45 liters$^{0.5}$ mole$^{-0.5}$ min^{-1}
PStL	Cyclohexane	0.35 liters$^{0.5}$ mole$^{-0.5}$ min^{-1}
PHL	Cyclohexane	200 liters mole^{-1} min^{-1}
PStL	+0.2 moles/liter THF	50 liters mole^{-1} min^{-1}

From the statistical data on the distribution of the structural units in these polymers, Matsuzaki et al. concluded that the phenyl group of the penultimate unit affects the growth reaction in anionic polymerization. A series of instances, including polystyrene obtained under the action of initiators containing potassium and cesium counterions, mentioned by the authors, were found to obey first-order Markov statistics [23]. On the other hand, some data on polystyrene formed in cationic systems show that the microstructure obeys Bernoulli statistics.

Considerable attention has been paid to the question of the mechanism in the anionic polymerization of α-methylstyrene (MSt). The dependence of the microstructure of MSt and similar polymers on the conditions of polymerization is not especially significant but is quite well marked. This is interpreted quite differently by the various groups of workers in the field.

The most important conclusion to be drawn from the research of Wicke and Elgert [31-33] is that multicenteredness can be ruled out

TABLE 5.4. The Microtacticity of Poly-α-methylstyrene Formed in THF under the Action of n-Butyllithium [32]

No.	Concentration, moles/liter		t, °C	Triad content, %			ρ
	monomer	initiator $\times 10^3$		I	H	S	
1	6.7	0.3	50	5	40	56	0.87
2	3.9	0.3	30	6	35	59	1.03
3	1.0	0.5	−50	11	44	45	1.00
4	0.6	6.7	−80	14	49	37	0.96
5	0.1	0.6	−49	6	39	54	0.96
6	0.1	1.5	−70	9	41	50	1.02
7	0.1	1.0	−101	14	50	35	1.25
8	0.01	0.2	−70	3	34	63	0.94

as a possible cause of the structural effects found. The basis for this is an analysis of the results of a study of the polymerization of MSt in THF under the action of n-butyllithium over a large range of temperature and monomer concentration, both with and without a buffer. Some of the results obtained are given in Table 5.4.

Apart from minor exceptions (Nos. 1 and 7) the structure of these polymers is in good agreement with Bernoulli statistics. This is shown by the values of ρ* given in the table. As the authors emphasize, the increase in the isotactic triads (I_T) with a decrease in temperature must be interpreted while taking into account the concentration of the monomer. This was chosen to be higher with increased temperature (because of the high equilibrium concentration of the monomer) and lower with low temperature (due to the freezing out of the monomer). In order to come to any conclusions regarding the structural differences which have been established for the polymers, it is essential that they do not depend either on the presence of a buffer ($LiB(Ph)_4$ was used) or on the concentration of the initiator.

Taking these circumstances into account and also the fact that log(m/r) varies linearly with inverse temperature (where m and r

---

*These are not given in [32]. In order to evaluate these we used the usual formula $\rho = 2(I_d \cdot S_d)/H_t$ in which I_d and S_d are the meso and racemic dyads and H_t are the heterotactic triads. The value $\rho = 1 \pm 0.1$ refers to the Bernoulli statistics.

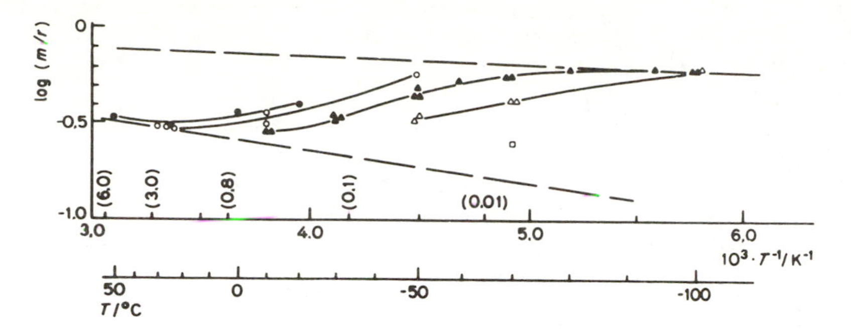

Fig. 5.2. The temperature dependence of the microtacticity of poly-α-methylstyrene as a function of the monomer concentration during polymerization under the action of n-butyllithium in tetrahydrofuran [32]. Monomer concentration, moles/liter: •) 6, o) 3, ▲) 0.8, Δ) 0.1, □) 0.01. The numbers in parentheses show the equilibrium concentration of the monomer in polymerization.

are the meso and racemic dyads in the polymer; see Fig. 5.2), the authors conclude that the observed variations in the structure of poly-MSt are the result of competition between propagation, depolymerization, and isomerization in the growing chains. Ideas were based on studies of the model compound $(CH_3)_3CCH_2C(Ph)(CH_3)Li$ [31]. ^1H- and ^{13}C-NMR data of this compound in benzene and THF are in agreement with the following structure for the end group:

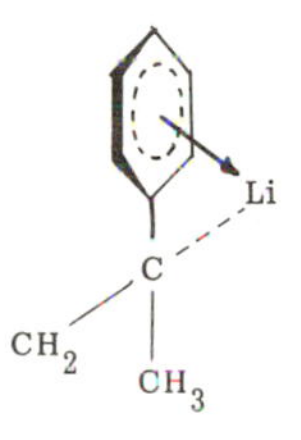

The authors propose that isomerization takes place via a stage of ionic dissociation:

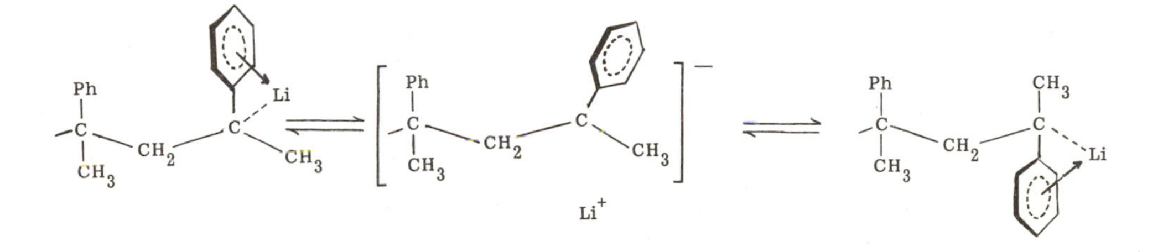

The effect of the competition of all these reactions on the microstructure of the growing chains is illustrated by the following scheme in which m* and r* denote the end dyads of the active sites (counterions omitted):

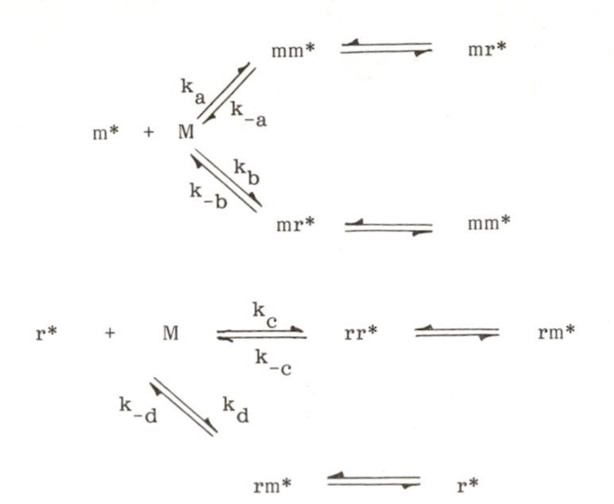

The fact that Bernoulli statistics apply enables rate constants $k_a = k_d$ and $k_b = k_c$ to be regarded as identical. The authors point out the complexity of the complete mathematical treatment of Scheme (5.10) and introduce simplifications in which depolymerization is neglected. The correspondence obtained between experimental data and the curves which are calculated using the simplified scheme is completely satisfactory (Fig. 5.2). This indicates that depolymerization does not contribute significantly to the total result.

The mean rate constants for iso- and syndiotactic growth were evaluated from the linear portion of the temperature plot for the overall constants in the range from −10 to −75°C.

The possibility cannot be excluded that this circumstance is connected with the relatively small difference between the depolymerization constants of the coexisting growth sites in comparison with the corresponding growth constants. The absence of compensation of the type $k_a/k_b = k_{-b}/k_{-a}$ which might be expected must be borne in mind.

Different conclusions are drawn by Malhotra [34-36] on the anionic polymerization of MSt in which the molar-mass characteristics of the polymers formed has been studied in the greatest detail. In the majority of cases a polymodal MMD has been established. For a number of systems these studies have been taken as far as a complete characterization of the fractions found. This has led to the discovery of the structural inhomogeneity of the macromolecules. Data on polymerization under the action of a metallic potassium–THF system at 55°C are given below as an example [35]:

	Content, %	DP	Triads, %		
			I	H	S
Low-molecular-weight fraction	52	14	19	51	30
High-molecular-weight fraction	48	40	8	40	52

The relation between fractions which differ in their degree of polymerization depends markedly on the conditions in which the reaction is carried out (mainly upon the temperature and concentration of the initiator) but structural inhomogeneity of the polymers is found for all the initiators used. This includes, apart from those already mentioned, potassium naphthalenide and potassium anthracenide in THF and a complex of n-butyllithium with tetramethylethylenediamine in bulk polymerization.

On the basis that the distribution of the structural units in the fractions obeys the Bernoulli statistics (deviations are found only for the low-molecular-weight fraction of the polymers formed in the temperature range −25°C and below), the authors give an explanation for the differences found by the coexistence of different active sites. These ideas have been defined by Malhotra in [36]. It is assumed that at the initiation stage, which takes place with excess metallic potassium or its derivative, two reagents are formed, each with considerably different rates. By analogy with the data in the literature (for bibliography see [36]), they ascribe to K–MSt–THF systems structures of the type (5.IX) and (5.X), and to the (NPhK)–MSt–THF system, the structures (5.XI) and (5.XII):

$$K^+ \; {}^-[\underset{Ph}{C}(CH_3)CH_2\underset{Ph}{C}(CH_3)CH_2CH_2\underset{Ph}{C}(CH_3)CH_2\underset{Ph}{C}(CH_3)]^- \; K^+ \tag{5.IX}$$

$$K^+ \; {}^-[\underset{Ph}{C}(CH_3)CH_2CH_2\underset{Ph}{C}(CH_3)\underset{Ph}{C}(CH_3)CH_2\underset{Ph}{C}(CH_3)]^- \; K^+ \tag{5.X}$$

$$NPhK^{+-}(A)^{-+}KNPh \tag{5.XI}$$

$$NPhK^{+-}(A')^{-+}KNPh \tag{5.XII}$$

Here A and A' are the dianions of the formulas (5.IX) and (5.X), respectively. In particular for the first of these systems the scheme

$$MSt + K \begin{cases} \xrightarrow{k'} (5.IX) \\ \xrightarrow{k''} (5.X) \end{cases} \tag{5.11}$$

is proposed, where k' ≪ k".

The individuality of the behavior of such reagents is expressed by the author through the following structural differences for these polymers:

Initiator	(5.IX)	(5.X)	(5.XI)	(5.XII)
Meso dyad content in the polymer, %	26	35	26	30

Such a concept demands an explanation of the reasons why the differences in the structures of the active sites compared at this stage of growth are conserved. This is not reflected in the above formulas. The affect of the difference between the initiating reagents (5.XI) and (5.X), or (5.XI) and (5.XII), on the microstructure of the polymers could be expected, for instance in the case of a predominant tendency of some of these compounds to form aggregates (ionic triplets, quadruplets, etc.), if such aggregates were to participate directly in polymerization. For the systems we are considering here, it is doubtful whether precisely such active sites coexist. However, the structural nonuniformity of the polymer fractions supports the concept of multicenteredness as the cause of the effects discovered by Malhotra et al.

It is necessary to note that in the work of Wicke and Elgert (op. cit.) the molar-mass characteristics of poly-MSt which would enable the mechanism of the processes they describe [see scheme (5.10)] to be approached from an analogous point of view are absent. This makes it difficult to formulate general conclusions about this cause, the more so since the bimodality of the MMD and the structural inhomogeneity of the polymer fractions seem to be possible also for the process of anionic polymerization of MSt which have not yet been studied from this point of view.

The marked variation in the microstructure of the macromolecules has been recently established by Matsuzaki et al. for the polymerization of methyl- [26] and methoxy-substituted styrene [27] in various anionic systems. Of the detailed information obtained we will give data for those polymers whose structure, expressed in dyads, is distinguished by the greatest homogeneity (see Table 4.5). Comparing these results with the characteristics of styrene polymers formed under similar variations in the conditions of polymerization (Table 5.3), these authors, depending upon the nature of the monomer, emphasize either the spatial influence of the substituent (in the case of o-methylstyrene) or its participation in the interaction with the counterion (in the case of o-methoxystyrene). In the discussion of the formation of the structure of chains, use is made of spectroscopic and quantum-chemical data from the study of compounds which model anionic active sites of o-methyl- and o-methoxystyrene [37]. An interesting effect has also been discovered for

TABLE 5.5. The Microstructure of the Polymers of Methyl- and Methoxystyrene Derivatives Formed in Anionic Polymerization in Toluene at −25°C [26, 27]

Monomer	Initiator	r, %
o-Methylstyrene	n-Butyllithium	81
	Potassium naphthalenide	38
m-Methylstyrene	n-Butyllithium	69
	Potassium naphthalenide	29
p-Methylstyrene	n-Butyllithium	70
	Potassium naphthalenide	32
o-Methoxystyrene	Sodium naphthalenide	33
	Potassium naphthalenide	84
p-Methoxystyrene	n-Butyllithium	71

[a]Data for polymerization in THF at −78°C.

the first of these models, namely the ability of the original compound (5.XIII) to undergo intramolecular regroupings to (5.XIV):

$$\text{o-}C_6H_4\begin{matrix}\diagup CH(Mt)CH_2CH_3 \\ \diagdown CH_3\end{matrix} \qquad (5.XIII)$$

$$\text{o-}C_6H_4\begin{matrix}\diagup CH_2CH_2CH_3 \\ \diagdown CH_2Mt\end{matrix} \qquad (5.XIV)$$

In particular for Mt = K, a transformation which has been noted at room temperature is not observed at −20°C. When Mt = Li the existence of each of the above compounds can already be seen in the NMR spectra at −30°C. The authors do not exclude the possibility of the occurrence of such events during polymerization and the inclusion of isomerized units in the structure of the growing chain.

In connection with the results from a study of compounds (5.XIII) and their methoxy analogs, we recall the data from the same authors characterizing the styrene model [28] and draw attention to yet another study from the same series on models of anionic α-methylstyrene active sites [38]. The dependence of the orientation of the components of the latter type of active site on the nature of the counterion is illustrated by the following structures which are based on NMR spectra (in the second case Mt is K or Cs):

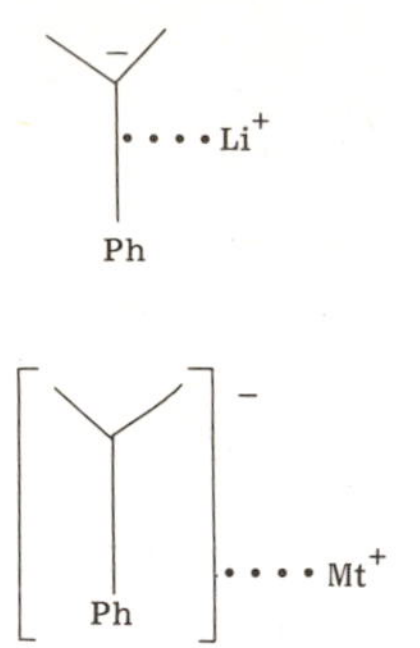

(5.XVa)

(5.XVb)

The results of quantum-chemical calculations lead to a qualitatively close geometry of the lithium derivatives (CNDO/2 method) [38]. Of the parameters of other models investigated using the same methods, the charge characteristics of some of the C atoms of the corresponding ion pairs and free ions are of interest:

Anion	Counterion	Charge in fractions of e		
		C_α	C^1	C^2
$o\text{-}C_6H_4(\bar{C}HCH_2CH_3)(CH_3)$	-	−0.56	+0.12	−0.18
	Lithium	−0.47	+0.12	−0.15
	Sodium	−0.43	+0.10	−0.14
$o\text{-}C_6H_4(\bar{C}HCH_2CH_3)(OCH_3)$	-	−0.52	+0.11	−0.16

As we can see there is not a great difference between the distribution of the electron density in ion pairs and free ions. In both cases about half of the negative charge is found to be on the phenyl group. The NMR data of [38] lead to a similar conclusion as also do the results of studies of the related aromatic reagents. See Chapter 2.

Of the structural effects characteristic of cationic polymerization of styrene monomers, the well known tendency of α-methylstyrene and its π-substituents to form syndiotactic polymers is remarkable. The recent results of Favier et al. [39] on the low temperature polymerization of π-isopropyl-α-methylstyrene in cationic and anionic systems are interesting examples illustrating the features of such macromolecules (data for −78°C):

System	Triad content in the polymer, %		
	I	H	S
Metallic sodium—THF	22	47	30
$TiCl_4$—CH_2Cl_2	0.5	13	86.5

The determining role of the separated ion pairs in the growth of cationic active sites and of contact ion pairs in the case of anionic polymerization may be the cause of such a difference. It is possible however that the features of the active sites under comparison are the result of the large tendency of anionic chains (as compared with cationic chains) to exist as "cissoid" forms, i.e., in a conformation which reduces its ability to form syndiotactic sequences. The results of theoretical calculations on free crotyl ions $C_4H_7^+$ and $C_4H_7^-$ indicate this difference betwen the cationic and anionic reagents. Schleyer et al. [40] showed, using nonempirical methods, that for the cation the most appropriate form is the trans form whereas for the anion it is the cis form. The corresponding values of the energy differences are found to be 3.36 and 1.48 kcal/mole. Carrying out parallel calculations for crotyl and α-methylstyrene ions is, of course, somewhat risky, but it is perfectly possible that the features which are a consequence of the sign of the charge on the end atom may give rise to similar conformational effects.

5.1.2. Polar Monomers

In regarding the structural effects in ionic polymerization which distinguish polar monomers from nonpolar, the ability of the former to form macromolecules of increased regularity in homogeneous systems is often regarded as a key feature. If, however, one proceeds on the basis of the total amount of information available on such polymerization, then the great limitation of systems in which this feature is manifested to a considerable degree becomes immediately obvious. It is not difficult to show the importance of the selection of initiating compounds, the reaction medium, and the conditions in which polymerization is carried out in the overwhelming majority of cases of synthesis of macromolecules which are really stereoregular. For a discussion of the mechanism of stereoregulation, the inability of some of these typical polar monomers to form stereospecific polymers is no less important.

Turning to information on structural effects during the polymerization of monomers of the general type $CH_2{=}C(R)X$ (in which X is a polar group and R is H, an alkyl, an aryl, or a second polar substituent), we will attempt to highlight the especially important features which are the subject of considerable discussion. These are based on facts which are difficult to interpret on the basis of generally accepted concepts.

5.1.2.1. Methacrylates and Related Monomers

Research into the microtacticity of polar molecules is concentrated to a considerable extent on polymethacrylates which are formed in anionic sytems and especially on poly(methyl methacrylates) (PMMA). By the sixties the high isotacticity of PMMA obtained under the action of lithium and certain magnesium initiators in nonpolar media at low temperatures had been established as had the enrichment of the polymers with syndiotactic dyads with changes in the reaction conditions (the transition to other initiators or to polar media, and increase in temperature). Later research broadened these concepts. In particular, it was shown that the actual type of structure and even the structure of the ligands in the coordination sphere of the counterion of the growing chains could be significant. Such variation of the active sites sometimes leads to unexpected effects.

The large number of anionic initiators and initiating systems which are used in the study of the dependence of PMMA structure on the polymerization conditions is covered quite fully in the review of Yuki and Hatada [41] which summarizes the literature up to 1978. The question of the statistical type of the distribution of structural sequences in these polymers (which is not touched upon in this review) has been given special attention more recently [22]. We will deal with those facts which are of more interest and which reflect the influence of the nature of the original reagents and conditions of polymerization on the formation of the polymer structure of different α-substituted acrylates.

We note initially the case which shows that the influence of electron donors on the structure of PMMA depends to a considerable extent upon the nature of the reagents chosen. As the data given below show, the well-known tendency towards the formation of predominantly syndiotactic sequences in such systems is not universal:

System	t, °C	Triads in PMMA, %			Ref.
		I	H	S	
sec-Butyllithium–toluene	−78°	90	10	0	42
sec-Butyllithium–tetramethylethylenediamine (1:4)–toluene	−78°	24	30	46	42
$CH_3O(CH_2)_2ONa$–dimethylformamide (1:4)–toluene	0°	79	12	9	43
t-Butylmagnesium bromide–toluene containing 2.6 moles/liter THF	−50°	84	6	10	44

The results of Okamoto et al. [45] are an interesting illustration of the dependence of the microtacticity of PMMA on the structure of a ligand in the coordination sphere of the same central atom of an active site under the same conditions. They relate to the polymerization of PMMA in toluene at −78°C under the action of various EtMgOR initiators. Several examples of these (about 35) show the possibility of a wide variation in structure as a result of changes in the nature of R in the initiator. The amounts of heterotactic triads in the polymers obtained vary from 3 to 33%.

The differences are brought about as a result of changes in the structure of the butyl group in ethylmagnesium butoxides, the action of which is compared below with the data from diethylmagnesium (toluene, −78°C) [45]:

Initiator	Triad content in PMMA, %		
	I	H	S
n-Butoxymagnesium ethyl (5.XVI)	24	26	50
i-Butoxymagnesium ethyl (5.XVII)	91	4	5
t-Butoxymagnesium ethyl (5.XVIII)	9	14	77
Diethylmagnesium (5.XIX)	6	25	69

In all these cases growth proceeds via the C—magnesium bond. Hence the cause of the individuality of the behavior of the growing chains corresponding to the initiators (5.XVI) to (5.XIX) must be looked for in the features of the influences of the alkoxy groups of different structure on the electronic and spatial characteristics of the active sites. Direct participation of the alkoxy ligands in the formation of the polymer chains seems unlikely.

For an explanation of the effects which are found the authors appeal to the concept of multicenteredness using the known equation of Schlenk and including the following complexes:

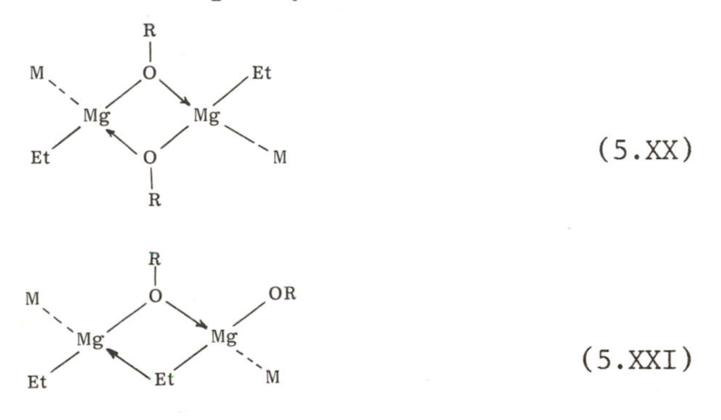

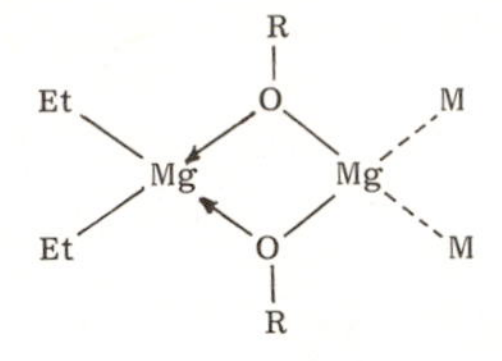

(5.XXII)

It should be borne in mind that both the stereospecificity of each of these participants in the equation and their individual contributions to the initial growth are determined by the structure of the substituent in the alkoxy ligand of the active site.

Such an assertion in no way reveals the essential features of the mechanism of stereoregulation. In order to form an opinion on this it is necessary to take into account the influence in these cases of the principle of the sequential distribution of different microtacticities. Essentially in all processes which are initiated by different EtMgOBu compounds there is an inherent deviation of the macromolecules formed from Bernoulli statistics. In the case of the smallest deviation the magnitude of ρ (see page 216) is 1.71 (data for the n-butoxy derivative). It is extremely difficult to reconcile this circumstance with the structures of the active sites corresponding to the formulas (5.XX) to (5.XXII). The intramolecular interaction of the counterion with a more distant group of the growing chain (which may cause the above deviation) would be natural for the coordination sphere of the magnesium atom with fewer number of ligands (see Section 2.3), i.e., at least when the structures proposed in [45] are partially dissociated.

The following transformation of one of the fragments of the original (5.XX) complex shows a possible variation of the structure of the active site corresponding to this condition:

$$\mathrm{{>}Mg{\genfrac{}{}{0pt}{}{\diagup Et}{\diagdown\cdot M}}} \longrightarrow \mathrm{{>}Mg{-}M{-}Et} \xrightarrow{\mathrm{M}} \mathrm{Mg{-}M_2{-}Et} \qquad (5.12)$$

The intramolecular cyclization, similar to that examined in Chapter 3 for growing chains of polar monomers, becomes realistic for the final product of this transformation. In this case an analogous situation is possible (see Section 3.1.4 for meaning of the symbols AS_ℓ and AS_i).

$$\underset{(AS_\ell)}{\mathrm{{>}Mg{-}M{-}M{-}Et}} \rightleftharpoons \underset{(AS_c)}{\mathrm{{>}Mg{\leftarrow}\overset{\diagup M}{\underset{}{M}}{-}Et}} \qquad (5.13)$$

We emphasize again that the very probable assumption that the reactivity of AS_ℓ is greater than that of AS_c does not exclude the probability of comparatively similar contributions of these and other forms of active site to the growth. At least the partial compensation of this inequality by the relation between the concentrations of these agents themselves is possible. The difference in the structural effects which is caused by the variation of the substituents in the initiators (5.XVI) to (5.XIX) can only be partly explained by this circumstance and by the dependence of the state of equilibrium (5.13) upon the nature of all the ligands near the Mg atom.

Nevertheless a direct connection between the changes in the structure of the butyl group in BuOMgEt initiators and the microtacticity of the corresponding polymers does not seem obvious on the basis of the above reports alone. The high isotacticity of PMMA which is formed by the isobutyl derivative does not extend to other related systems. Such a result, by analogy with the extremely rare alternative cases of the synthesis of PMMA of a similar structure, can be taken as an indirect indication that in the polymerization by initiating compounds (5.XVII), the role of intramolecular complex formation as a stereoregulating factor is an important one. This conclusion is essentially in accord with the notation of the predominant, if not exclusive, growth of the chain under the action of the active sites AS_c. This could mean that the counterions in the given system differ by a smaller coordination saturation by the "external" ligands. The apparent contradiction between these conclusions and the structural variations of the initiators proposed by Okamoto et al. [45] disappear immediately if one turns to the question of the genesis of the structural differences being discussed. When applied to growing chains this may be shown by a simple example which considers not so much the difference in the structures of the associates, but the state of equilibrium (5.14):

$$2\ M_nMgOBu \rightleftharpoons M_n{-}Mg\langle {}^{O(Bu)}_{O(Bu)} \rangle Mg{-}M_n \qquad (5.14)$$

The specific action of the iso-butyl derivatives is easily explained by its monomeric form which creates the optimum conditions for the formation of an active site. The transition $AS_\ell \longrightarrow AS_c$ is not excluded for associates of the type (5.XXII) either, but is energetically less favorable. This follows from the experimental and theoretical data on the reduction in the energy of complex formation in the RMt—D systems when the number of D ligands in the coordination sphere of the central atom increases [46, 47]. On the same basis it is possible to connect the tendency of the active site AS_ℓ to convert to AS_c in the presence of small quantities of D agents in a nonpolar environment and the resultant sensitivity of the microstructure of the polymers to the formation of active site complexes with the independent electron donors. A trivial example in this respect is the change of structure of PMMA when going from n-butyllithium to its complex with TMED. However, contradictory situations are also known, in particular, the polymerization of MMA under the action of t-butylmagnesium bromide in a solution of a mixture of toluene and THF; see page 224.

It is especially interesting that here, as in the above systems with butoxymagnesium alkyls, there appears a specific feature of the initiators of the general type BuMgBr which depends upon the structure of the butyl group. The individuality of their actions deserves attention in connection with the different features found in the research of Allen et al. [44, 48-52]. We note first of all that in polymerizations initiated by sec-butyl magnesium bromide, even smaller concentrations of THF influence the microstructure of PMMA markedly when simple experimental conditions are maintained such as those used in the case of t-butyl derivatives (toluene, −50°C; data for the fraction insoluble in CH_3OH) [49]:

THF concentration in solution, moles/liter	Triad content in PMMA, %		
	I	H	S
0.8	67	17	16
1.3	48	23	29
3.6	28	26	46

The study of the MMD of PMMA formed under the action of different butylmagnesium bromides in the same mixed solution has led to some very important conclusions. Polymers obtained under the action of n- and isobutyl derivatives are characterized by a symmetrical unimodal MMD. (Data on the fractions insoluble in CH_3OH are given here and also below. There are no data in the cited works on their content in the polymers investigated.) On the other hand, sec- and tert-derivatives produce a polymodal MMD for PMMA. This effect is preserved during changes in concentration of the THF in the system and with increased conversion. Both this and other events are accepted as evidence for the absence of exchanges between the coexisting active sites. This circumstance in its turn is used by the authors for their conclusions on the covalent nature of the organomagnesium active sites which are assigned structures of strong coordinating complexes (see [51, 52] for further details). The authors have noted the extremely strong influence of the conditions chosen for the initiation on the occurrence of multicenteredness. In particular, results obtained when these conditions are varied as follows have been compared. In one of these experiments, solutions of initiator and monomer were mixed at a temperature at which polymerization occurred (−43°C). In another experiment the monomer was condensed onto a solid solution of initiator at the temperature of liquid nitrogen, after which the temperature was raised to −43°C. The time required for this rise in temperature (several minutes) was very much less than that taken for polymerization (hours). The characteristics of the corresponding polymers are shown below (data for t-butylmagnesium bromide) [50]:

Conditions	Type of MMD	Ratio I:H:S
Simultaneous combination of the components of the system	Trimodal	99:0:1
Condensation of the monomer to a solid solution of initiator	Unimodal	1:1:1

On the basis of these results only it is difficult to stretch the assumption that the second of these cases is characterized by the coexistence of rapidly exchanging active sites of different types rather than by monocenteredness. It is easier to fit Bernoulli statistics to this case (the value $\rho = 1$ corresponds to the ratio 1:1:1), the more so since for organomagnesium initiators a statistically similar type of structure for MMA polymers is an extremely rare exception. Apart from the data given on page 225, this also follows from the fact that there are a large number of polymers which are related to the BuMgX compounds by butyl groups of different structure and by different halogens. The value of ρ for the corrresponding polymers obtained at −50°C in toluene–THF of different compositions ranges from 1.23 to 3.77 [44]. We note also that such a complete equalization of the fractions of all types of triad* is very rarely found and in no way corresponds to the "ideal" atacticity of a polymer, which is distinguished by an I:H:S ratio of 1:2:1, and which obeys Bernoulli statistics.

It follows therefore that the statistical type of chain structure is just one of the parameters necessary for formulating the mechanism of formation of macromolecules. A sound basis for the connection between the type of distribution of structural units in the polymer and the features of the growth reaction requires additional information on the active sites responsible for growth. The

*A result similar to such a distribution is also found for the polymerization of MMA under the action of the initiator $Ph_3CCaCl \cdot 2THF$ at 15°C with no solvent present; the triad content I, H, and S in PMMA is found in this case to be 30%, 35%, and 35% with an error of 5% [53].

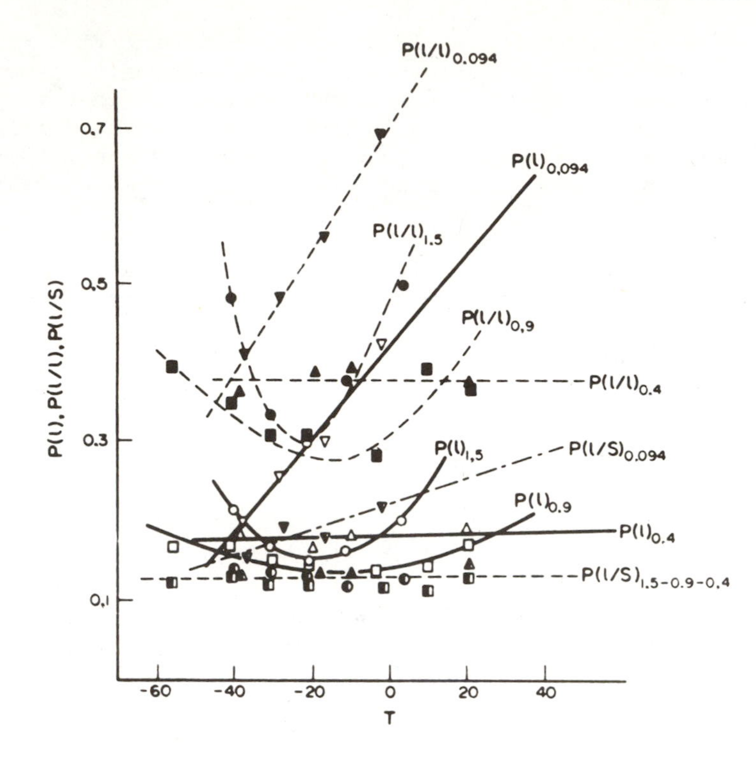

Fig. 5.3. The temperature dependence of the microtacticity of poly(methyl methacrylate) as a function of the ratio $MgCl_2$/t-Bu_2Mg (Q) [58]. $P(\ell)$ is the fraction of isotactic dyads, $P(\ell/\ell)$ is the ratio of isotactic triads to isotactic dyads, $P(\ell/S)$ is the fraction of heterotriads. Q: 1.5 (●, ○, ◑), 0.9 (■, □, ◧), 0.4 (▲, △, ◭), 0.094 (▼, ▽, ⧨).

amount of such information in such systems based on organomagnesium compounds is clearly insufficient to enable conclusions to be drawn about the structure of the active sites. For this reason conclusions amount to stating that in each system there exist active sites of different stereospecificity, that limiting cases can be distinguished (which is sometimes justified), and that such active sites act independently and their exchanges proceed at a rate faster than that of the growth reaction. In a paper by Allen et al. [52] these reports are amplified by ^{1}H-NMR data according to which the original monoesters RMgBr·THF form, in toluene solutions, mixtures of RMgBr, THF, and R_2Mg·2THF components with the latter predominating. The influence on the effects of multicenteredness of varying the experimental conditions does not extend to the possibility of controlled regulation of the relative contributions of the parallel-acting active sites.

Pham et al. [54-58] adopted a different approach to the problem of possible structural variants of organomagnesium active sites by using magnesium chloride as a model reagent. The system $MgCl_2$–MMA–THF was chosen for direct investigation. ^1H-NMR spectroscopy data led to the conclusion that there exists an overall equilibrium given by the general equation

$$2M + MgCl_2 \cdot 4THF \rightleftharpoons MgCl_2 \cdot 2M + 4THF \qquad (5.15)$$

Together with the discovery of $MgCl_2 \cdot xTHF$ and $MgCl_2 \cdot yM$ complexes of different stoichiometry, the ^1H-NMR spectra point to the probability of formation of more complex aggregates in which THF and monomer are present simultaneously as ligands.

As has been shown, the microstructure of PMMA depends upon the ratio $t\text{-}Bu_2Mg$–$MgCl_2$ (see Fig. 5.3), which is in agreement with the conclusion of the influence of the nature and concentration of the above complexes on the stereochemistry of the growth reaction.

It must be noted that Allen came to the conclusion, on the basis of the kinetics of the polymerization of MMA in mixed THF–toluene solvents under the action of various butylmagnesium halides and also butylmagnesium [48, 51], that growth proceeded via a stage in which the active sites formed complexes with the monomer.

The ability of initiators based on other alkaline earth metals to form isotactic PMMA has not as yet been observed. The possibility of obtaining PMMA with a high content of syndiotactic triads has been established for one of the calcium compounds while carrying out polymerization in dimethoxyethane. The data of Lindsell et al. [53] show the dependence of the microstructure of PMMA on the nature of the alkaline earth metal in initiators of a closely similar type:

Initiator	Solvent	Triads, %		
		I	H	S
$Ph_3CCaCl \cdot 2THF$	THF	51	34	15
$Ph_3CSrCl \cdot 2THF$	THF	25	30	45
$Ph_3CBaCl \cdot 2THF$	THF	23	46	31
$Ph_3CCaCl \cdot 2THF$	DME	0	12	88

The change in the structure of the macromolecules on transition from THF to DME for the same initial calcium initiator is very marked. This is obviously caused by the exchange between the ligands in the coordination sphere of the counterion with molecules of the solvent and the formation of growing chains of the separated ion pair type. It is also known that DME, to a considerably greater extent than THF, brings about the ionic dissociation of the active sites. This is illustrated by the characteristics of certain living chains with alkali metal counterions (see Section 2.1).

At the same time the polymerization of MMA by strontium and barium initiators shows that changes in the nature of the solvent have a much lesser effect on the structure of the polymers. There is, however, some tendency in these cases to an increase in the isotactic triads in THF which is not observed in DME. A maximum content of isotactic triads of 60.5% has been established for the initiator $Ph_2CBr \cdot 5THF$ (as compared to 10% for DME). These effects do not lend themselves to exact interpretation and it is possible only to assume that the determining role is played on the one hand by the state of the active site in the sense that they are of the IP_s type and on the other hand by the intramolecular complex formation of the growing chains. The situation regarding the establishment of the formation of the chain structure for systems which include alkali metals as counterions is somewhat better.

The smaller variety of "ligand environments" for counterions and the availability of data obtained in polar media using buffer compounds must be borne in mind. This circumstance makes it possible to compare the stereospecificity of the related IP_c, IP_s, and also free ions.

As yet all the data necessary for such a comparison is not available but a considerable part of this material is contained in the studies of Schulz et al., some of which have already been cited in Chapter 4.

We will give some examples which illustrate the structural effects in an MMA–cumylcesium–$Cs(BPh_3CN)$ system over a wide range of temperature [59]:

t, °C	Triads, %			
	I	H	S	ρ
20	10	55.5	34.5	0.85
−11	8	52	40	0.86
−80	0.5	52	47.5	0.75
−100	0	40	60	0.80

In interpreting these results the authors emphasize the deviations from the Bernoulli statistics which they found in the structures of these polymers when multicenteredness in its generally accepted sense is absent. The presence of a buffer precludes the existence of growing chains in the form of free ions, but a linear Arrhenius graph for the growth rate constants over the same temperature interval (see Section 4.1.1.3) enables the ion pairs to be taken as uniform. Assuming Markov statistics to apply the authors propose a mechanism which includes the active sites as intramolecular complexes of one of the following structures which may possibly be in equilibrium:

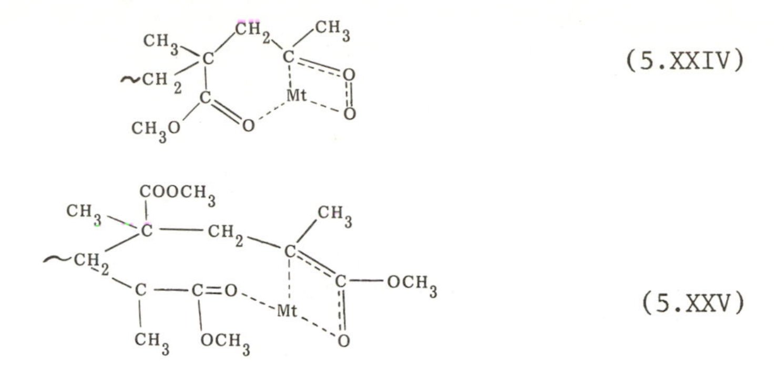

On the basis of the results obtained for the sodium counterion under the same conditions and their comparison with data for both counterions in the polymerization of MMA in a dimethoxyethane medium, the notion that the active sites were similar was somewhat modified. See Chapter 4 structures (4.XII) and (4.XIII).

Among the forms in which PMMA active sites can exist and which are recognized by Schulz et al., the intramolecular associates of dianionic growing chains must be mentioned [60, 61]:

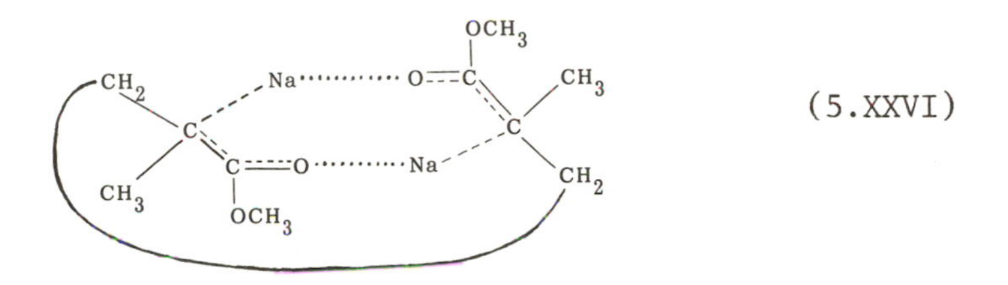

The possibility of forming similar structures in a polar medium is of sufficient general interest to deal with this aspect in somewhat greater detail.

In discussing the stability of the associated forms of organometallic compounds, the extremely large difference in the behavior of such aggregates with the nature of the initial reagents and the polarity of the medium has already been noted (see Sections 2.1.1 and 4.1.1.1). The information available in this field did not allow, for a system such as Na–M_n–Na–THF (for which the structure (5.XXVI) is proposed), the assumption of equilibrium between the linear and cyclic forms of disodium chains to be taken further, this equilibrium being markedly displaced towards the former. There are results quoted which form the basis of a contrary conclusion. The structure (5.XXVI) is taken as the only form for those chains whose intramolecular associates become more unstable with an increase in the degree of polymerization. For a DP of about 200 the tendency of these chains to transfer to the cyclic form practically disappears.

In order to evaluate the correctness of these conclusions it is necessary to mention the most important of the facts upon which they are based.

The features of the fractions of PMMA which are formed in THF under the action of α-methylstyrene disodium tetramer (MDT) are of primary importance. With low conversion PMMA has a well-marked bimodal MMD, the fractions corresponding to the various values of DP being quite strongly differentiated with respect to tacticity. With increased conversion the bimodality of the MMD gradually smooths out and the microstructure of the fractions becomes markedly similar. Of the great quantity of results given in [60] we limit ourselves to the following (data for −75°C):

	Conversion, %	DP	Triads, %			ρ
			I	H	S	
High-molecular-weight fraction (HF)*	25	670	5	33	62	1.02
Low-molecular-weight fraction (LF)	16	68	39	41	20	1.18

*Data for −73°C; the exact conversion for this value of DP not given. As pointed out by the authors, the tacticity of the high-molecular-weight fraction does not depend upon conversion.

The authors consider the partial deactivation of the initial initiator at the first stages of polymerization (additions, side reactions with the monomer) and the resulting presence of monofunctional M_nNa chains in the system to be the source of this bimodality. It is accepted that their fraction is about 10% of the total concentration of active sites.

The use of labelled termination reagent CH_3COOT and radiochemical analysis of the fractionated polymer permitted the discovery of double T labels on the chain in the LF in comparison with their content in the HF. Hence an unequivocal conclusion on the genesis of each fraction was drawn:

$$Na{-}M_n{-}Na \longrightarrow LF \text{ and } M_nNa \longrightarrow HF$$

In its turn such a conclusion may be taken as evidence of the reduced growth rate which takes place in bifunctional active sites. This has also led to a concept of their structure which is described by formula (5.XXVI) for which the reduced activity as compared with linear chains seems perfectly natural.

Ideas on the reasons for smoothing out of the differences between the MM and the structural characteristics of the PMMA fractions with increased conversion also agree with the kinetic data. Together with rapid initiation, which has been confirmed by independent observations, a gradual increase in the polymerization rate is noted. A marked increase in this rate is observed when low-molecular-weight fractions with a DP of about 200 are obtained.

In order to ascertain the role of DP more precisely for dianionic growing chains, the polymerization of PMMA under the action of high-molar-mass disodium-poly-α-methylstyryl (DP = 70) was studied. As a result of this it was established that there was no induction period and that PMMA with a bimodal MMD was formed whose character did not depend upon the conversion; the GPC peak stays at the same retention volume. Here the radiochemical characteristics of the fractions which were separated after termination by labelled acetic acid are seen to be the inverse of the above. The T-label content in the LF is found to be only half that in the HF. Thus both the kinetic and radiochemical data show that there are no differences in the reactivity of the active sites in the case of mono- and bifunctional growing chains obtained using disodium initiator with a high initial DP.

We now turn to the structural features of PMMA formed in systems based on MDT. It is obvious that the counterions play a part in determining the structure of the chains growing on dianionic active sites (this is shown by the increased isotacticity of the LF and the deviation of the corresponding value of ρ from unity). If the associated state of the active site is regarded as the reason for these effects then it is not difficult to ascribe the features found to the difference in the C_α—Na distances of the growing chains being compared. For monofunctional compounds of the general type $RCH_2-\underset{R'}{\underset{|}{C}}(Mt)COOR'$, the considerable stretching of this bond as a result of the approach of the Mt atom to the group follows from both the experimental and calculated data (see Chapter 2). Essentially, the ideas of Schulz et al. on the structure of the end group of the ion pairs of living PMMA chains which they regard as being intermediate between carbanionic and enolated are also based on these data. The occurrence of intramolecular associates of dianionic chains of this type must have as its consequence the reduction in the distance between the counterions and their C_α atoms.

Taking this circumstance into account, any doubts as to the correctness of the nontrivial conclusions drawn from these facts discovered in studies of MDT—MMA—THF system almost completely disappear. There remains a certain vagueness only in the reasons for the stability of the associates (5.XXVI), which is conserved up to a high DP. For this reason the following may be proposed.

If the coexisting forms of active sites are regarded as the system

$$Na{-}M_n{-}Na \rightleftarrows (5.XXVI) \qquad (5.16)$$

then obstacles to the intramolecular cyclization must be expected at much smaller DP than would seem likely from the experimental data. Therefore it mut be assumed that there is a great advantage in forming (5.XXVI) structures at small values of n and the virtual irreversibility of cyclization when n is increasing, at least up to several tens of units. Such a situation seems all the more correct since under these conditions the rate of polymerization is quite high. A conversion of about 20% is attained at −75°C in about ten seconds [60]. This permits one to believe the system to be far from equilibrium, a situation which, for small values of n, could correspond to a smaller concentration of active sites of the type (5.XXVI) than is in fact the case. The functional relationship $AS_\ell/AS_c = f(DP)$ cannot be realized since the growth rate is greater than that for the establishment of equilibrium (5.16) which, if a structure of the type (5.XXVI) is taken for AS_c, depends upon the degree of polymerization.

We will supplement these results of the study of the microstructure of PMMA formed in various anionic systems by some remarks on processes initiated by organolithium compounds in nonpolar media. The degree of isotacticity of the macromolecules in these cases depends considerably upon temperature and the selection of the actual initiator. It is not difficult to associate this with the degree to which side reactions occur and with the modifying influence of the transformation products (see Section 4.1.1.3) on the active sites. Obviously the great difference in the structure of PMMA found by Trekoval et al. [62] while carrying out polymerization of MMA in the same conditions of temperature and concentration under the action of various organolithium initiators and some systems based on them, result from just this circumstance. We give as an example the difference in the content of isotactic triads in PMMA obtained in toluene at −40°C using n-butyllithium and $(CH_3)_3C(C{=}O)\cdot CH_2Li$; these are 75.2 and 86%, respectively.

There is still one interesting effect which also relates to the polymerization of MMA in a nonpolar medium and this is the dependence of the structure of the macromolecule on monomer concentration. This is illustrated by the following (toluene–n-butyllithium, 70°C [63]):

	Triads in PMMA, %	
MMA, moles/liter	I	S
4.70	45	30
0.21	77	9

This difference may be ascribed to the fact that mechanism (4.1) is realized in the purest form at minimum monomer concentration.

As is well known, the structural effects which are observed in the polymerization of unsaturated monomers can sometimes be connected with the volume of the substituent on the C_α atom. This increase may facilitate syndiotacticity of the growth reaction. In the case of the polymerization of methacrylates in anionic systems the appearance of the role of the steric factor in the variation of the substituent in the ester group is by no means obvious. The data of Hatada et al. [64] on the polymerization of methacrylates with $-COOCH_2R$, $-COOCHR_2$, and $-COOCR_3$ under standard conditions (octylpotassium, THF, −79°C), where R is methyl or phenyl, enable the changes in the structure of these macromolecules to be traced. Of nine monomers in this series, significant increases in syndiotactic triads as compared to MMA were found only in the polymerization of ethyl methacrylate. In all other cases the triad content either differs very little from that of PMMA (substituent $-COOC(CH_3)_2Ph$) or is considerably less:

		Substituent (R in −COOR group)				
		CH_2	$CHCH_3$	$CH(CH_3)(Ph)$	$C(CH_3)_3$	CPh_3
Triads, %	I	11	18	20	15	90
	H	53	29	64	60	2
	S	36	53	16	25	8
	ρ	0.86	1.08	0.78	0.82	0.82

It is possible in principle to interpret these differences from the point of view of changes in the relative role of the electronic and spatial factors in each individual case [22], but there is no substantial basis for this approach. We refer for example to the structure of the polymers of the trimethyl and triphenyl derivatives of this series. The usual spatial concepts make it possible to expect the appearance of qualitatively similar structural effects in the polymerization of both monomers. However, the total absence of such a similarity may easily be ascribed to the fact that the high isotacticity of the trityl methacrylate derivatives is the result of the regulating influence of the counterion. Hence, bearing in mind that the participation of the counterion in the formation of polyligand stereospecific active sites is facilitated by the phenyl groups, it must be emphasized that the question must arise as to the direct influence of the aromatic nuclei on the fixing of the conformation of the growing chains which favor isotactic sequences in the macromolecules. On the other hand, in any attempt to develop this hypothesis further it is necessary to take into account that, according to Okamoto et al. [65], the marked tendency to isotactic growth in triphenylmethylmethacrylate has also been found in the radical polymerization of this monomer. The isotactic triad content in this case reached 64% at 60°C (as compared to 4% in PMMA) [65]. Consequently the actual conformation of the growing chains with this substituent creates conditions for a predominantly isotactic structure for this polymer.

At the moment only suppositions are possible on the nature of the additional advantages which are in this sense inherent in anionic systems. In any case the possibility that the regulating action of the counterion at reduced temperature plays some role here cannot be excluded.

General considerations cannot explain the strong dependence on the nature of R [41] of the structure of macromolecules obtained by anionic polymerization of a related series of monomers, CH_2=C(R)$COOH_3$. The behavior of alkyl derivatives (R = ethyl or t-butyl) is rather similar to that of MMA. In both cases, n-butyllithium in toluene yields highly isotactic macromolecules, while in THF a significant shift to syndiotactic sequences is observed. The phenyl derivative shows neither a tendency to form stereoregular polymers nor solvent sensitivity. Macromolecules obtained from this compound with various anionic initiators and media are structurally similar to atactic polymers.

α-Chloroacrylates, CH_2=C(Cl)COOR, where R = CH_3, C_2H_5, or C_3H_7, show some interesting features. These compounds also fail to yield isotactic polymers under conditions that often produce this effect (e.g., n-butyllithium in hydrocarbon media) [41]. However, the use of heterogeneous catalysts, obtained by reacting RMgX with α,β-unsaturated carbonyl compounds (the so-called Breslow—Kutner catalysts), results in structurally heterogeneous fractional composition of the polymers and high isotacticity of THF-insoluble fractions (polymerization in toluene at 30°C) [66]:

	Triads	Gross structure	Insoluble fraction
Polyethyl-α-chloroacrylate	I	68%	80%
	S	21%	12%
Polyisopropyl-α-chloroacrylate	I	51%	90%
	S	24%	4%

Date on the relative amounts of the isotactic fractions in these polymers are not available from the above literature. The given structural differences enable it to be assumed that the insoluble fraction constitutes a large part of these only in polyethyl-α-chloroacrylate.

Polymerization processes in these systems are not very efficient (a conversion of the order of 10% is attained under normal conditions after many hours) but they are of interest as examples of the coexistence of independently functioning active sites. Furthermore, some of the polymers obtained have been more fully described than have the alkylmethacrylates, the information on which is limited to data on the content of the various triads. Therefore conclusions which go beyond the limits of the deviations from Bernoulli statistics are in these cases conjectural. The content for the various tetrads has been established for a number of poly-α-chloroacrylates (by high resolution NMR data for the CH_2 lines relating to the ester groups) and a detailed structural analysis of the individual members of this series has been carried out. It is shown that the structures of the methyl and butyl derivatives formed in homogeneous anionic polymerization are in good agreement with the Bernoulli statistics and deviate from them only in the case of the isopropyl derivative; for these polymers the values of ρ are 0.09, 1.00, and 1.35. Acccording to their overall characteristics all of these polymers must be regarded as atactic. The isotactic fractions isolated from polymers obtained with Breslow–Kutner catalysts obey first order Markov statistics. This is well shown by the fact that there is a good correspondence between the contents of the majority of the possible types of tetrads evaluated experimentally (T_E) and theoretically (T_T) on the basis of the above statistics. We will show this for examples of the above fractions [66] (where R is a radical in the ester group of the monomer, tetrad content in %):

R		mmm	mmr	rmr	mrm	mrr	rrr
Ethyl	T_E	79	2	3	1	7	8
	T_T	76	8	0	1	6	9
n-Propyl	T_E	92	1	2	1	4	1
	T_T	92	5	0	1	2	1

On the basis of the statistical characteristics, the authors assign the dominant role to the asymmetry of the active sites.

We emphasize that the Breslow–Kutner catalysts are distinguished by a still higher stereospecificity when polymerizing the common alkylacrylates [41]; however, in these cases they are similar to other anionic initiators. For α-chloroacrylates, as has already been pointed out, the action of different anionic reagents is not uniform. Consequently the formation of stereoregular fractions in the polymerization of the α-chloroacrylates under the action of these catalysts is not the result of the presence of a halogen in the monomer molecule, but due to the special conditions which arise when growth proceeds at the phase boundary. On the other hand the structure of the macromolecules formed from the same monomers in homogeneous conditions using initiators which give rise to stereospecificity in alkylacrylates is an indication of the negative effect of the chlorine atoms as substituents in the vicinity of the C_α atoms of the monomer on the spatial selectivity of the growth reaction. The reason for this may be competitive complex formation of the counterion with the Cl atoms in the ester groups of the polymer chain, i.e., in the reduction of the stability of the forms of active site which are distinguished by increased stereospecificity. Apparently the intramolecular complexes of active sites AS_c, in which the carbonyl groups closest to the end units of the growing chain function as electron donors, are such forms.

5.1.2.2. Vinylpyridines

Of all the monomers of this series, 2-vinylpyridine (2-VP) is distinguished by its particular ability to form isotactic polymers, and several of the organomagnesium compounds are among the corresponding best anionic initiators. Their high stereospecificity with respect to 2-VP was discovered by Natta et al. in the sixties, but detailed studies in this field have been carried out in recent years by Soum and Fontanille [67-69].

With the objective of creating optimum conditions for the physicochemical characteristics of the stereospecific active sites these authors directed their efforts to finding an initiator which excluded side reactions which usually accompany anionic polymerization of 2-VP and its isomers. These reaction products considerably complicate the study of such reactive mixtures. In a series of experiments they established that benzylpicolylmagnesium (BPM) possessed the necessary properties. This may be regarded as the simplest single-unit model of the active site of 2-VP with a magnesium counterion. This compound affords practically 100% initiation efficiency and the absence of side products.

Isotacticity of poly-2-VP formed with BPM practically coincides with the data for the usual R_2Mg initiators, but the latter are distinguished by a low initiation efficiency and do not eliminate side reactions. The structural and MM characteristics of these polymers are compared below (data for complete conversion in benzene at 25°C) [67]:

Initiator	$M_n \times 10^{-3}$	$M_{theo} \times 10^{-3}$	Triads, %		
			I	H	S
Di-sec-butylmagnesium	145	17.5	92	7	1
Dibenzylmagnesium	3	1	89	9	2
BPM	49	50.1	93	(7 total)	
Oligomer-2-VP, obtained under action of BPM	230	240	91	7	2

The difference in the values of M_n are caused by variations in the monomer/initiator ratio.

In order to explain the structure of the active sites which are responsible for the initiation of BPM it is essential to assert that the initiator is initially in the dimer state (cryoscopic data), whereas growing chains do not exhibit a tendency to dimerization (viscometric data characterizing living and deactivated polymers) [68]. This is explained by the authors by the fact that in contrast to the single-unit models whose dimerization leads to structure (5.XXVII), for real growing chains the formation of intramolecular complexes (5.XXVIII) is preferred; here Bz is a benzyl group:

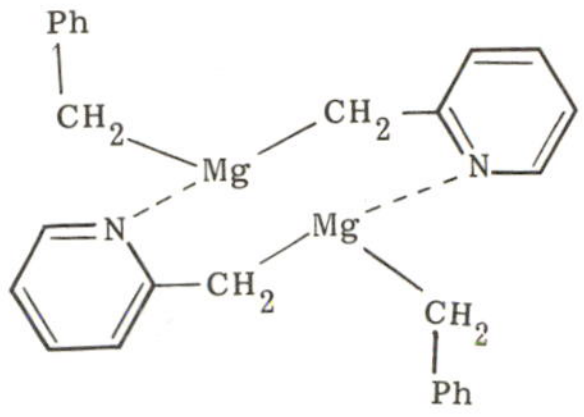

(5.XXVII)

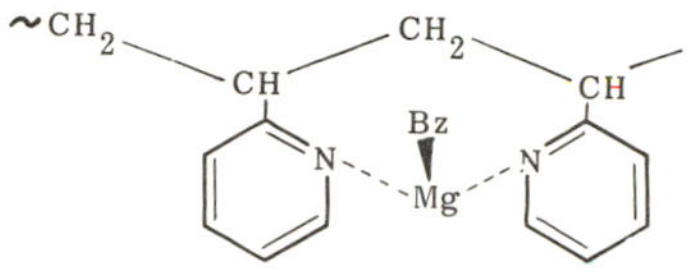

(5.XXVIII)

Such a structure of the active site may be responsible for the construction of a polymer chain obeying first-order Markov statistics. There is a fairly satisfactory correspondence of the content of the various pentads (P) in the macromolecules (evaluated from ^{13}C-NMR spectra using the lines of the C_2 atom of the pyridine nucleus) with the calculated values. Instead of the structural composition we give values which characterize the degree of correspondence of the experimental values (P_E) with the theoretical values (P_T) assuming first-order Markov statistics for poly-2-VP formed under the action of BPM in benzene at 20°C [69]:

	mmmm	mmmr	mmrm	rmrr	rrrr
P_E/P_T	1.01	0.89	0.89	0.75	15.0

The remaining pentad content is given in the form of totals and not individually.

The greatest deviation in the ratio P_E/P_T from unity occurs for those structural sequences which are poorly represented in the macromolecules, which is perfectly natural. This is particularly so for the pentads rmrr and rrrr for which the experimental values are 0.3 and 1.2%. Nevertheless, a growth mechanism is possible which includes the participation of not only the penultimate group but also more distant units in the formation of the chain. This may manifest itself at the very least in the form of some contribution from higher-order Markov statistics. In the general case such a situation is more natural than "pure" statistics of one type. This question may amount in principle to a degree of participation of AS_c which have different cyclic forms; see Section 3.1.4.

There are as yet no data which could serve as a starting point for an informed discussion of this question. The situation exists in which the general form of the growing chains does not vary (i.e., remains as AS_c) and there is a variety of active sites present. It is only possible to assume that some indication of the effects produced by such a situation is afforded by the degree of regularity of the microstructure of such polymers. In the cases which we are considering, I_T is of the order of 90% but the highest known content of isotactic pentads is 78% [69]. The very modest amount of information on the structural characteristics of other stereoregular polymers beyond the triad level does not give grounds for a general discussion of whether the difference in value given for mm and mmmm is typical, or exclusive to macromolecules which are taken to be stereoregular.

Regarding the point of view of Soum and Fontanille, they regard structure (5.XXVIII) as applicable to active sites of high stereoregularity on the basis of their data on the conservation of Bz—Mg groupings which do not vary throughout the growth. In order to explain the latter situation the distribution of the electron density on BPM is used. This is calculated from characteristics obtained from ^{1}H and ^{13}C-NMR studies [structures (5.XXIX) and (5.XXX), respectively]. Here Py is the pyridine ring in the BPM compound.

(5.XXIX)	(5.XXX)
−0.032 −0.012	−0.0092 −0.008
$Py{-}CH_2{-}Mg{-}CH_2{-}Ph$	$Py{-}CH_2{-}Mg{-}CH_2{-}Ph$

Without attaching special significance to the absolute values based on data from these and other spectra (the conditionality of the charge characteristics based on NMR spectra has already been pointed out in Section 2.1.3) it is possible to assert the qualitative agreement of the differences between the electron density on the C_α atoms attached to the Py- and Ph-rings in the structures compared above. The considerably increased affinity for carbanions of the picolyl group modelling the end units of the growing chains is obvious.

These authors regard the end and penultimate units as the groups which actually have a stereoregulating function, i.e., the fragment with two asymmetric C atoms. They also take into account four diastereoisomeric pairs; isotactic growth is possible for two

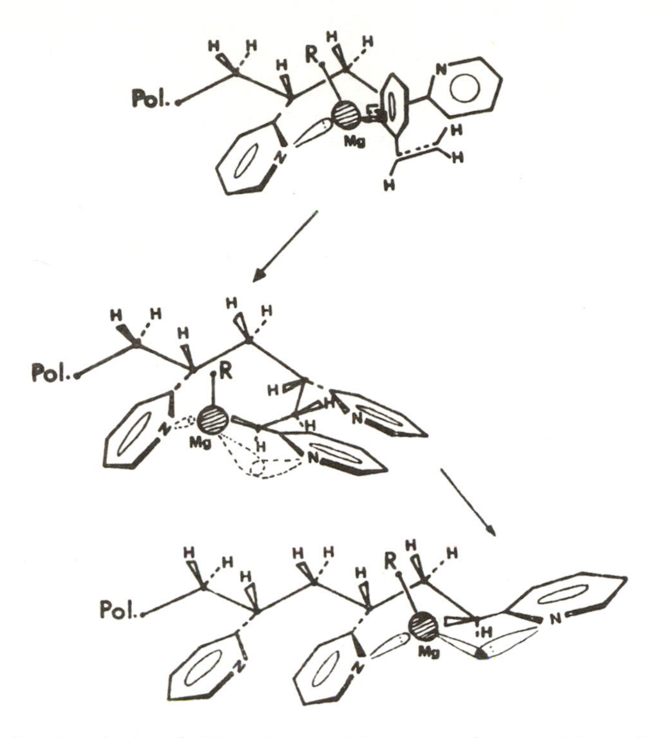

Fig. 5.4. Mechanism of the isotactic growth reaction during the polymerization of 2-vinylpyridine under the action of organomagnesium initiators (from Soum and Fontanille [69]).

of these and syndiotactic growth for the other two. It is not possible to substantiate the stereospecific effects observed on the basis of the dominant role of the diastereoisomers because in terms of stability they do not differ essentially from one another.

This circumstance served to bring about the formulation of a mechanism corresponding to an equation of the general type (4.1). For the intermediate complex AS·M a structure is proposed in which the monomer molecule, without hampering the interaction of the counterion and the penultimate unit of the growing chain, becomes the ligand in the coordination sphere of the magnesium instead of the pyridine ring of the end unit:

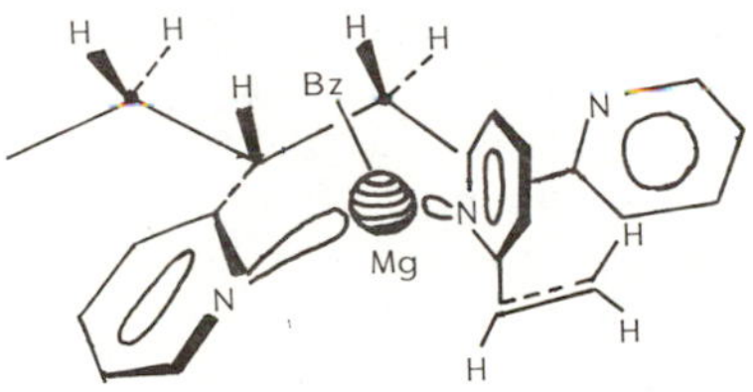

The insertion of the monomer into growing chains from within this complex is regarded as a concerted 4-center mechanism with conservation of the initial configuration. The authors have shown quite convincingly that the transition states, which, due to the insertion of the monomer, cause the transformation of the end dyads into triads according to Eqs. (5.17, 18), are sterically superior to those of pathway (5.19, 20);

$$\mathrm{m} \xrightarrow{\mathrm{M}} \mathrm{mm} \quad (5.17)$$

$$\mathrm{r} \xrightarrow{\mathrm{M}} \mathrm{rm} \quad (5.18)$$

$$\mathrm{m} \xrightarrow{\mathrm{M}} \mathrm{mr} \quad (5.19)$$

$$\mathrm{r} \xrightarrow{\mathrm{M}} \mathrm{rr} \quad (5.20)$$

On the basis of the microstructure of the polymers, Eq. (5.17) may be regarded as the dominant event in growth; for this the scheme given in Fig. 5.4 is proposed.

In conditions which usually favor the maximum formation of isotactic sequences in poly-2-VP, organolithium initiators are considerably inferior to organomagnesium initiators with respect to stereospecific action. The highest known value of I_T for the first of these is 85% [70]. On the other hand, in the synthesis of oligomers of 2-VP a significant tendency to isotactic growth is observed with organolithium active sites even in THF. The stereochemistry of these reactions which has been studied in detail by Hogen-Esch et al. for several model active sites is formulated very precisely in [70-73] which also contain references to earlier work.

As has already been shown, the interaction of 2-VP with the compound

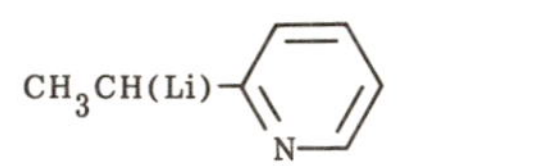

(5.XXXI)

proceeds with extremely high spatial selectivity in the formation of the dimer. The meso-dyad content of this oligomer is more than 99% (THF, −78°C). The stereospecificity of subsequent growth events falls sharply although the isotactic direction remains markedly superior. Thus, in trimers and tetramers of 2-VP obtained under the same conditions, the I_T content is somewhat greater than 60%. The authors ascribe such a tendency to the participation of the penultimate units in the formation of complexes with the counterion. This is shown in their proposed structure

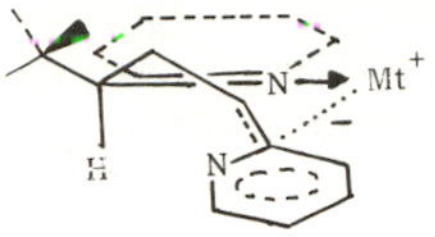

(5.XXXII)

This is confirmed by the difference in the character of the interaction of 2-VP with the single-unit models (5.XXXIII) and (5.XXXIV):

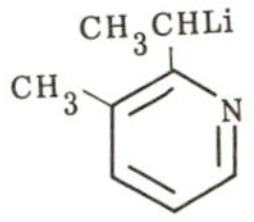

(5.XXXIII)

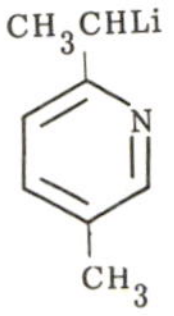

(5.XXXIV)

The 3-CH_3 derivative (5.XXXIV) causes the formation of oligomers which are differentiated according to structural characteristics and are very similar to those already mentioned for the reaction products of 2-VP with the compound (5.XXXI). On the other hand, the results corresponding to the 5-CH_3 derivative indicate that there is no similar tendency. The relevant data are shown below (THF, −78°C) [72]:

Initiator	Structural units, % (in dimers)	(in trimers)	(in tetramers)
(5.XXXI)	99.6	63.8	62.5
(5.XXXIII)	24	27	5
(5.XXXIV)	99.3	70.0	56.3
	(m)	(mm)	(mmm)

It is possible that in the case of initiator (5.XXXIII), the CH_3 group hampers the formation of structure (5.XXXII) which favors isotactic growth.

Turning to the possible causes which limit the high spatial selectivity of the compounds (5.XXXI) and (5.XXXIV) at the stage of dimer formation, the following features of anionic polymerization of monomers with polar substituents must be borne in mind. First, side products, which are typical for the majority of such processes, are capable of affecting the stereochemistry of the growth reaction. Secondly, the change in the detailed structure of the end unit may play some role as the very first monomer molecules are added to the primary site. We explain this using the data of [74] in which the geometry of the compounds (5.XXXI) is described. According to the ^{13}C-NMR spectra, in the equilibrium between isomers

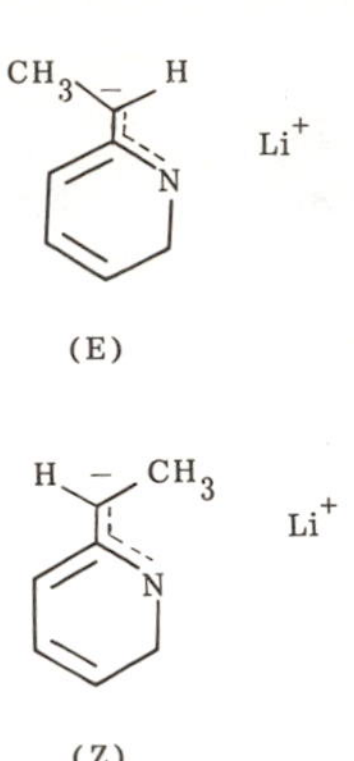

(E)

(Z)

the (E) form, whose content in THF is 94% at 25°C, is responsible for the main contribution. The authors have sufficient grounds for ascribing the stereochemical selectivity of this compound to the fact that it exists mainly in one of the two forms. However, it does not follow from the structure of the single-unit model of the active site (AS^1) that the same ratio E/Z is conserved by the end unit of the two-unit model (AS^2). A change in this ratio on transition from AS^1 to AS^2 may be the cause of the above difference in the structure of the oligomers of 2-VP.

Third and finally, the increase in length of the polymer chain increases the number of actual forms of active site of the general type AS_c. For the counterions of these the possibility of intramolecular complex formation with a polar group of any unit of the growing macromolecule is not excluded. To regard this feature of anionic active sites as dominant is hardly reasonable because the synthesis of polar macromolecules of high stereoregularity is possible in a large number of cases. We note nevertheless that the cause of the increased stereospecificity of heterogeneous initiators often found cannot be ascribed to only one preorientation of the monomer at the interface. Apart from this, for growing chains in such systems the number of possible real structures for AS_c must decrease sharply in comparison with related homogeneous initiators. This fact, in itself indisputable, must be considered among other circumstances when interpreting the features of the stereochemistry of the growth reaction in systems of different phase state.

The work of Hogen-Esch et al. also contains physicochemical features of anionic models of the pyridine series obtained using IR spectroscopy and conductometric and quantum-chemical methods. These results relate to a single research program and we considered it more appropriate to deal with some of them here at this point rather than in Chapter 2.

A considerable volume of information is obtained from studies of the complex models (5.XXXV) (in which Mt is lithium, sodium, or cesium and Py is 2-, 3-, or 4-pyridyl). The results obtained are interesting in comparison with those from the (5.XXXVI) compounds which are used in the synthesis of reagents (5.XXXV) [75]:

$$\underset{\displaystyle Py}{\underset{|}{MtC}}(Ph)CH_2C(Ph)_2(CH_2)_2C(Ph)_2CH_2\underset{\displaystyle Py}{\underset{|}{C}}(Ph)Mt \qquad (5.XXXV)$$

$$MtC(Ph)_2(CH_2)_2C(Ph)_2Mt \qquad (5.XXXVI)$$

We will limit ourselves to giving the characteristics of lithium members of these series; here Py is 2-pyridyl:

Compound	λ_{max}, nm		$K_{diss} \times 10^8$, moles/liter
	toluene	THF	(THF, 25°C)
(5.XXXV)	435	443	28
(5.XXXVI)	450	485	400

They will illustrate the strengthening of the interaction between the components of the ion pair which results from the insertion of the pyridyl substituent into the original compound.

A similar study was carried out for the simpler 2-pyridyl derivative $C_5H_4N{-}CH(Na)CH_3$ (5.XXXVII), in which it was shown that there was a considerable difference in the dependence of the UV and electrochemical characteristics of the ion pair on the actual type of polydentate electron donor [73]. Thus, dibenzo-18-crown-6 [in about 10% excess over (5.XXXVII) in a solution of THF] does not affect the UV spectrum of the compound and has comparatively little effect on the corresponding K_{diss}. More of an effect is seen with the presence of cryptand [222], which in the same concentration produces a noticeable shift in λ_{max} in the long-wave region and a considerable increase in the electrolytic dissociation constant of the (5.XXXVII) compound. The values of the latter obtained in various systems are shown below (25°C):

Systems	THF	THF—crown	THF—cryptand
K_{diss}, moles/liter	$1.4 \cdot 10^{-10}$	$1.5 \cdot 10^{-9}$	$5.5 \cdot 10^{-7}$

Quantum-chemical calculations were carried out for the lithium model (5.XXXI) and free picolyl anion (CNDO/2 method) [73]. In the first case the optimum geometry was found when the counterion is situated above the plane of the pyridine ring equidistant from the C_α and N atoms. A similar positioning of the components of the ion pairs is typical for a number of other anionic active sites with polar or aromatic substituents (see Chapter 2). In the second case the distribution of the electron density is described. As has been shown the negative charge of the anion is concentrated mainly on the C_α and N atoms (0.36 and 0.32 e, respectively).

Additional conformational features of two- and three-unit lithium models and their deactivation products are given in [76].

In this series there are also data on the structure of the 4-VP oligomers. The polymerization of this isomer of the series is characterized by the formation of atactic macromolecules: this has been known since Natta's earliest work in which this feature was associated with considerable separation of the N atom of the monomer from its vinyl group. As can be seen from the features of the models studied by Hogen-Esch et al., no less important is the mutual orientation of the C_α and N atoms in the end unit which excludes the possibility of the occurrence of stereospecific active sites which are typical of the polymerization of 2-VP.

5.1.2.3. Alkenyl Ethers

The discovery of the stereoregularity of macromolecules formed from vinyl monomers was made long before stereoregular polyalkyl ethers were synthesized; the history of this problem has been discussed many times (see, for example [77]). At the same time studies of the mechanisms of the corresponding active sites has not been particularly systematic. Over the last decade only uncoordinated works have appeared, more often than not repeating the general features of Kunitake's scheme which was proposed as long ago as 1970 [78].

This scheme consists in essence of concepts which are similar to those which are accepted for anionic sites. The idea that the dependence of the stereochemistry of the growth reaction on the nature of the counterion decreases with an increase in its size and with the stretching of the active bond (this is caused by the increase in the permittivity of the medium), i.e., with the decrease in the "degree of contact" of the ion pair, must be borne in mind.

Recently Kunitake and Takarabe [79] investigated the applicability of this concept of the polymerization of t-butylvinyl ether (t-BVE) using a series of uniform initiators of the general type $Ph_3C(EL_p)$ in which E is the central atom of the counterion (B, Al, Sn, Ti, etc.) and L is a ligand in its coordination sphere (basically Cl and F) in media of permittivity from 1.5 to 10.0 (mixtures of toluene–dichloromethane of different compositions).

The conclusions drawn by these authors are based on the features given in Fig. 5.5 which show the variation of meso-dyad content with the nature of the counterion and the permittivity of the

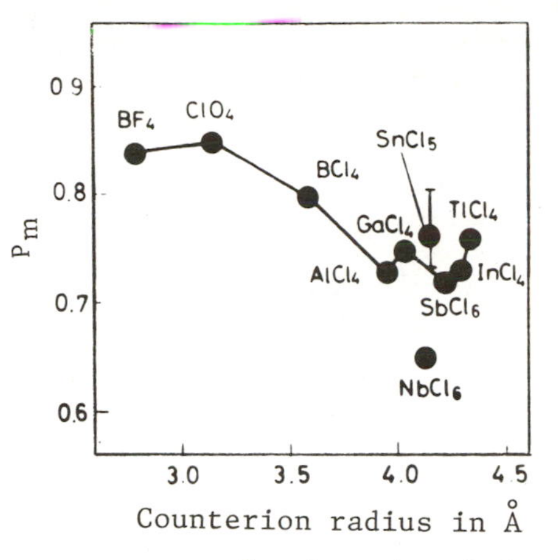

Fig. 5.5. The microstructure of poly-t-butylvinyl ether as a function of the radius of the counterion during polymerization under the action of $Ph_3C^+Y^-$ initiators [78]. Temperture, −76°C. Solvent: a mixture of methylene chloride and toluene (7:3 by volume). P_m is the meso-dyad fraction.

medium. The maximum value (m = 88%) corresponds to an isotactic dyad content of about 65%. This applies to polymers obtained in media of low polarity (the ratio toluene/CH_2Cl_2 = 7:3 by volume; $\varepsilon = 2.5$) under the action of $Ph_3C(GaCl_4)$. It will be recalled that the highest content of isotactic dyads noted so far has been found for polyisobutylvinyl ether synthesized in toluene under the action of diethylaluminum chloride at −78°C (81%) [80] and for polybenzylvinyl ether obtained in the same medium and at the same temperature with BF_3 etherate (89%) [81].

The dependence of the meso-dyad content on the dimensions of the counterion agrees well, within certain limits, with the concepts put forward by Kunitake. In particular, in a medium with CH_2Cl_2/toluene = 7:3 by volume, it has been found that for a number of counterions with ionic radii from 2.4 (BF_4^-) to 4.0 Å ($AlCl_4^-$) there is a gradual reduction in the meso-dyad content from 85% to 72%; data for −76°C. In the transition to larger counterions no clear dependence of the structure of the macromolecules on this parameter of the active sites is found. In discussing the possible role of other factors the authors associate the observed structural effects with the spatial features of the substituents in these monomers (this determines the conformation of the growing chains) and with the ability of certain counterions to form aggregates. The latter concept is based on the quite considerable dependence of the structure of poly-t-BVE which is formed under the action of Ph_3CSnCl_5 on the concentration of the initiator.

The mechanism of stereoregulation in the polymerization of these monomers has recently been discussed again for allylvinyl ether. The possibility of obtaining highly isotactic polymers (95% meso-dyads) has already been noted in an earlier work of Yuki et al. in 1970 [82]. In particular, just such a result was obtained for polymerization in toluene under the action of BF_3 etherate at −78°C. The more recent studies of Sikkema and Angad-Gaur [83 and 84] are interesting in that they contain triad characteristics obtained using high-resolution ^1H- and ^{13}C-NMR spectra. For the majority of these calculated values good agreement was found between the data based on the relevant lines in these and other spectra. Significant differences between the ^1H- and ^{13}C-NMR data appear only when attempts are made at a precise characterization of the structure of monomers synthesized at relatively high temperatures (15°C).

The majority of experiments were carried out within the temperature range from −75° to −70°C in a hydrocarbon medium under the action of BF_3 and $SbCl_5$ without deliberately introducing co-catalysts. The authors were much more concerned with the efficiency of the processes as a function of the experimental conditions than with their effect on the microstructure of the polymers. Therefore it is difficult to assert that the results which they obtained are optimum (up to 80% meso-dyads for each initiator). Nevertheless, we will give as an example data on one of the polymers which has been studied in the most detail, i.e., information obtained at −76°C in toluene under the action of antimony pentachloride:

Meso dyads, %		Triads, %					
		^1H-NMR data			^{13}C-NMR data		
^1H-NMR data	^{13}C-NMR data	mm	mr	rr	mm	mr	rr
76	78	64	24	12	61	33	6

We now turn our attention to features which distinguish the approach to the interpretation of the mechanism of stereoregulation in cationic systems from the formulation of chain structures in homogeneous anionic polymerization. A stage in which intermediate counterion–monomer coordination compounds are formed is natural for anionic systems but for understandable reasons is not involved when considering cationic polymerization. This, however, does not exclude the notion of the preorientation of the monomer which precedes its insertion into the growing chain. From the point of view of Heublein [85] such an orientation is possible at the stage at which the monomer squeezes out the solvent molecule from the solvation sheath of the cation of the separated ion pair or the free cation. According to the scheme proposed in [85] the monomer competes in the donor–acceptor interaction forming a π-complex with the cationic center. This conclusion is based on the results of a study of model

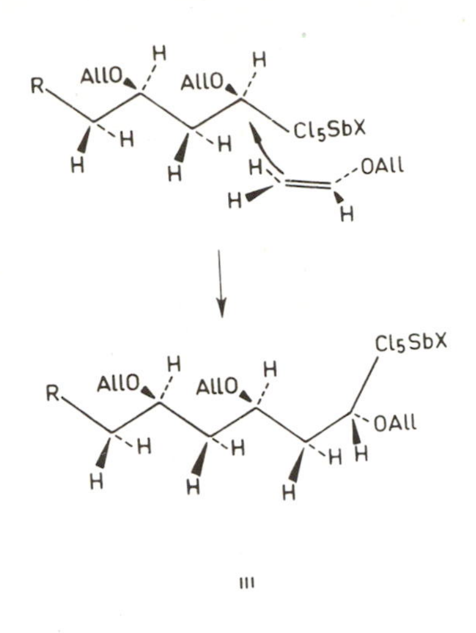

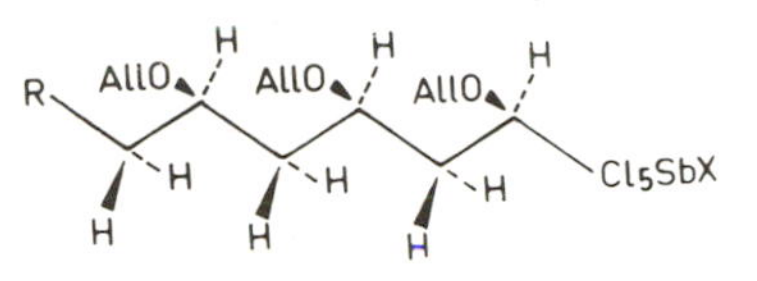

Fig. 5.6. The stereoregulating mechanism in the cationic polymerization of allylvinyl ether (from Sikkema and Angad-Gaur [83]).

systems in which benzene and its methyl substituent are used as π-donors and $Ph_3C(SbCl_6)$ and $Ph_3C(FeCl_4)$ as cationic reagents. In particular the constants of complex formation evaluated for the first of these salts range from 0.35 (for benzene) to 35.0 liters/mole (for hexamethylbenzene); data obtained in CH_2Cl_2 at 0°C.

On the other hand, Sikkema and Angad-Gaur [83] have formulated a mechanism in which there is weak association of the monomer with the counterion resulting from the coulombic interaction which is caused by the positive polarization of the oxygen atom in the monomer. It is assumed that this leads to orientation of the monomer which favors the formation of isotactic sequences; see the scheme in Fig. 5.6.

It is difficult to agree with the assumption that the oxygen atom is positively polarized. There are no experimental data on this, but it certainly does not follow from theoretical calculations on the model system CH_3^+–vinylmethyl ether which is fairly similar to the active sites under consideration here. The changes in the distribution of the electron density when the monomer interacts with the cation which is expressed in the total charges is illustrated

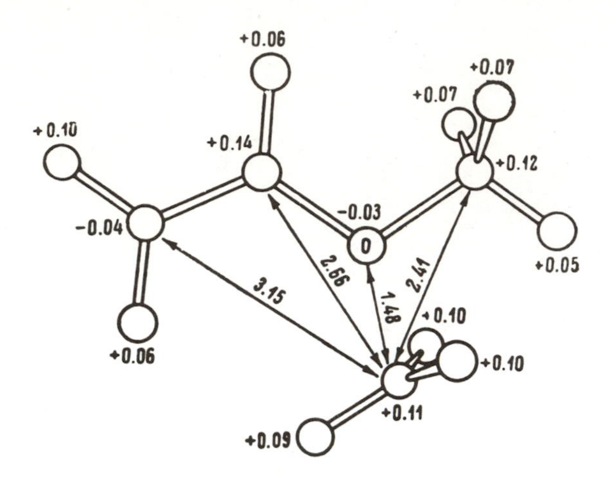

Fig. 5.7. The optimized geometry of the methylvinyl ether–methyl cation system and its electronic structure [87]. Calculations carried out using the CNDO/2 method.

by structure (5.XXXVIII) [86] and Fig. 5.7 [87]; data were obtained by the CNDO/2 method:

```
+0.03 H                                          H  0.00
        \       +0.17      -0.19               /
  -0.12  C ═══════ C ──────── O ──────── C ── H  -0.01
        /                             +0.13   \
+0.02 H             H                          H  -0.01
                  -0.02
```
(5.XXXVIII)

In the presence of the counterion the degree to which the charge on the atom of the monomer is reduced when it interacts with the cation would be still smaller.

There is another point which is worthy of attention; this is the possibility of realizing the transition AS_{ℓ} to AS_c for cationic growing chains.

In the preceding sections of this chapter such events have been quoted in discussing the stereoregulation in anionic systems. For the corresponding growing chains, particularly for those containing polar substituents, the donor–acceptor interaction of the positive charge of the counterion with the functional groups of the elementary units of the macromolecule is accompanied by a considerable gain in energy.

It is possible to propose the formation of the AS_c form in the case of cationic growing chains only if it is accepted that a special type of exchange of ligands takes place in the coordination sphere of the counterion $(EL_p)^-$ which corresponds to a scheme of the general type (5.21):

The possibility of such events is rather doubtful. To a greater extent the possibility of intramolecular interactions during cationic polymerization must be taken into account, bearing in mind that it is the end atom of the growing chains which takes part in these events and not the counterion. Reoxonization, mainly the splitting off of various "foreign" monomers during the polymerization of formaldehyde, cyclic oxides, and acetals, is a likely basis for this (see Section 4.1.4). Free ions, but not ion pairs, may exhibit a tendency to the transformation $AS_\ell \longrightarrow AS_c$ via a mechanism of this type. This circumstance does not, however, reduce the reality of the transformations under consideration. For the carbenium active sites which can appear in the polymerization of O-containing heterocycles, event (5.22) is obviously comparable energetically with the initial interaction with the monomer [Eq. (5.23)], being independent of the existence of such active sites in the form of free ions or of ion pairs. In the reactions shown below the counterions have been omitted for simplicity:

(5.22)

(5.23)

We have dealt with this in connection with the questions concerning the applicability of the interpretation of the nature of stereospecific effects, which is often used for anionic systems, to the cationic polymerization of alkenyl ethers. The question arises of the possibility of connecting the known facts of the dependence of the structure of polyalkenyl ethers on the nature of the counterion with the formation of intermediate AS_c forms, bearing in mind that their geometry, stability, and reactivity determine the structure of the macromolecules. There is nothing concerning this in the literature but the idea that events of the type (5.23) took place on growing chains of these monomers was presented as far back as the sixties. Thus Imanishi [88], in discussing the different variations of the termination reaction, came to the conclusion that the most probable mechanism is that shown on the next page using, as an example, methylvinyl ether:

$$\sim CH_2CH(OCH_3)CH_2CH(CH_3)CH_2\overset{+}{C}H(OCH_3)\cdots\cdot X^-$$

(5.XXXIX)

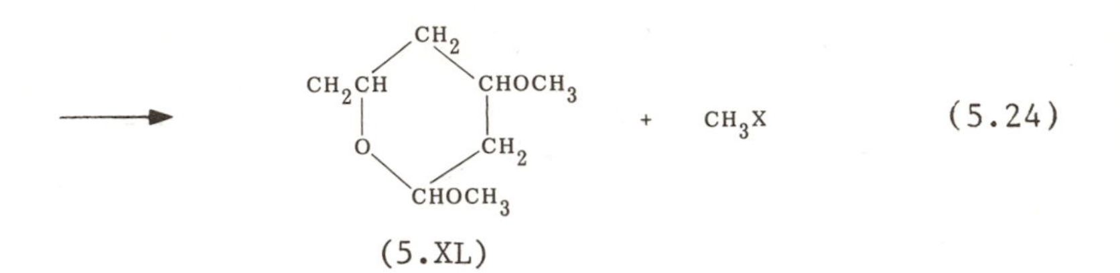

(5.24)

(5.XL)

Naturally reaction (5.24) must proceed via an intermediate stage in which trialkyloxonium ions are formed and which is assumed to be reversible. This situation corresponds to scheme (5.25):

$$(5.XXXIX) \underset{b}{\overset{a}{\rightleftharpoons}} CH_3O-CH \frown CH_2-\overset{+}{O}CH_3\ ,\ X^- \qquad (5.25)$$

(5.XL)

The considerable difference in the fate of the trialkyloxonium ions which are formed in this system and during the polymerization of O-containing heterocycles deserves attention. In the second case such a structure is taken for the active sites themselves and the competitiveness of reactions (5.22) and (5.23) is of second-order significance for polymerization. The reason for this, when intermolecular cyclization of carbocation active sites is not energetically superior to attack of the monomer by these active sites (or if the cyclic form of the end unit is taken as basic), lies in the great ease of reoxonization. Such a relationship between the events may be connected with the electronic characteristics of the O atoms which are well known for cyclic ethers, for example, THF. From data calculated using the CNDO/2 method, the total charge on the O atoms of this monomer is −0.27 e; see [87]. It is natural to take roughly the same value for the charge on the O atoms of the polyalkylene oxides. On the other hand, the characteristics of the monomers of the alkenyl ether series and the alkoxy groups in these macromolecules must be very different. Hence, for the active sites of the alkenyl vinyl ethers, reaction (5.25a) may be energetically superior to the proposed but rarely occurring formation of the oxonium ion at the expense of the O atom of the monomer.

These ideas enable one to assume the possibility of the formation of the distinctive AS forms of the cationic chains under discussion at the expense of the independence of the efficiency of such events on the nature of the counterion. The appearance of such forms is most natural for free ions, but is also absolutely real for IP.

5.2. DIENE MONOMERS

In discussing the problem of stereospecificity, the polymerization of dienes is a particularly rich source of information. This gives grounds for attempts to find the reasons for the connections between the varied parameters and the structural effects. The first of such facts was discovered at an early stage in the study of polymerization in diene—alkali metal systems, i.e., about fifty years ago. Towards the end of the fifties a large volume of work was carried out in this field. However, this still did not lead to a generally accepted and undisputed scheme for chain formation in the polymerization of dienes initiated by ionic reagents.

By combining all that is known about formation of diene polymers by non-free-radical methods, it is possible to assert that there are various circumstances which make the interpretation of the mechanism of stereoregulation difficult in such systems. The most important of these are the following: 1) the restriction of information on the nature of active sites by the statistical characteristics obtained in the absence of the monomer and mainly in a concentration range which differs sharply from that which prevails in real polymerization conditions (for example, anionic living chains); 2) the restriction of such information by the data of initial initiating systems, i.e., on reagents which may be taken as the active sites for initiation but not growth (for example, Ziegler—Natta catalysts); 3) the course of side reactions which lead to considerable expenditure of unsaturated bonds in the macromolecules and thereby to the masking of initial structural effects (cationic systems).

From the aspect of the discussion of the mechanism of stereoregulation, of which reactions of formation of polymers of the diene series are of fundamental interest, for instance, the role of the variable parameters is especially obvious. Nevertheless, we will deal first with processes of a different kind, namely the polymerization of dienes under the action of cationic initiators.

5.2.1. Polymerization in Cationic Systems

We have chosen to discuss these systems first for two reasons. One is that cationic polymerization of dienes is usually accorded much less attention than other processes which include the same monomers and ionic reagents. This situation can be found in both original papers and reviews. Secondly, there are in the polymerization of dienes features which somewhat simplify the approach to the mechanism of formation of the structures of polydiene chains.

The peculiarity of the cationic polymerization of dienes which distinguishes it from ionic processes consists of the comparatively low sensitivity of the structure of the polymers formed to the nature of initiator and the conditions of polymerization (solvent and temperature). Thus, according to review [89], a quite general feature of butadiene and isoprene polymers formed in different cationic systems is the higher content of 1,4-trans links (75-90%) and the complete absence of 1,4-cis links.

The small variations in the above limits are quite clearly insufficient for finding the connection between the conditions in which the polymerization is carried out and the structure of the final products. It is more appropriate to ask the opposite question to that usually asked when discussing the mechanism of stereospecific polymerization of dienes. It must be borne in mind that an explanation is needed of the causes which lead to a comparatively narrow range of structural features of diene polymers which are formed by cationic initiation. It is also necessary to emphasize that the corresponding macromolecules usually have a low value of MM, reduced branching, and reduced content of unsaturated bonds (about 25-50% of the theoretical value); see [89].

The opening of unsaturated bonds under the action of cationic reagents with the formation of ring structures is also known for "dead" polydiene chains. These reactions have recently been studied for polybutadiene by Priola et al. [90], who also give references on earlier work in this field. Apparently side reactions of cationic active sites which accompany the growth of diene chains proceed in a similar way, both uni- and bimolecular mechanisms being possible. Taking into account the increased reactivity of the outer bonds (1,2 and 3,4) of the macromolecules as compared with internal bonds, the participation of side groups of the growing chains in such events must be regarded as advantageous. This is shown for examples of butadiene [Eq. (5.26)] and isoprene [Eq. (5.27)], counterions omitted:

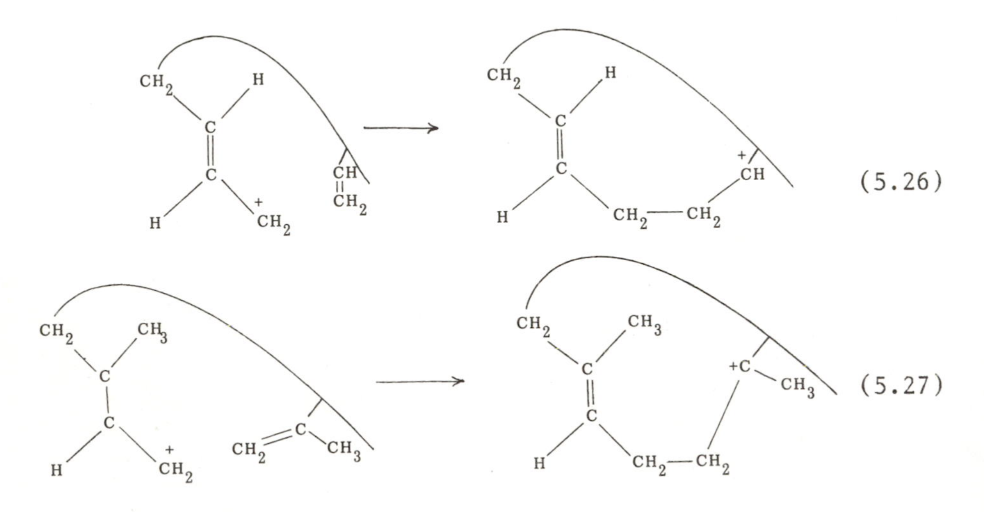

Apparently the products of such reactions preserve the ability to participate in growth. Only under this condition and with the periodic repetition of reactions of type (5.26) and (5.27) (and/or the corresponding bimolecular reactions) with the participation of the same macromolecule can such a significant reduction in the unsaturation of the polymer chains be realized.

For cyclic active sites of the butadiene type, isomerization with the transition of secondary to tertiary carbocations is possible. Such transitions were examined in Section 5.1.1. For this case an example can be given of a cyclic active site which is formed by reaction (5.26) by the interaction of the final 1,4 group of the growing chain with the penultimate 1,2 group:

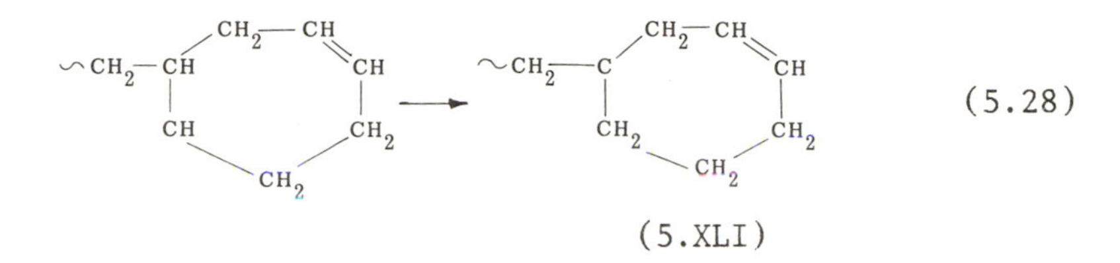

For the strongly shielded cation (5.XLI) a decrease in reactivity is possible and consequently the tendency to stabilization is accompanied by a reduction in unsaturation:

(5.XLI) $\xrightarrow{-H^{+}}$ CH_2—C(=CH—CH_2—CH_2—CH=CH—CH_2—) (5.29)

It is extremely difficult to discover such effects directly. It seems possible to evaluate them indirectly by methods based on the NMR spectra of the macromolecules which take part in reactions (5.28) and (5.29), but as yet there are no such data available. We therefore draw upon the results of Cesca et al. [91] who carried out detailed ^{1}H- and ^{13}C-NMR and IR studies on isobutylene copolymers with different dienes obtained in a CH_2Cl_2–pentane solution at −70°C under the action of $EtAlCl_2$. The bibliography for this series through 1982 is given in [91]. Of particular interest in the present discussion is the almost complete absence in these molecules of ring structures which are typical for diene homopolymers formed in such conditions. The insignificant content of such structures which are clearly defined (but without quantitative evaluation) are determined only in the case of copolymers of isobutylene with butadiene [92].

This fact, together with certain structural features of this polymer given in [92], may be used in an attempt to determine details of the mechanisms of secondary reactions of the type (5.26).

The form proposed above raises questions as to the dependence of the efficiency of the irreversible interaction of such a type upon the position of the attacking unit relative to the attacking C atom of the active site. Not having any additional information it is possible to assume comparative probability of the active site reacting with the penultimate unit as with any other. It is possible to draw more definite conclusions on this by examining the reasons for the sharply reduced effects of cyclization in the isobutylene—butadiene system.

Bearing in mind the average characteristics of butadiene polymers of cationic origin (about 25% 1,2 units with the loss of about 25% of the double bonds) and assuming a mechanism of the type (5.26), it is easy to accept that cyclization concerns about half the total number of vinyl groups. When no cyclization takes place, the content of 1,2 units would apparently approach 50%.

The structure of the diene components of these copolymers essentially depends upon the nature of the diene. Thus, the content of 1,2 and 3,4 units in the isoprene component is found to be negligible [93]. In copolymers of isobutylene with butadiene it is found that about 20% of the butadiene forms 1,2 units [92].

Unfortunately it is not shown in [92] whether or not the 1,2 unit content depends upon the total composition of the polymer which varies within the range 10-77% butadiene. Nevertheless, taking into account the insignificant role of cyclization (and consequently the amount of unsaturated bonds in the polymer which is close to that calculated), it is possible to accept that the "dilution" of the diene sequences by isobutylene sequences reduces the relative content of 1,2 units.

Of the possible suggestions as to the causes of the difference in the structural effects which are inherent to cationic homo- and copolymerization of dienes, the following appears to be preferable. The reduction in the vinyl groups in the diene units of the macromolecules on transition to copolymers (which is especially significant in the case of isoprene) may be ascribed to the fact that the C_α atom of the end diene unit of the growing chain is attacked mainly (and in the case of isoprene practically exclusively) by the isobutylene molecule. For the homopolymerization of butadiene and isoprene such selectivity is manifested to a much smaller degree. Obviously an analogous situation must also hold at the stage of interaction of the diene with the end diene unit of the growing copolymer. It would be more interesting to judge to what extent this perfectly feasible assumption is realized in practice on the basis of the succession of the isobutyl and diene units of different structures in polyisobutylene, which contains practically no 1,2 or 3,4 units; see [93]. However, there are coexisting qualitative features only for copolymers of isobutylene with butadiene. Some of these obtained using ^{13}C-NMR for copolymers of different composition are shown below (here T and V are the 1,4-trans and 1,2-units, respectively); see [92].

Butadiene content in the copolymer, %	Content of sequences, %	
	TTT	VTT + TTV
44	39	31
37	38	27
14	31	26

In mixed triads, i.e., those which also contain isobutylene units (I), there are found only structural units in which the I units lie next to the T units, but not butadiene V units: for example, ITT, ITI, and TTI. This is in agreement with ideas on the point of attack of the butadiene end unit by the isobutylene molecule and, very importantly, is a serious argument in favor of the 1,4-trans structure of the end unit of the growing chain.

In contrast to anionic diene active sites which are characterized sufficiently well using various physicochemical methods, cationic active sites which are similar in their monomers have not so far been studied successfully. This is due to their extremely low stability. Therefore, we are compelled to base our conclusions as to the forms of such active sites only on indirect data or on theoretical results. Such calculations have been carried out for the crotyl cation [40], which is as yet the only model of a cationic diene active site investigated in this respect. We have already had cause to mention that according to these calculations the trans configuration is the most stable form of this reagent (see page 223).

Nevertheless, among the facts which enable one to take for the end unit of the growing cationic butadiene chain only the 1,4-trans structure, the features of the structures of the copolymers of butadiene with isobutylene considered above occupy an extremely important position. They may be regarded as a basis for excluding the 1,2 end unit as a structural variation of the active site under discussion.

We now turn to the question of the precise definition of reaction mechanism (5.26). For this question particularly we draw upon the structural features of the copolymers of butadiene with isobutylene. The triad composition of these copolymers enables one to assert quite definitely the absence of a clear tendency of the end butadiene T unit to interact with the penultimate V unit of the growing chain and with its nearest neighbor. This follows first from the existence of sequences VT and VTT and, second, from the absence of ring structures in the composition of the copolymers. Obviously, rings in homopolybutadiene are formed as a result of the 1,2 units which are more distant from the end of the chain. In the case of copolymers this may be hampered by isobutylene sequences either due to an increase in the distance between fragments which can interact, or as a consequence of additional steric hindrance caused by the I units. The lower flexibility of these copolymer chains as compared to those of butadiene may also have some significance.

In discussing the mechanism of chain formation in cationic polymerization of dienes, relatively more attention is paid to the lack of a role for the counterion as a regulating factor. Conclusions regarding this cause more often than not are based on the fact that the dimensions of the counterions are typically comparatively large and do not differ much from one to the other. We note that such an interpretation does not contradict the opposing character of the polymerization of alkenyl ethers for which the dependence of the structure of the macromolecule on the nature of the counterion is obvious (see Section 5.1.2.3). In the latter instance the stereoregulation manifests itself in the selectivity of the iso- and syndiotactic directions of the growth reaction whose mechanism we simply do not compare with the competition of the 1,4 and 1,2 opening of the unsaturated bonds in the molecules of monomers of the diene series. We note nevertheless that the specific features of the cationic polymerization of dienes, namely the complete absence of competition in the 1,4 opening of the double bonds (i.e., these events lead exclusively to the formation of trans units) limit the number of possible structural variations to a certain range of the ratio T/V. These limits are small but they are considerably greater than the possible experimental errors.

Examining the formation of the structure of polydiene chains from the point of view of the competition of attack of the C_α and C_γ atoms of the active sites by the monomer we make use of the generally accepted concepts of the anionic polymerization of dienes; for more detail see Section 5.2.2. The correctness of such a transfer of the "anionic concept" to cationic systems requires some foundation. In the short term it may amount to the following. From the above data and concepts it may be concluded that, together with side reactions which complicate the course of cationic polymerization of dienes, these processes have a feature which simplifies the mechanism of stereoregulation. The question arises of the absence of isomerization of the end units which is so typical for anionic active sites. Although, as has already been noted, this has not as yet been successfully demonstrated directly, the indirect data available (the structure of homo- and copolymers) and quantum-chemical calculations are completely sufficient for such a conclusion.

We emphasize that the backbone of the trans structure of the 1,4 units of polydiene chains is much more favorable from the energy point of view. However, in the case of terminal units of anionic active sites the interaction of the C_α atom of negative charge with one of the remote CH_2 groups leads to an increase in the stability of the cis structure. According to theoretical calculations carried out in [40] for the crotyl anion which models the butadiene active site, the function of acceptor is fulfilled by the "quasi-π-orbitals" of the hydrogen atoms connected to the C_δ atom.

For free cations there is no tendency towards analogous intramolecular interaction, and for ion pairs, taking into account the charge characteristics of their components, only an increase in the advantage of the trans forms of the end units may be expected.

Hence it follows that each growth event may be taken as the entry of a monomer molecule into an active site in the form of a trans-1,4 structure. If in this event the C_α atom of the end unit of the initial active site is the attacking center, then the same structure is formed on the penultimate unit.

It would seem possible to contrast such a concept with the probability of the existence of active sites with vinyl end units. Arguments against such a suggestion have already been presented (see page 259). From our point of view these exclude any other mechanism for the formation of 1,2 (or 3,4) units in the cationic polymerization of dienes apart from attack of the C_γ atom of the final unit of the active site (which has 1,4-trans structure) by the monomer molecule.

The type of structure of the polydiene chains in the various processes point to a much smaller difference in the efficiency of the C_α and C_γ directions of the growth events during cationic polymerization as compared to anionic polymerization. The question is whether this is caused by the general type of corresponding active sites or by narrower constraints of variable parameters in the cationic processes which have so far been studied. It is all the more appropriate that if, for example, organolithium compounds are excluded when comparing polydienes which are formed under the action of alkali metal derivatives, then the limits of corresponding structural differences become extremely narrow and a basic shift takes place towards the reduction in the 1,4-unit content. This leads to the conclusion that the reason for the low selectivity of these directions in the known cases of cationic polymerization of butadiene is largely due to the properties of the actual reagents (i.e., in addition to being cationic). It must be borne in mind that ion pairs $M_n^+Y^-$, in which Y^- is a general counterion of the type $[E^nX_n]^-$, are comparatively close "geometrical analogs" of anionic active sites with higher alkali metals as counterions. At the same time the difference in the electron type of these and other active sites appear quite clearly in the structure of the corresponding nonstereoregular macromolecules.

5.2.2. Polymerization in Anionic Systems

Reviewing the literature of the last 25-30 years on the mechanism of anionic polymerization and, in particular, the spatial effects which are typical of these reactions, it is not difficult to assert that the formation of diene polymers plays a leading role in this research. If here certain important details must still be regarded as open to discussion, then deficiences relating to anionic systems are less significant than in other processes for the ionic polymerization of dienes. This is largely the result of the great developments in physicochemical studies of living oligo- and polydienyl chains which we have discussed in previous chapters.

In approaching the discussion of the mechanism of formation of chain structures in anionic polymerization of dienes in the present section, we emphasize first of all the following points which have served as a basis for selecting materials and the order in which they are dealt with. In the first place, this problem forms the subject of an extremely large number ot reviews and specialized chapters which have been written in various monographs. This excludes the appropriateness of the new systematization of well-known phenomena and the striving for bibliographical completeness when compiling a list of references. It is our purpose to attempt to describe the existing situation concerning this question. The history of the formulation of concepts is touched upon only if it is necessary for explaining the significance of conclusions which result from new experimental facts or from the results of theoretical studies.

In the second place, the greatest volume which comprises the basis for a detailed description of these processes is concentrated on systems which include butadiene and isoprene. Therefore we will deal first of all with just these monomers. Thirdly and finally, it must be recalled that the subdivision of the processes of polymerization under the action of organic compounds of nontransition metals as anionic—coordination or anionic was not used by us. Of course such a subdivision is not difficult to substantiate, but from the point of view of our exposition it can be avoided.

5.2.2.1. Butadiene and Isoprene

Many years of discussion on the mechanism of stereoregulation in anionic polymerization of dienes to a considerable extent are concentrated around the question of whether the structure of the macromolecule is fixed at the end unit or at the penultimate unit of the growing chain. According to the interpretation most widely held at present, the answer to this question depends upon the actual structural type of the unit and this is reflected in the following scheme (here Y is the counterion):

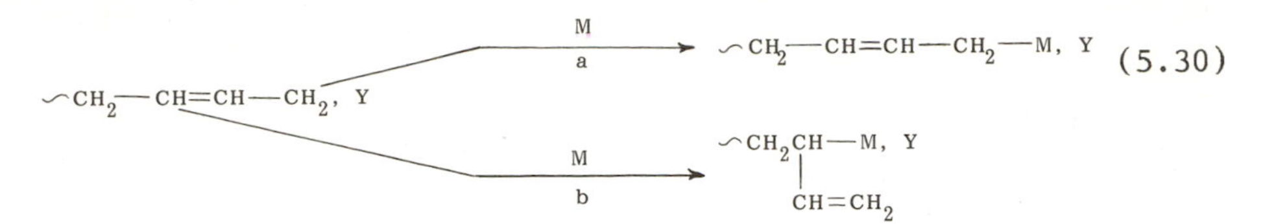

(5.30)

Individual opinions (which still exist sometimes) on the possibility that the end units of anionic diene active sites have a structure of the type $-CH_2CH(CH=CH_2)$ are based only on the failure to prove their absence and not on any serious experimental data of a contrary nature. Hence scheme (5.30) may be regarded as sufficiently universal for explaining the genesis of alternative structures: the fixing of 1,4 units (without additional details) on the end and, in this case, of 1,2-units on the penultimate units of the growing chains.

Ideas on the factors which control competition of reactions (5.30a) and (5.30b) and the formation of cis and trans structures in reactions (5.30a) are less common. The reactivity of C_α and C_γ atoms of the end unit as direct growth centers is evaluated from the point of view of their charge characteristics and their spatial accessibility, but opinions vary as to the relative roles of these factors. Concepts on the connection between the vulnerability of the C_α atom of the active site and the associated state of the growing chain (there are a few extremely convincing facts which favor this; see Sections 2.3 and 4.1) have still not become widely accepted. Changes in the contributions of Eqs. (5.30a) and (5.30b) are generally regarded as differences in the diene active sites which are the cause of the variation in the parameters of the end unit–counterion system, irrespective of the effects of association.

On the other hand, associates responsible for the isomerization of the end units of the polydienyllithium chains have been discussed ever since the work of Sinn [94] starting from the early seventies. We note, in presenting the somewhat earlier hypothesis of Dolgoplosk et al. [95], that the isomerization of diene active sites is a factor which determines the structure of the macromolecule formed and this is not connected with the state of association of the growing chains.

According to Sinn's scheme the initial event of growth in diene–$(M_nLi)_m$–hydrocarbon medium systems takes place only on monomeric forms of growing chains with the formation of end cis units; this may be expressed in condensed form as follows (here A and A' are polydienyllithium associates):

$$(M_nLi)_m \rightleftharpoons mM_nLi$$

$$M_nLi + M \longrightarrow M_nM_{cis}Li$$

$$M_nM_{cis}Li + A \rightleftharpoons M_nM_{tr}Li + A' \qquad (5.31)$$

$$M_nM_{cis}Li \downarrow M_nM_{cis}M_{cis}Li \qquad M_nM_{tr}Li \downarrow M_nM_{tr}M_{cis}Li$$

In the more general system of complex equilibria it is also envisaged that a certain fraction of the monomeric forms with vinyl end groups leave the associates.

In evaluating these or any other premises upon which attempts to substantiate the existence of anionic diene active sites with 1, 2 (3, 4) end groups are based, it must be remembered that it is necessary to delocalize the electronic structure of all such fragments as compared with the corresponding internal units. In the limit it may be imagined, using butadiene as an example, that there is a hypothetical symmetrical structure (5.XLII) to which the concept of an isomeric form in the sense being discussed here is not generally applicable:

$$\sim CH_2 - CH \overset{CH}{\cdots} CH_2 \quad Mt \qquad (5.XLII)$$

Such a symmetry is not found in reality (with the exception of some allyl compounds of the type $CH_2{=}CR{-}CH_2{-}Mt$) and the structure of real active sites does not lose the definition necessary for it to be classified. At the moment, of all the cases studied, this definition is sufficient for the 1,4 (or 4,1) conformation to be regarded as the only structure which has been firmly established.

Although the subject of the present section does not often arise nowadays, it still remains unresolved and in any new approach to this question it must be borne in mind that the problem can be expressed in a number of ways. Among these may be included, in the first place, the question as to whether it is necesary to go beyond the framework of scheme (5.30) which enables all the experimental facts to be interpreted. Of course the answers to this question will vary, but our answer would be in the negative. Secondly is included the question of whether there are methods which are unused (or which have not been exploited sufficiently) which can be applied to obtain more accurate experimental data which can be used to decide which of the existing points of view is correct. To this an affirmative answer is appropriate. Studies on the structure of the various single-unit RM—Mt compounds (in which M is the diene molecule) and their hydrolysates show that the ratio of the cis and trans conformations of the 1,4 units varies with the nature of M, Mt, and even R. However, the probability that for the attacking diene molecule the 1,4 opening of the conjugate double-bond systems is not only the most favorable but also the only direction of the reaction using anionic reagents cannot as yet be regarded as a firmly established fact. The information available on the structure of such compounds is not sufficiently exhaustive for conclusions which go beyond the bounds of individual materials which have already been studied.

We have paid special attention to the above, because in recent studies, which have concentrated on the stereospecificity of anionic diene active sites, the genesis of 1,2 (3,4) groups is often ignored. This refers particularly to the work of Worsfold and Bywater [96], in which the scheme, which is in a sense similar to that of Sinn [see Eq. (5.31)] served as a basis for the determination of several quantitative features.

It is most convenient to commence an explanation of the above work by giving the following scheme (in which R is t-Bu):

$$\underset{\text{(5.XLIII)}}{\overset{M'}{RCH_2CH{=}C(CH_3)CH_2Li}} \xrightarrow[-LiCl]{(CH_3)_3SiCl} \underset{\text{(5.XLIV)}}{RM'{-}Si(CH_3)_3} \qquad (5.32)$$

$$\xrightarrow{M''} \underset{\text{(5.XLV)}}{RM'M''Li} \xrightarrow[-LiCl]{(CH_3)_3SiCl} \underset{\text{(5.XLVI)}}{RM'M''{-}Si(CH_3)_3} \qquad (5.33)$$

Here M'' is the second isoprene molecule.

Taking the ratio of the cis and trans forms in the (5.XLIV) compounds, the authors traced the changes in this value in the deactivation products of the transformation of (5.XLIII) to (5.XLIV) at various stages of reaction, i.e., characterizing single- and two-unit compounds on the basis of the consumption of the former. This in fact compares the structure of the M' unit in the (5.XLIV) compounds with the "gross-structure" [(5.XLIV) + (5.XLVI)] when the transformation of the initial model to the final product is incomplete. All the results obtained were examined from the point of view of the difference in the reactivities of the end cis and trans units according to the following scheme (counterions omitted):

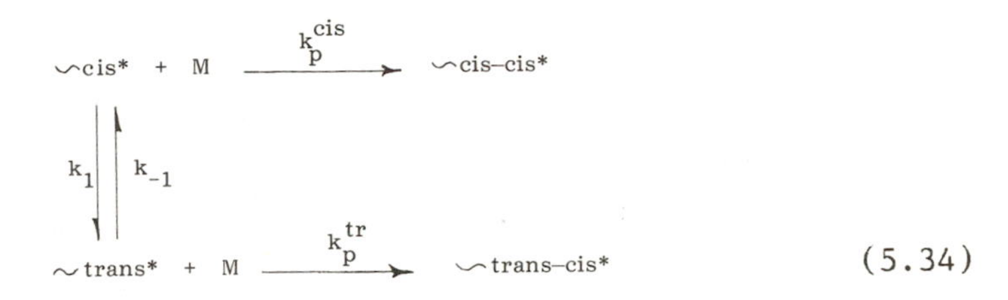

(5.34)

TABLE 5.6. Equilibrium Fractions of the cis–trans Forms of $RCH=CHCH_2Mt$ Compounds Based on the Features of Their Deactivation Products in Hexane with Oxirane and in Tetrahydrofuran (in brackets) [97]

	Mt		
R	Lithium[a]	Sodium[b]	Potassium[b]
CH_3	67/33 [85/15]	93/7	96/4 [99.2/0.8]
$i\text{-}C_3H_7$	14/86	38/62	56/44 [78/22]
$t\text{-}C_4H_9$	3/97 [4/96]	7/93 [8/92]	8/92 [13/87]

[a]Equilibrium was established at 0°C.
[b]Equilibrium was established at −48°C.

where k_p^{cis} and k_p^{tr} are the growth-rate constants for the corresponding isomeric forms of the active sites.

The ratio cis/trans which decreases as the (5.XLIII) compounds are consumed is interpreted as a consequence of the higher value of the constant k_p^{cis}. The structural changes found are used to calculate the ratio k_p^{cis}/k_p^{tr} which for this approach is equal to 8.4.

In order to discuss the significance of this value for a real polymerization process, it must be mentioned that Worsfold and Bywater [96] compared the isomerization rate constants for compound (5.XLIII) and its linear isomer n-Bu—M—Li (5.XLVII) in which M is the isoprene molecule. For this purpose both the above compounds were synthesized by interacting the corresponding organomercury derivatives with dispersed lithium. The structure of the original Hg derivatives is characterized from ^1H-NMR data by a cis-form content of 65%. The isomerization rate constants for these compounds proved to be very different. For (5.XLIII) this value is 10^{-2} sec^{-1}. It is difficult to obtain similar characteristics for compound (5.XLVII) because its isomerization rate is too high. Only an approximate value of the order of 0.75 sec^{-1} has been estimated for this. The authors were therefore forced to take compound (5.XLIII) as a model active site although it is a less close analog of the actual polyisoprene chains than compound (5.XLVII). It is also impossible to ignore the highly probable dependence of the position of the equilibrium [see Scheme (5.34)] on the actual hydrocarbon substituent in the C_γ atom of the end unit. This concept, which is based on the data of Schlosser and Hartmann [97] (see Table 5.6), is designed to emphasize the necessity of great care when making use of conclusions based upon low-molar-mass models for real polymer chains.

TABLE 5.7. The Effect of the Deactivator on the Structure of Single-Unit Dienyllithium Compounds Synthesized in Benzene at Room Temperature

Initial compound	Terminating reagent	Unit content (%) in the deactivation product				Ref.
		1,4-cis	1,4-trans	1,2	3,4	
t-Bu–C_4H_6–Li	H_2O	23	70	7	-	98
	CH_3OH	18	70	12	-	
t-Bu–C_5H_8–Li	CH_3OH	87		0	13	99
	C_2H_3I	73.2		0	26.8	
	CH_3I	66.4		0	33.6	

Great care is also needed when simply transferring structural features established for deactivated compounds to the original organometallic reagents. The idea that the structure of the end unit does not change when the next monomer molecule is added or when termination takes place under the action of some deactivating agent is contradicted by a large number of experimental facts. Here we limit ourselves to those which illustrate the role of the nature of the deactivating agent (see Table 5.7).

Although aggregates do not figure in scheme (5.34), they are mentioned as a reservoir of "dormant" dienyllithium chains in [96]. Obviously opinions regarding their relative roles (those of Sinn et al. on the one hand and Worsfold and Bywater on the other) coincide.

The results of Garton and Bywater [100, 101] which attempt to evaluate the effect of the polarity of the medium on the equilibrium between cis and trans conformations in the end groups of diene active sites are extremely interesting. They studied particularly the temperature dependence of the UV spectra of polybutadienyllithium–THF systems in the range from −40°C to 22°C. On the basis of the considerable displacement and great breadth of the values of λ_{max} found, the authors considered it possible to interpret the features observed as a superposition of bands corresponding to the cis and trans forms of the active site. They take into account the known NMR data for polybutadienyllithium [102] and assume that the relative proportions of the isomeric forms remain constant over a wide range of concentrations. By separating the bands of the UV spectra observed, values of λ_{max} of 285 nm and 325 nm were obtained for the cis and trans conformations, respectively. These data are illustrated by Fig. 5.8 which shows a marked preponderance of cis active sites at low temperatures. This is due to the equilibrium which results from the absence of monomer. In low-temperature polymerization the trans form is dominant.

We emphasize that these changes in the end unit have a very modest effect on the polymer structure which is basically a 1,2 structure (80% at 0°C and 86% at −70°C); see [101]). From this it can only be concluded that the trans structure of the end group is to be included in the factors which favor C_γ attack [see Eq. (5.30)].

Taking the concepts of Bywater et al. to be perfectly probable we turn our attention to the conclusion regarding the inverse activity relation between cis and trans forms for models of polyisoprenyllithium in nonpolar media and for polybutadienyllithium in a polar medium. In comparing these conclusions it is necessary to take into account that in fact what is being compared are the actions of reagents of the different types M_nLi and $M_nLi \cdot D$. Furthermore, due to the mainly C_α attack in the first case and the C_γ attack in the second, the effects found must be regarded separately. The actual ki-

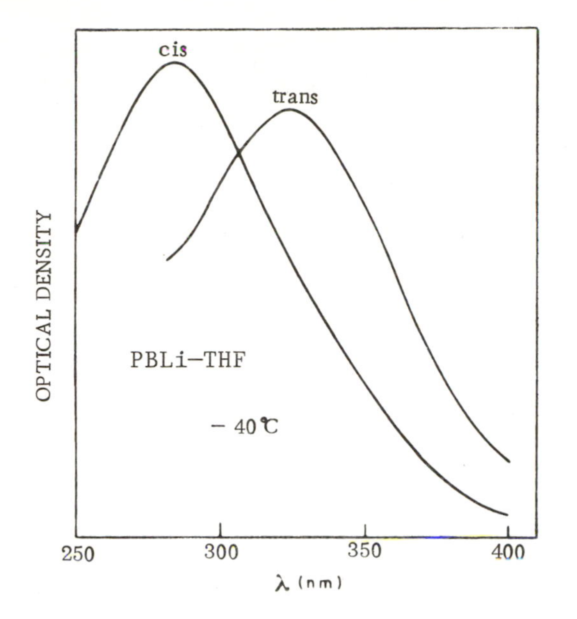

Fig. 5.8. Extrapolated UV characteristics of polybutadienyllithium obtained from the total spectrum [100]. Temperature, –40°C. Solvent, tetrahydrofuran.

netic characteristic of the corresponding growth events is a complex constant which averages to different extents and directions (5.30a) and (5.30b). Of course it may depend upon the nature of the end group itself and its corresponding monomer, but obviously in the case of the simplest dienes the nature of the central atom of the counterion and its surroundings play the decisive role.

Before entering a more detailed discussion on the polymerization of dienes in polar media, we recall that the structure of the unit which is added during growth depends to a considerable extent upon the nature of the adjacent groups. Facts which indicate the existence of such a dependence have already emerged from the structure of the butadiene sequences in styrene and butadiene copolymers obtained with organolithium initiators in hydrocarbon media; see [103, 104]. As was shown later, the enrichment of the butadiene components by trans-1,4 units with increased styrene content (St) in the copolymer is brought about by both the attack of the end butadiene (Bt) unit by styrene and the presence of St as the penultimate unit of the growing chain with a Bt end group. Studies of the hydrolysates of "living" oligomers with single Bt groups which are situated in an environment of deuterobutadiene (Bt-D_6) or styrene (St) point to this conclusion. The original compounds $RM_nC_4H_6Li$ (in which M is Bt-D_6 or St and n = 8) were synthesized under standard conditions in a nonpolar medium. The structure of the end products $RM_nC_4H_6M'_mH$ has been established by Melenevskaya et al. [105] using ^{1}H-NMR and IR spectra. The structural features of the butadiene unit which is "sandwiched" between M_n and M'_m were determined by assigning bands; this was carried out independently for poly-(D_6)-butadiene:

M	M'	Structure of C_4H_6 unit, %		
		1,2	cis-1,4	trans-1,4
Bt-D_6	Bt-D_6	49	21.5	29.5
St	Bt-D_6	19	43	38
St	St	17	0	83

The qualitative agreement of the structural effects in both cases of deviation from homopolymerization conditions (i.e., in growth events Bt—Bt—Li + St and St—Bt—Li) could not have been foreseen. However, it has a simple explanation for which it is sufficient to accept: 1) the cis-trans ratio in the end group is dependent upon the nature of the penultimate unit, and 2) styrene mainly reacts with the trans form and as the latter is used up this is compensated (albeit partially) by a shift in the equilibrium $AS^{cis} \rightleftarrows AS^{tr}$.

Pham et al. [106-108] have studied the structural dependence of these polymers on the nature of the alkali metal counterion and the reactive medium. Other publications in this series are cited in [106] and [107]. Before turning to some of the results obtained, we note the particular combination of the various points of view in the authors' interpretation of the mechanism of formation of the chain structure. On the one hand, they clearly formulate a principle which corresponds with scheme (5.30). At the same time, in discussing the processes which take place in electron-donor media, they use another approach which was applied widely in the sixties. In an extremely brief form this concept may amount to the formation of intermediate complexes by diene molecules with the counterions of growing chains. These intermediate complexes have a bidentate structure in nonpolar media and a monodentate structure in polar media. An interpretation developed on this basis contains, however vaguely, the thought that in the second case the end group of the active site may have a 1,2 (or 3,4) structure. Nowadays there are no very good reasons for appealing to such a concept which occupied a leading position twenty years ago, at least for the active sites which we are considering.

Of the large number of structural features (which are given most fully by Salle [108]) we will make use of the data for the polymerization of isoprene in benzene and dioxane (DO). It is interesting to compare these with the results of Dyball et al. [109] which were obtained under similar conditions in diethyl ether (DEE);

TABLE 5.8. The Dependence of the Microstructure of Polyisoprene on the Nature of the Counterion in Polymerization in Benzene [108], Diethyl Ether (DEE) [109], and Dioxane (DO) [108] (Temperature of polymerization in benzene and dioxane, 15°C, and in DEE, 20°C.)

Counter-ion	Contents of the units, %								
	1,4[a]			1,2			3,3		
	benzene	DEE	DO	benzene	DEE	DO	benzene	DEE	DO
Lithium	92	35	3 cis 11 trans	0	13	18	8	52	68
Sodium	35	17	10	10	22	12	55	61	78
Potassium	54	38	4 cis 32 trans	4	19	14	42	43	50
Cesium	56	52	51	11	16	13	33	32	36

[a]Data on the isomeric forms of the 1,4 units are absent in the majority of cases; obviously the content of cis form in DEE and DO does not exceed a few percent.

see Table 5.8. It will be recalled that the ratio 3,4/1,2 > 1, which is characteristic for the majority of the results given in Table 5.8, and indicates that the basic composition of the end group of the active sites in these circumstances is the 4,1, and not the 1,4, configuration.

The permittivities of DEE and DO are practically equal and not large (about 2). Hence the structural differences of the corresponding polymers which are especially noticeable in the case of the lithium counterion must be ascribed to the greater ability of dioxane to form complexes. The properties of the electron donors which are given in Table 5.8 enable the possibility of a real contribution from the free ions and separated ion pairs to be excluded from these discussions. They also make it possible to ascribe the individual features of real systems which are observed to the characteristics of the contact ion pairs M_nMt and $M_nMt \cdot xD$ in benzene and in a polar medium, respectively. This circumstance, while limiting the number of factors whose variation affects the microstructure of the polymers, makes the evaluation of their relative values easier. In the cases we are examining stoichiometric variations of the $M_nMt \cdot xD$ chain are possible, but, since the reagent D is at the same time the reaction medium, the main type of functioning active sites must be the growing chains with a maximum ratio D/Mt for the given conditions. The extremely low value of K_{diss} in the above solvents enables the systems discussed to be taken as practically single-centered.

If the data given in Table 5.8 are analyzed, taking all the above features into account and selecting the most illustrative of these, the following features appear quite remarkable: 1) the deviation from the uniform structure of the polymer (with M_nLi in C_6H_6 as a standard) is practically the same when either benzene is substituted for DEE or lithium for sodium; 2) the structural features of polyisoprene which is formed under the action of M_nNa chains occupy an extreme position. This is unquestionably due to the total content of 1,4 groups (these have minimum values in all three solvents) and to the 3,4 groups (maximum values). Such regularity is less clearly expressed in the 1,2 groups.

It is easy to ascribe the first of these points to the great similarity between the stereochemistry of the growth reactions which take place in systems in which the active sites are M_nNa and $M_nLi \cdot xDEE$ (where x seems to be greater than 2). The reason for such a similarity may be explained by turning again to scheme (5.30). Essentially the question is whether the increase in the role of "C_γ attack" as compared to direction (5.30a) is caused by steric hindrance or the redistribution of the electron density. It seems to us that the answer depends upon the material chosen for examination and therefore may vary. In this case it is more correct to regard the former as playing the more decisive role for $M_nLi \cdot xDEE$ chains and the latter factor for the M_nNa active sites. The competitive role of C_γ attack is further strengthened by steric hindrance. This can be seen in the growth reaction of polyisoprenylsodium when it is functioning in the form of $M_nNa \cdot xD$ complexes. This is supported by the overall content of 1,2 and 3,4 groups in polymerization in polar media.

If the problem of how the corresponding active sites function is approached taking into account the parameters of the C—Mt bonds, then it is possible to make certain assumptions on the extreme structural features which are typical of polymers which are formed in Na systems. The weakening of the "degree of contact" of the ion pairs on transfer from sodium to its higher analogs, i.e., the stretching of the active bond and the weakening of the screening effect of the C_α atom, must be borne in mind. Changes of this type in a number of Na, K, and Cs derivatives may somewhat lower the basic tendency of the corresponding active sites to growth in direction (5.30b) and cause the second of the alternative growth directions to gain somewhat.

Turning to the isomerization of the end groups of active sites in the polymerization of dienes in polar media in which the effects of association may, in the majority of cases, be neglected, we recall the scheme given in Chapter 3 (page 85) in which active sites participated. This scheme, which reflects the view of Dolgoplosk [95] on the existence of intermediate highly reactive active sites, cannot simply be rejected since this state may appear in the course of cis-trans transformations. However, the question remains as to whether it is a real intermediate state or a transitional one. It is appropriate here to recall the possibility of structural effects which are capable of polymerizing dienes in polar media by insertion of extremely small quantities of a strong electron donor into the nonpolar solvent or by using previously prepared $RMt \cdot D$ complexes as initiators. The results which in this respect characterize real systems have been discussed so often (in [110-112] for instance) that there is no need to dwell on them here. From the possibility of forming a polydiene structure which has been synthesized for example under the action of M_nLi chains in THF, by initiating polymerization by stable $M_nLi \cdot D$ complexes in a hydrocarbon medium, it follows that the mechanisms by which the chains are formed in both cases may be identical. Such a result is most easily explained by ascribing the dominant role to C_γ attack in each case [Eq. (5.30b)].

Recent data on complexes of lithiumdienyls with 1,2-dipiperidinoethane (DPPE), which, according to Halasa et al. [113], favor the highest content of 1,2 groups in polybutadiene when polymerized in a hydrocarbon medium (up to 100%), provided additional grounds for this conclusion. Results obtained from spectroscopic studies of complexes of single-unit models of lithiumdienyls with DPPE which were carried out by Worsfold et al. [114] point to a direct link be-

TABLE 5.9. The Dependence of the Structural Features of the Diene Unit (M) in t-BuMLi Compounds on the DPPE/Lithium Ratio (Q) (Solvent, cyclohexane; temperature, 0°C)

M	Q	Trans form, %	Chemical shift of the C_γ atom[a]	
			in the trans form	in the cis form
Butadiene	0	77	101.6	102
	0.15	77	98 (broad signal)	
	1.09	86	74.0	85.4
	2.10	93	70.9	81.9
Isoprene	0	66	104	102
	0.13	60	100 (broad signal)	
	1.80	20	72	90
	3.65	32	69.9	89.7

[a]In ppm with respect to TMS.

tween the structure of diene groups in these complexes (which model the end groups of growing chains) and the selectivity of reactions (5.30a) and (5.30b). An increase in the relative role of reaction (5.30b) is favored by an increase in the trans content of the active end group. This attains the highest value for the butadiene model. Some data established for t-BuMLi complexes with different DPPE/lithium ratios (Q) are given in Table 5.9. Together with the results of other physicochemical measurements (low-temperature UV spectroscopy) they enable one to conclude that when $Q < 0.5$ there exist different forms of polydienyllithium, including uncomplexed tetramers and complexes of tetramers and dimers with DPPE. The authors emphasize the considerable difference between the characteristics which are given in Table 5.9 and the structure of complexes of the same models with TMED. In the latter case with the ratio TMED/lithium = 2, the trans content of the butadiene derivative is about 60% whereas in isoprene it is completely absent; see [114]. It will be recalled that the growth of polybutadienyllithium chains (PBL) in a hydrocarbon medium in the form of complexes with TMED also leads to a high content of 1,2 groups (up to 85%), but these are inferior in this respect to the active sites of PBL·DPPE.

Butadiene systems containing DPPE become single-centered when $Q = 2$ (this follows from the first-order growth reaction for the active sites). For isoprene this occurs at higher values of Q.

Changes in the structure of polybutadiene with Q at PBL concentration of the order of $1 \cdot 10^{-3}$ moles/liter are illustrated by the following data (cyclopentane, 0°C); see [114]:

Q	0	0.1	0.5	1.2	2.0
1,2 units, %	15	62	90	97	99

An approximately 20-fold increase in the polymerization rate is noted at Q = 2 as compared with a system without DPPE.

The general conclusion of [114] is that for the trans form of the active site, the direction of growth reaction (5.30b) is the only one, whereas for the cis form there is competition between both directions of reaction (5.30). Furthermore, for the complexes under discussion the condition $k_p^{tr} > k_p^{cis}$ is taken to hold true since active sites which contain some cis forms ensure the almost complete absence of cis-1,4 forms in polybutadiene.

These ideas which agree quite well with the data of [114] and with those of earlier studies remain open to question upon at least two points. These are the cause of the considerable difference in the nature (butadiene) and degree (isoprene) of the effects of DPPE and TMED on the cis/trans ratio in the end units of the active sites under discussion, and the cause of the difference in the effects which are observed under the same conditions for the complexes of DPPE with the model active sites of butadiene and isoprene. The difference in the degree of the effect while its direction remains constant can be easily explained by the stability of the corresponding complexes. However, in order to interpret the contrary nature of the influence of DPPE and TMED on the cis–trans ratio of the active sites, it is necessary to assume the existence of an additional structural feature for the materials being compared. As regards the conditions in which the monomeric forms of these reagents are the only ones present (or are dominant), the difference between them must be limited to stoichiometry or whether they are monodentate or bidentate complexes. Obviously, similar differences apply to complexes of DPPE with butadiene and isoprene active sites.

The connection between the selectivity of the reactions (5.30) and the spatial configuration of the 1,4 end unit seems not to be unexpected. The increased accessibility of the C_γ atom in the end trans unit (as compared with that of its cis isomer) to attack by the monomer seems obvious. Nevertheless, this situation does not enable such a clear similarity as has been established by Worsfold et al. [114] between the fractions of the trans configuration and the content of vinyl groups in polybutadiene to be foreseen.

TABLE 5.10. The Microstructure of Polymers Formed in Anionic Polymerization of 1-Phenyl-, 2-Phenyl-, and 1-Methylbutadiene

No.	M	Initiator	Solvent	Temperature, °C	Units, %				Ref.
					1.4-trans	1,4-cis	1,2	3,4	
1	1-PB	sec-BuLi	Toluene	20	58	25	0	17	116
2	"	sec-BuLi	THF	0	77	13	0	10	"
3	"	1,1-Diphenylhexyllithium	Diglyme	−78	84	8	0	8	"
4	"	Sodium naphthalenide	THF	−78	79	12	0	9	"
5	2-PB	sec-BuLi	Toluene	30	0	92	8	0	117
6	"	t-BuLi	Diglyme	50	0	97	3	0	"
7	"	t-BuLi	THF	0	90	10	0	0	"
8	"	Sodium naphthalenide	THF	50	0	97	3	0	"
9	"	Sodium naphthalenide	THF	0	0	90	10	0	"
10	1-Methyl Butadiene	n-BuLi	Heptane	80	60	25	15	0	121
					42	26	25	7	122

5.2.2. Diene Phenyl Derivatives

Of the vast quantity of information covering the behavior of other dienes in anionic systems we make use mainly of those results which are known for the corresponding phenyl derivatives.

The systematic studies of Suzuki et al. concentrate on the polymerization of 1-phenyl- and 2-phenylbutadiene (1-PB and 2-PB) [116, 117] in different anionic systems and on the physicochemical charateristics of the corresponding lithium oligodienyls (1-PBL and 2-PBL) [118-120]. Data on the influence of the conditions under which the polymerization is carried out on the structure of poly-phenylbutadiene polymers is particularly valuable for the comparison of butadiene and isoprene polymers (Table 5.10). This also contains data on piperylene [121, 122] which is similar to that for the alkyl analog 1-PB.

As we can see, the behavior of 1- and 2-substituted butadiene in a nonpolar medium in the cases of methyl and phenyl derivatives, judging from the overall content of 1,4 units and from the distribution of the cis-trans forms (see also Table 5.8), is quite similar. Clearly marked differences in the structures of the same polymers appear on transfer to polar media. For isoprene this leads to the total disappearance of the 1,4-cis units whereas for poly-2-PB the effect on the structure of such a change in the reaction medium is insignificant. The content of these units falls markedly only when the polymerization temperature is reduced, i.e., at −78°C it is 73% (t-BuLi—THF); see [117]. It is impossible to form a soundly based opinion as to the reasons for this effect because there is a lack of data on the polymerization of 2-PB at this temperature in hydrocarbon media. In the corresponding 1-substituted butadienes different changes in the structure of the polymers are also noticeable in polar media.

We will deal in more detail with the phenyl derivatives drawing upon the features of the corresponding model lithium dienyls. The paths used by Suzuki et al. [116, 117] for their synthesis are illustrated by the following schemes:

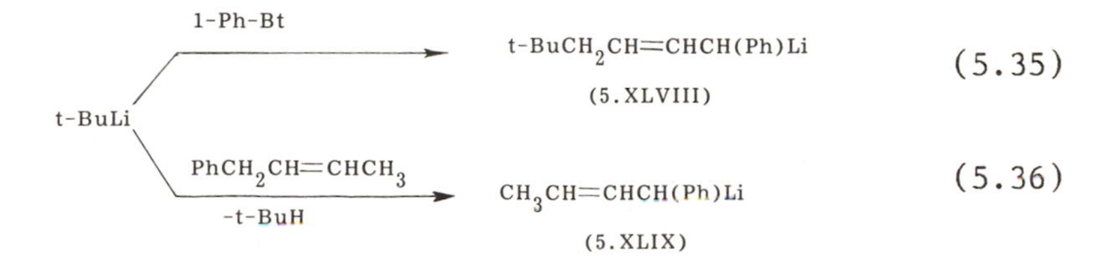

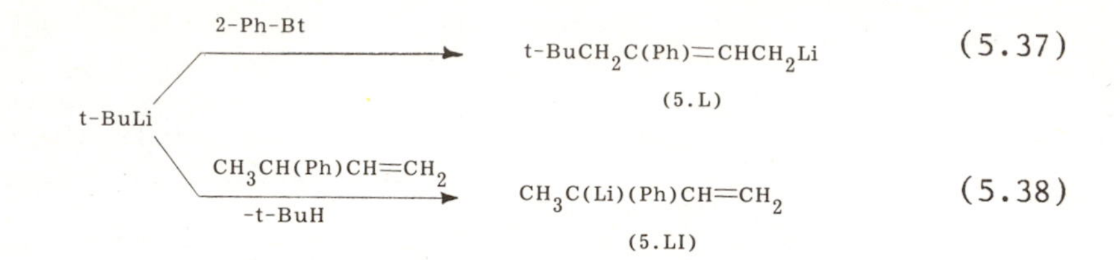

The compound (5.LI) which models the hypothetical 1,2 end group of the 2-PB growing chain was included among the material studied because of the impossibility of synthesizing the 1,4 model by a reaction of the type (5.36). Some features of the reaction products of Eqs. (5.35) to (5.38) and their derivatives obtained by deactivation with CH_3OH are given in Table 5.11.

The data of Tables 5.10 and 5.11 reflect the marked lack of correspondence between the structure of the model active sites and polymers. In particular this is so for 2-PBL which contains basically the 1,4 unit, with some contribution from the 4,1 structure. At the same time poly-2-PB contains a small portion of 1,2 groups while 3,4 structural groups are absent entirely. Consequently the end groups of real growing 2-PB chains, in contrast to 1-PB chains, exist practically exclusively in the 1,4 form.

Turning to the electronic characteristics of 1-PBL and 2-PBL we note that although the authors have differentiated them in their nomenclature, i.e., have assigned the former to π-benzyl and the latter to π-allyl, they differ very little with respect to the π-electron density on their C atoms. In both cases it is essentially the concentration of a significant fraction of the charge on the phenyl nucleus which apparently causes the absence of an appreciable sensitivity of the characteristics of the C_α and C_γ atoms to changes in the polarity of the medium. As a result, the C_α atoms of both active sites preserve their advantage over the C_γ atoms in any variation in the experimental conditions.

Regarding the difference in the geometry of the 1,4 groups in these polymers, the existence of 2-PBL (5.L) in THF in the form of the (E) conformation may explain the spatial selectivity of the growth reaction which is reflected in the superiority of the formation of cis-1,4 units. When examining the growth mechanism in detail the above authors prefer a scheme of type (3.10), taking into account, however, only the mutual transformations $A \rightleftarrows C \rightleftarrows B$. This concept is based on the difference in the geometry of the 1,4 (or 4,1) active end units and the internal units of the macromolecule. The cis-trans isomerization is regarded as a consequence of the formation of an intermediate state C. As has already been emphasized, it is by no means compulsory to interpret such an intermediate state as an independent active site (i.e., capable of growth). It is perfectly possible that the lifetime of reagents of type C

TABLE 5.11. Features of Model Lithium Phenyl Dienyls and Their Alcoholysis Products [118-120]

Compound	Solvent (t, °C)	cis/trans ratio	π-Population of the C atoms[a]					Solvent (t, °C)	Units (%) in the protonated products					
			C_α	C_γ	C_1	C_{ortho}	C_{para}		1,4 cis	1,4 tr	4,1 cis	4,1 tr	1,2 cis	4,3 tr
1-PBLi (5.XLVIII)	Toluene	2/3	1.41	1.24	0.91	1.12	1.17	Toluene	–	–	23	–	–	12
	THF	only trans	1.42	1.28	0.90	1.11	1.17	THF	–	–	100	–	–	–
1-PBLi (5.XLIX)	THF	3/1	1.30	1.31	0.90	1.12	1.17	THF	–	–	19	–	–	46
2-PBLi (5.L)	THF (24°C)	only trans[b]	1.47	1.29	0.91	1.11	1.16	Toluene	13	45	4	17	10	11
								THF (25°C)	1	69	0	23	3	4
								THF (-78°C)	0	86	0	8	6	0
3-Phenyl-3-lithium-1-butene (5.LI)	THF	–	1.49	1.32	0.91	1.11	1.14	–	58	–	–	–	–	42

[a]Calculated from ^{13}C-NMR data at 24°C, according to O'Brien et al. [123].
[b]Assumption based on the absence of signs of the existence of isomeric forms and on the spatial concepts of [119].

TABLE 5.12. Methoxy Derivatives of 1-Phenylbutadiene (CH_3O–1-PB). Features of Li AS Models and Polymers Obtained under the Action of t-BuLi at 0°C [124]

M	Solvent	π-Population of Li AS					Microstructure of the polymers, %			
		C_α	C_γ	C_1	C_2	C_4	solvent	1,4-trans	1,4-cis	3,4-
o-CH_3O-1-PB	THF	1.43	1.26	0.96	0.90	1.18	Toluene	-	-	98
							THF[a]	-	-	98
p-CH_3O–1-PB	THF	1.43	1.31	0.96	1.11	0.91	Toluene	-	-	98
							THF[a]	79	11	10

[a]Reducing the temperature of polymerization in THF to −78°C does not affect the structure of the polymers.

(the formation of which is not excluded at an intermediate stage of polymerization) is very much less than is required for participation in growth. Competition between C_α and C_γ attack may be decisive in these cases.

Fresh obstacles arise when evaluating the relative roles of the spatial and electronic features of the active sites in determining the actual direction of growth reaction (5.30). For this question we draw upon data for the methoxy derivatives of 1-PB. According to the results of the same groups of researchers [124], in the anionic polymerization of the ortho- and para-monomers of this series, particular features appear which are illustrated in Table 5.12. This also contains electronic characteristics for the coresponding active sites in THF. To obtain similar data in nonpolar media proved not possible due to the insolubility of the single-unit models and the lack of definition in the NMR spectra of the active oligomers.

On the basis of the microstructure of o-methoxy polymers, the authors conclude that the spatial effect of the substituent is more significant than its effect on the π-electron density of the active site. In order to explain the considerable difference in the structure of polymers of the p-derivative which are formed in polar and nonpolar media, they accept in the first case that stereoregulation is determined by associated growing chains.

It is interesting to supplement the data obtained by these authors on the dependence of the C_α and C_γ reactions for 1-PBL and 2-PBL on the nature of each compound and on the reactive medium by information on deactivation products of the above model active sites obtained by various terminating reagents. Even if we limit ourselves to the overall "C_γ yield" (see Table 5.13) without subdividing it into (E) and (Z) forms (which we will deal with later), the complexity of the correlation of the specificity of the anionic active sites and of the electrophilic reagents used is very clearly manifested. The greatest difference was found in the interaction products of acetone and the (5.L) and (5.LI) compounds. Tsuji et al. [120] regard these materials as π-allyl systems (5.La) and (5.LIa), and considering the small difference in their π-electron structure (see Table 5.11), place special significance upon the t-Bu group which screens the C_γ atom.

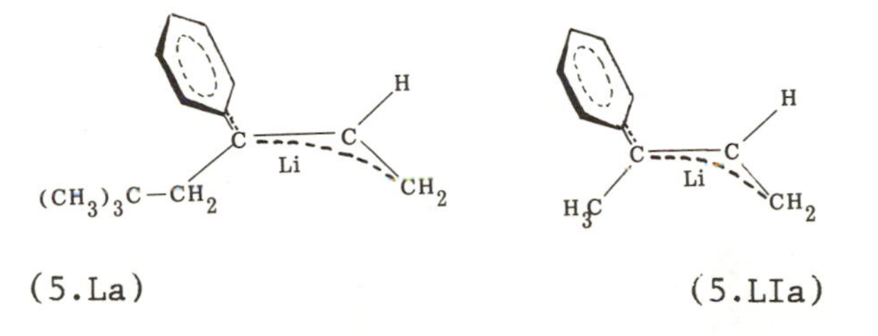

(5.La) (5.LIa)

TABLE 5.13. Deactivation Products of Model Lithium Phenyl Alkenes in THF at 0°C [120]

Initial compound	Direction	Deactivators				
		CH_3OH	CH_3I	$(CH_3)_3SiCl$	$(CH_2)_2O$[a]	$(CH_3)_2C=O$
t-$BuCH_2CH=CHCH(Ph)Li$ (5.XLVIII)	C_α	78	96	95	100	62
	C_γ	22	4	5	0	38
$CH_3CH=CHCH(Ph)Li$ (5.XLIX)	C_α	54	96	67	100	55
	C_γ	46	4	33	0	45
t-$BuCH_2C(Ph)=CHCH_2Li$[b] (5.L)	C_α	87	82	100	25	100
	C_γ	13	18	0	75	0
$CH_3C(Ph)(Li)CH=CH_2$[c] (5.LI)	C_α	61	45	100	0	0
	C_γ	39	55	0	100	100

[a]Data for −78°C.
[b]Total yield of deactivation products of this compound is about 91-93%; here the values are normalized to 100%, see [120].
[c]This system of labelling the C atoms cannot be regarded as successful but we have retained it as being the system used by the authors of the references cited.

However it hardly follows that the role of the electron structure and the geometry of the models (5.L) and (5.LI) [which are masked in such "smoothed out" formulas as (5.La) and (5.VIa)] should be completely excluded.

The superiority of C_α attack is also seen to a marked degree in the reactions of the compounds (5.L) with CH_3OH and CH_3I, although in these cases the differences between the active sites being compared are somewhat "smoothed out" as compared to the action of acetone. The cause of the latter situation can be seen in the fact that the greatest steric hindrance to the reaction is to be found in the reaction of the active sites with just this deactivator.

It is more difficult to understand why ethylene oxide (EO) is left out of the deactivator series. For this the C_γ reaction is the sole or principal one in the case of both structural variations of 2-PBL. Any attempt to interpret this effect would be risky since data for EO relates to a considerably lower temperature than the other results given in Table 5.13. Therefore, we emphasize that the only circumstance when EO exhibits high selectivity is when 1-PBL and 2-PBL are involved (C_α and C_γ attack respectively).

The structural features of the deactivation products of the model active sites also deserve attention from the point of view of their degree of correspondence with the microstructure of the polymers formed in conditions coinciding with those given in Table 5.13 (THF, 0°C). In confining ourselves to the overall reactions of C_α and C_γ attack, it is possible to see quite similar values of the structural indices of the polymers and the model active sites which are deactivated by the majority of the compounds used. Notable exceptions are the data for EO–2-PBL (5.L) and the acetone–1-PBL (5.XLVIII) systems. However, conclusions on such a correspondence vary considerably on comparing the contributions of the cis and trans forms to the overall 1,4 structure. Some examples are given below (THF, 0°C) [120]:

Deactivator	Structure of 1,4 units in the deactivation products of the compounds, %			
	1-PBL		2-PBL	
	cis (Z)	trans (E)	cis (E)	trans (Z)
CH_3OH	8	70	81	0
CH_3I	11	85	78	0
$(CH_3)_3SiCl$	4	91	91	0

It will be recalled that for these polymers (see Table 5.10) the following characteristics are obtained:

poly-1-PB		poly-2-PB	
cis (Z)	trans (E)	cis (E)	trans (Z)
13	77	91	0

Only the data for the product of the interaction of 2-PBL (5.L) with $(CH_3)_3SiCl$ yield a good correspondence between the structure of the polymer and the model active site. The structural characteristics do not depend upon the nature of the deactivating reagent in any regular fashion. For example, in the case of 1-PBL, trimethylchlorosilane leads to the formation of a product which is structurally different from poly-1-PB to a greater extent than is found for other terminating reagents which we have mentioned. Therefore we must still limit ourselves to asserting that it is necessary to take these facts into consideration, particularly when attempting to draw qualitative conclusions regarding a polymerization mechanism based on the structural features relating to model compounds.

These remarks must be supplemented by the possible role of the difference in the rates of reaction of these deactivators with organolithium compounds. Systematic studies in this field have not apparently been carried out, but there are examples known which point to the comparatively slow course of such reactions when trimethylchlorosilane or acetone are used as deactivators; see [127]. In such cases the advantageous choice of one isomeric form of the active reagent which is compensated by the rapid reestablishment of equilibrium is perfectly possible.

In discussing the causes of the differences between the structure of the polymers formed in anionic systems and the deactivated model reagents, it is usual to refer to a cis-trans equilibrium which may not be attained in real polymerizations. The role of this is significant when the difference between the relative reactivities of the cis and trans forms is considerable. Obviously the different "rate of selection" of one or another of these forms is also inherent in the deactivation reaction, where the initial position of the cis-trans equilibrium may be destroyed and partially restored. The degree to which this takes place depends upon the nature of the deactivating reagent. The latter can determine the overall rate and selectivity of the corresponding reactions of the model active sites. This affects not only the deactivation or selection of the cis and trans forms but also the competition of the C_α and C_γ reactions between the active sites and the terminating reagents.

In summarizing the above it is possible to conclude that the study of model compounds and their reaction products with different reagents leads to results which are seldom amenable to unequivocal interpretation. The simple transfer of experimental data which characterize such materials to real active sites can often lead to qualitatively false conclusions.

The available data on the structural characteristics of polydienes of anionic origin indicate the variety of the effects which dominate in systems which are fairly similar. Only a partial analysis of the connection between the variable parameters and the final results is possible. It is therefore natural that those phenomena which are difficult to interpret sometimes become the centers of "micro-discussions" which have fundamental significance for the mechanism of stereoregulation. From this point of view the opinions of the different authors on the geometry of the end groups of diene active sites are significant. That these concepts are contradictory becomes particularly obvious if the above results are supplemented with the data of other studies.

As can be seen from the characteristics obtained from RCH=$CHCH_2Mt$ compounds, the increase in the cis fraction while the R substituent remains constant proceeds in parallel with the increase in the electropositive nature of the Mt atom; see Table 5.6. In [97] this fact is taken as evidence favoring the tendency of the carbanion fragment of the compounds studied to form intramolecular complexes which favor the cissoid form. The calculation on the crotyl anion which was carried out by Schleyer et al. [40] using a nonempirical quantum-chemical method leads to qualitatively similar conclusions.

Bartmess et al. [125], on the basis of the acidity of the cis- and trans-2-butenes evaluated using pulsed-ion cyclotron resonance spectroscopy,* came to the opposite conclusion regarding the geometric isomers of the crotyl anion. The results of a study of the transfer of the proton from these monomers in the gas phase to H_2O and C_3H_6 acceptors are interpreted by these authors as facts which point to the somewhat lower stability of the cis anion in comparison with the trans. This conclusion was recently confirmed by the theoretical studies of Dewar et al. [126], who applied the semiempirical MNDO method to this anion and to the compound $CH_3CH{=}CHCH_2BeH$ chosen as a model of the organometallic crotyl agent. In both cases it was found that the trans conformation was the most advantagous (0.36 and 0.80 kcal/mole, respectively). Dewar et al., noting the impossibility of a rigorous discussion with such a small value of ΔE, gave preference to the results obtained by the MNDO method which in their view were the more correct with respect to the materials studied than the nonempirical method used in [40].

It is quite clear that this question is still not exhausted and discussion cannot yet be regarded as complete.

---

*The method is based on following the motion of charged particles in magnetic and electric fields. For references see [125].

5.2.3. Polymerization in Systems with Transition Metals

While limiting ourselves to only a brief treatment of the processes of the polymerization of dienes under the action of active sites based on transition metals (which we will designate AS–TM), we recall first of all the recent work of Burford [128] which concentrates mainly on butadiene but which contains the majority of the summaries in this field. The bibliography of [128] must be supplemented by the reviews of Dolgoplosk et al. [129, 130], Lobach and Kormer [131], and Monakov et al. [132]. It may be asserted that the wide use of Ziegler–Natta catalysts and related systems for the synthesis of stereoregular polydienes has not led to an especially notable development of theoretical studies in recent years. In particular, schemes of stereoregulation which are contained in [128] in fact describe the level of ideas reached about ten years ago. The absence of fundamentally new ideas does not reflect only the situation connected with the "classical" Ziegler–Natta catalysts, i.e., with insoluble complexes which are difficult to study from the physicochemical point of view. The advantage which the soluble π-allyl complexes of transition metals enjoyed in this respect had already been reduced considerably in the first half of the seventies. The use of these complexes for the synthesis of low-molar-mass adducts which model to some extent the diene AS–TM (details of these are given in [131]) led in these years to the formulation of concepts on the microstructure of the polymer chain as a function of the structure and the kinetic and thermodynamic parameters of the end groups of the model compounds.

Detailed factual material contained in [128-132] and the stereoregulation schemes presented in them exclude the advisability of a full examination of these questions. We note only some features of these systems which enable parallels to be drawn between the diene AS–TM and the anionic growing chains of the same monomers.

The similarity between these and other systems can be seen in the initial formation of cissoid end groups whose fate is determined by the competition of cis-trans isomerization and the growth reaction. Obviously such a concept is sufficiently general whereas the possibility of applying scheme (5.30) to AS–TM, although not completely excluded, is less obvious. Dolgoplosk [129, 130] uses for this the existence of intermediate σ-forms at the stage of anti-syn isomerization, regarding them as responsible for the formation of 1,2 or 3,4 groups in the polydiene macromolecules. There is no firm basis for the choice between this interpretation and the competition of the C_α and C_γ attack of the end group by the monomer. The results obtained from a study of structure of the deactivation products of model AS–TM which are sometimes used for this purpose are difficult to regard as sufficiently informative in this respect. As has already been noted (see Tables 5.4 and 5.13), the nature of the deactivating reagent often determines the structure of the end groups in the passivated products. It is essential here that the monomer can also be regarded as a reagent which exhibits stereochemical individuality at the stage of "deactivation" of the end group, i.e., when it becomes the penultimate group. On the basis of this feature it is difficult to regard as successful the view that the reaction on the C_γ atom is a consequence of the "transfer of the center of reaction" (see [129]). Essentially the question must concern the selectivity of the attack which depends, for any given active site, conditions being equal, on the nature of the attacking molecule including the nature of the monomer. This has been so far shown especially clearly in processes initiated by organolithium compounds (see page 278), but analogous effects are also probable for AS—TM.

There is also a definite similarity in the behavior of AS—TM which are capable of selective formation of macromolecules of 1,2 or 3,4 structures and anionic active sites of the types RMT·D. Both these and others, as a rule, are not distinguished stereospecifically in the strict sense of the term. The parallel combination of the C_γ attack with the fixing of the sequence of groups of a certain high atacticity is a distinguishing feature only for some heterogeneous systems based on transition metals. Obviously, the formation of isotactic macromolecules in the polymerization of any nonpolar monomer is possible only with heterogeneous catalytic complexes. It is not possible to illustrate this for polydienes using a large number of facts. There is quite a large number of examples of the formation of 1,2 or 3,4 groups as basic structural units for diene polymers but a detailed evaluation of the microtacticity of the corresponding studies has been comparatively rare. Bearing in mind the synthesis of iso- and syndiotactic polydienes, data which mark the beginning of the work of Natta in the sixties have been augmented only to a small extent in recent years. It is quite clear that the majority of polymers with high content of vinyl (or vinylidene units) must be regarded as atactic structures.

We note that the discussions of the stereoregularity of some similar polymers are based on crystallinity data without introducing quantitative structural features; see [128, 129]. Hence the possibility of a statistical analysis of the structure of the macromolecules, the results of which possibly giving additional information on the mechanism of such processes, is excluded. At the same time polymers of this type which are regarded as amorphous must dffer in their degree of atacticity, i.e., according to their absolute microstructure. As has been shown for a number of other polymers (see Section 5.1), such data often prove to be very useful for determining the details of the mechanism of formation of the structure of macromolecules.

REFERENCES

1. W. Cooper, "Stereospecific polymerization," Progress in High Polymers, J. C. Robb and F. W. Peaker (editors), Vol. 1 (1961), pp. 279-340.
2. A. Zambelli and C. Tosi, "Stereochemistry of propylene polymerization," Adv. Polym. Sci., 15, 31 (1974).
3. J. A. Licchelli, A. D. Caunt, R. N. Haward, and J. W. Parsons, "The polymerization of propylene using titanium-based catalysts by halogen-free magnesium alkyls," IUPAC Symposium on Macromolecules, Florence (1980), Vol. 2, pp. 48-50.
4. Gy. Sárosi and A. Simon, "Some peculiarities of the $MgCl_2$-supported catalyst systems for stereoregulation of propene polymerization," IUPAC Symposium on Macromolecules, Florence (1980), Vol. 2, pp. 44-47.
5. S. Kvisie, O. Nirinsen, and E. Rytter, "Structural studies of the supported catalyst $MgCl_2$—$TiCl_4$—PhCOOEt used for the polymerization of propene," IUPAC Symposium on Macromolecules, Florence (1980), Vol. 2, pp. 32-35.
6. N. Gaylord and H. Mark, Linear and Stereoregular Additive Polymers, Interscience, New York (1959).
7. B. L. Erusalimskii, Wang Fo-Sung, and A. P. Kavunenko, "Investigation of the reaction of organomagnesium compounds with the salts of heavy metals and the use of organomagnesium compounds and their complexes for polymerization," J. Polym. Sci., 53, 27-32 (1961).
8. Y. Doi, E. Suzuki, and T. Keii, "Stereoregularities of polypropylenes obtained with highly active supported Ziegler—Natta catalyst sytems," Makromol. Chem. Rapid Commun., 2, 293-297 (1981).
9. Y. Doi and T. Keii, "The stereoregulating ability of isotactic-specific sites in heterogeneous catalyst systems $TiCl_3$—$Al(C_2H_5)_2X$ ($X = C_2H_5$, Cl, Br, or I) for propene polymerization," Makromol. Chem., 170, 2117-2119 (1978).
10. U. Giannini, "Polymerization of olefins with high-activity catalysts," Makromol. Chem. Suppl., 5, 216-229 (1981).
11. P. Galli, L. Luciani, and G. Cechin, "Advances in the polymerization of olefins with coordination catalysts," Angew. Makromol. Chem., 94, 63-89 (1981).
12. J. P. Kennedy and A. W. Langer, Jr., "Recent advances in cationic polymerization," Adv. Polym. Sci., 3, 508-580 (1964).
13. G. Ferraris, C. Corno, A. Priola, and S. Cesca, "On the cationic polymerization of olefins and the structure of the polymers, 3. Polypropylene," Macromolecules, 13, 1104-1110 (1980).
14. C. Corno, G. Ferraris, A. Priola, and S. Cesca, "On the cationic polymerization of olefins and the structure of the polymers, 2. Poly-1-butene," Macromolecules, 12, 404-411 (1979).
15. A. Priola, C. Corno, and S. Cesca, "On the polymerization of olefins and the structure of polymers, 4. 1-Butene selectively ^{13}C-enriched in positions 3 and 4," Macromolecules, 13, 1110-1114 (1980).

16. J. Boor, Jr., "Ziegler-Natta Catalysts and Polymerizations, Academic Press, New York (1979).
17. P. Corradini, G. Guerra, R. Fusco, and V. Barone, "Steric effects and stereospecific polymerization on Ziegler-Natta heterogeneous catalysts," IUPAC Symposium on Macromolecules, Florence (1980), Vol. 2, pp. 1-4.
18. R. Zanetti, G. Guidetti, D. Ajo, and A. Martorana, "Polymorphism of $TiCl_3$ and the structural disorder in the σ-form," IUPAC Symposium on Macromolecules, Florence (1980), Vol. 2, pp. 5-8.
19. L. Locatelli, M. C. Sacchi, E. Rigamonti, and A. Zambelli, "Stereospecific polymerization of propene. Some evidence for monomer insertion on σ-metal-carbon bond," IUPAC Symposium on Macromolecules, Florence (1980), Vol. 2, pp. 28-31.
20. Y. Doi, "Structure and chemistry of atactic polypropylenes. Statistical model of chain propagation," Makromol. Chem. Rapid Commun., 3, 653-641 (1982).
21. K. H. Reichert, "Mechanismus der Olefinpolymerisation mit Ubergangsmetallkatalysatoren," Angew. Makromol. Chem., 94, 1-23 (1981).
22. B. L. Erusalimskii, "Über einige Besonderheiten der anionischen Polymerization polarer Monomerer," Plaste Kautsch., 15, 788-792 (1968).
23. T. Uryu, T. Seki, T. Kawamura, A. Funamoto, and K. Matsuzaki, "Synthesis of polystyrenes with different stereoregularities by anionic polymerization," J. Polym. Sci., Polym. Chem. Ed., 14, 3035-3044 (1976).
24. T. Uryu, T. Kawamura, and K. Matsuzaki, "The stereoregularities of polystyrene obtained by different ion pairs of polystyryl alkali salts," J. Polym. Sci., Polym. Chem. Ed., 17, 2019-2029 (1979).
25. T. Kawamura, T. Uryu, and K. Matsuzaki, "Analysis of isotactic polystyrenes obtained with butyllithium and olefin catalysts by carbon-13 nuclear magnetic resonance spectroscopy," Makromol. Chem., 180, 2001-2008 (1979).
26. T. Kawamura, T. Uryu, and K. Matsuzaki, "Stereoregularity of polystyrene derivatives, 2. Poly(methoxystyrene)s obtained by anionic catalysts," Makromol. Chem., 183, 143-151 (1982).
27. T. Kawamura, T. Uryu, and K. Matsuzaki, "Stereoregularity of polystyrene derivatives, 3. Poly(methylstyrene)s obtained by anionic catalysts," Makromol. Chem., 183, 153-162 (1982).
28. K. Matsuzaki, Y. Shinohara, and T. Kanai, "Nuclear magnetic resonance studies on polymer carbanions. 1. Living polystyrene and its model compounds," Makromol. Chem., 181, 1923-1934 (1980).
29. J. Dils and M. van Beylen, "Penultimate unit effects in anionic polymerization. The anionic polymerization of 1,1-diphenylalkali salts with styrene," Int. Symp. on Macromolecules, Dublin (1977), 1, pp. 69-76.
30. E. Walckiers and M. van Beylen, "Penultimate effect in anionic polymerization of styrene," Int. Symp. on Macromolecules, Boston (1977), Vol. 2, pp. 1199-1206.

31. R. Wicker and K. F. Elgert, "Zur Taktizität von Poly-α-methylstyrol, 4. Das ^{1}H- and ^{13}C-NMR-Spektrum des 1,3,3-Trimethyl-1-phenylbutyllithiums," Makromol. Chem., 178, 3063-3073 (1977).
32. R. Wicker and K. F. Elgert, "Zur Taktizität von Poly-α-methylstyrol, 5. Die Abhängigkeit der Taktizität von der Reaktionstemperatur und der Anfangsmonomerkonzentration," Makromol. Chem., 178, 3075-3084 (1977).
33. R. Wicker and K. F. Elgert, "Zur Taktizität von Poly-α-methylstyrol, 6. Zum stereoreguleierenden Mechanismus der Polyreaktion des α-Methylstyrols mit Butyllithium in Tetrahydrofuran," Makromol. Chem., 178, 3085-3099 (1977).
34. S. L. Malhotra, "Polymerization of α-methylstyrene in tetrahydrofuran with potassium as initiator, V. NMR analysis of the reaction products," J. Macromol. Sci., Chem., A12, 73-101 (1978).
35. S. L. Malhotra, "Polymerization of α-methylstyrene in p-dioxane and cyclohexane. VI. NMR analysis of the reaction products," J. Macromol. Sci., Chem., A12, 883-908 (1978).
36. S. L. Malhotra, "Anionic polymerization of α-methylstyrene. IX. Nature of propagating species," J. Macromol. Sci., Chem., A15, 533-552 (1981).
37. K. Matsuzaki, T. Iwamoto, and T. Kanai, "Nuclear magnetic resonance studies on polymer carbanions, 3. Model compounds for poly(o-methoxystyryl) and poly(o-methylstyryl) carbanions," Makromol. Chem. Rapid Commun., 2, 187-192 (1981).
38. K. Matsuzaki, Y. Shinohara, and T. Kanai, "Nuclear magnetic resonance studies on polymer carbanions, 2. Model compounds for α-methylstyrene and p-methylstyrene tetramer dianions," Makromol. Chem., 182, 1533-1540 (1981).
39. J. C. Favier, "Regularities of polv-p-isopropyl-α-methylstyrene," Polymer, 23, 1501-1507 (1982).
40. P. v. R. Schleyer, J. D. Hill, J. A. Pople, and W. J. Hehre, "Geometrical preferences of the crotyl anion, radical, and cation," Tetrahedron, 333, 2497-2501 (1977).
41. H. Yuki and K. Hatada, "Stereospecific polymerization of α-substituted acrylic acid esters," Adv. Polym Sci., 31, 1-45 (1979).
42. B. L. Erusalimskii (Erussalimsky), B. G. Belen'kii (Belenkii), A. A. Davidyan (Davidjan), V. D. Krasikov, V. V. Nesterov, N. I. Nikolaev, V. N. Zgonnik, and V. M. Sergutin, "Subcatalytic effects in anionic polymerization processes," 27th Int. Symp. on Macromolecules, Strasburg (1981), Vol. 1, pp. 182-185.
43. M. Tomoi, K. Sekiya, and H. Kaliuchi, "Anionic polymerization of methyl methacrylate by alkali metal alkoxides," Polym. J., 6, 438-444 (1974).
44. P. E. M. Allen and B. O. Bateup, "Kinetics of the polymerization of ·ethyl methacrylate initiated by butylmagnesium bromides a dibutylmagnesium in tetrahydrofuran + toluene," J. Chem. Sc. Faraday Trans., I, 71, 2203-2212 (1977).

45. Y. Okamoto, K. Urakawa, and H. Yuki, "Stereospecific polymerization of methacrylates with ethylmagnesium alkoxides," Polym. J., 10, 457-464 (1978).
46. Yu. E. Eizner and B. L. Erusalimskii, The Electronic Aspect of Polymerization Reactions [in Russian], Nauka, Leningrad (1976).
47. V. N. Zgonnik, E. Yu. Melenevskaya, and B. L. Erusalimskii, "The study of active centers in anionic polymerization using spectroscopic and quantum chemical methods," Usp. Khim., 47, 1479-1503 (1978).
48. P. E. M. Allen and B. O. Bateup, "Kinetics of the polymerization of methyl methacrylate initiated by organometallic compounds. VIII. Initiation by n-, i-, s-, or t-butyl magnesium compounds. Evidence for an enieidic pseudoanionic mechanism," Eur. Polym. J., 13, 761-768 (1977).
49. B. O. Bateup and P. E. M. Allen, "Polymerization of methyl methacrylate by organometallic compunds. VII. The n-butylmagnesium bromide—di-n-butylmagnesium system in THF + toluene solution," Eur. Polym. J., 13, 762-768 (1977).
50. P. E. M. Allen, "Active centers for the stereospecific polymerization of methyl methacrylate by organomagnesium compounds," J. Macromol. Sci., Chem., A14, 11-21 (1980).
51. P. E. M. Allen, M. C. Fisher, C. Mair, and E. H. Williams, "Polymerization of methyl methacrylate initiated by t-butyl and phenylmagnesium compounds. Factors influencing the nature of the active centers," ACS Symposium Series 166, J. E. McGrath (editor), 185-197 (1981).
52. P. E. M. Allen, C. Mair, M. C. Fisher, and E. H. Williams, "Some problems concerning the mechanism of the isotactic polymerization of methyl methacrylate initiated by organometallic compounds in toluene solution. Application of modified dilatometer and NMR method," J. Macromol. Sci., Chem., A17, 61-67 (1982).
53. W. E. Lindsell, F. C. Robertson, I. Sontar, and V. H. Richards, "Polymerization by alkaline earth metal compounds. I. Studies on the polymerization of methyl methacrylate by triphenylmethyl derivatives and related compounds of calcium, strontium, and barium," Eur. Polym. J., 17, 107-113 (1981).
54. J. Kawak, Q. T. Pham, C. Pillot, and J. P. Pascault, "Polymérisation anionique du méthacrylate de méthyle par les organoalcalins. II. Polymérisation en milieu solvent aprotonique et l'effet de la température sur la tacticité des polymères, Eur. Polym. J., 10, 997-1003 (1974).
55. J. P. Pascault, J. Kawak, J. Gole, and Q. T. Pham, "Polymérisation du méthacrylate de méthyle par les organoalcalins. III. Polymérisation en milieu solvent aprotonique, études des tacticités," Eur. Polym. J., 10, 1007-1114 (1974).
56. R. Pétiaud, P. Ciaudy, and Q. T. Pham, "Polymérisation du méthacrylate de méthyle par le chlorure de tertio-butyle-magnésium en solution dans le tétrahydrofuranne. I. Étude de l'équilibre de dissociation du chlorure de tertio-butyle-magnésium," Polym. J., 12, 441-447 (1976).

57. R. Pétiaud and Q. T. Pham, "Polymérisation du méthacrylate de méthyle par le chlorure de tertio-butyle-magnésium en solution dans le tétrahydrofuranne, II. Comportement de la polymérisation en fonction de la température et de la composition magnesienne," Eur. Polym. J., 12, 449-452 (1976).
58. R. Pétiaud and Q. T. Pham, "Polymérisation du méthacrylate de méthyle par le chlorure de tertio-butyle magnésium en solution dans le tétrahydrofuranne, III. Paramètres thermodynamiques des propagation iso- et syndiotactique en fonction de la composition magnesiennes, Eur. Polym. J., 12, 455-461 (1976).
59 A. H. E. Müller, E. Höcker, and G. V. Schulz, "Rate constants in the anionic polymerization of methyl methacrylate in tetrahydrofuran with cesium as counterion," Macromolecules, 10, 1086-1089 (1977).
60. V. Warzelhan, H. Höcker, and G. V. Schulz, "The anionic polymerization of methyl methacrylate with a bifunctional initiator," Makromol. Chem., 181, 149-163 (1980).
61. A. H. E. Müller, "The present view of the anionic polymerization of methyl methacrylate and related esters in polar solvents," ACS Symposium Series 166, J. E. McGrath (editor), 441-461 (1981).
62. L. Lochman, R. L. De, J. Janča, and J. Trekoval, "Some organolithium compounds as initiators of the anionic polymerization of methyl methacrylate," Collect. Czech. Chem. Commun., 45, 2761-2765 (1980).
63. V. N. Krasulina, A. S. Khachaturov, N. V. Mikhailova, and B.L. Erusalimskii, "On the mechanism of stereoregulation in the anionic polymerization of polar monomers," Vysokomol. Soedin., 12, 303-307 (1970).
64. K. Hatada, H. Sugino, H. Ise, T. Kitayama, Y. Okamoto, and H. Yuki, "Heterotactic polymers of α-substituted acrylic acid esters," Polym. J., 12, 55-62 (1980).
65. Y. Okamoto, K. Ohta, K. Hatada, and H. Yuki, "Anionic polymerization of triphenylmethyl methacrylate," ACS Symposium Series 166, J. L. McGrath (editor), 353-365 (1981).
66. G. R. Dewer, F. E. Karasz, W. J. McKnight, and R. W. Lenz, "Poly(alkylchloroacrylates). V. Preparation and properties of methyl, ethyl, and i-propyl polymers of varied tacticity," J. Polym. Sci., Chem. Ed., 13, 2151-2179 (1975).
67. A. Soum and M. Fontanille, "Living anionic stereospecific polymerization of 2-vinylpyridine, 1. Initiation of polymerization and stereoregularity of polymers," Makromol. Chem., 181, 799-808 (1980).
68. A. Soum and M. Fontanille, "Living anionic stereospecific polymerization of 2-vinylpyridine, 2. Kinetics of polymerization and nature of active centers," Makromol. Chem., 182, 1743-1750 (1981).
69. A. Soum and M. Fontanille, "Living anionic stereospecific polymerization of 2-vinylpyridine, 3. Structure of active centers and mechanism of polymerization," Makromol. Chem., 183, 1145-1159 (1982).

70. T. E. Hogen-Esch, W. L. Jenkins, C. F. Tien, and R. Smith, "Ion pair structure and stereochemistry in anionic oligomerization of some vinyl monomers," Polym. Prepr., 21, 13-14 (1980).
71. T. E. Hogen-Esch and C. F. Tien, "Oligomerization stereochemistry of vinyl monomers, 7. Diastereomeric ion pairs as intermediates in stereoregular anionic oligomerization of 2-vinylpyridines. A proposed mechanism," Macromolecules, 13, 207-216 (1960).
72. C. S. Huang, C. Mathis, and T. E. Hogen-Esch, "Oligomeriza- of vinyl monomers, 9. ^{13}C-NMR and chromatographic studies of oligomers of 2-vinylpyridine," Macromolecules, 14, 1802-1807 (1981).
73. T. E. Hogen-Esch and W. L. Jenkins, "NMR and conductometric studies of 2-pyridyl-substituted carbanions, 2. Effect of cation size and coordination," J. Am. Chem. Soc., 103, 3666-3672 (1981).
74. W. F. Jenkins, C. F. Tien, and T. E. Hogen-Esch, "Oligomerization stereochemistry of vinyl polymers, IV. Ion pair structure and β-carbon stereochemistry in anionic oligomerization of 2-vinyl pyridine," J. Polym. Sci., B16, 501-506 (1978).
75. C. J. Chang, F. F. Kiesel, and T. E. Hogen-Esch, "Ultraviolet spectroscopic and conductometric studies of pyridine-type carbanions," J. Am. Chem. Soc., 97, 2805-2810 (1975).
76. C. Mathis and T. E. Hogen-Esch, "Evidence for a conformationally restrained six-membered ring containing lithium ion," J. Am. Chem. Soc., 104, 634-635 (1982).
77. C. E. Schildknecht, "Vinyl ether stereoregulated polymerization," High Polym., C. E. Schildknecht and I. Skeists (editors), 29, 325-329 (1977).
78. T. Kunitake and C. Aso, "A proposal on the steric course of propagation in the homogeneous cationic polymerization of vinyl and related ethers," J. Polym. Sci., Part A1, 8, 665-678 (1970).
79. T. Kunitake and K. Takarabe, "The counterion effect on the steric course of the cationic polymerization of tert-butyl vinyl ether, Makromol. Chem., 182, 817-824 (1981).
80. K. Matsuzaki, M. Hamada, and K. Arita, "Stereoregularity of poly(vinyl ether)," J. Polym. Sci., Part A1, 5, 1233-1243 (1967).
81. H. Yuki, K. Hatada, K. Ohta, I. Kinoshita, S. Y. Murahashi, K. Ono, and Y. Ito, "Stereospecific polymerization of benzyl vinyl ether by $BF_3 \cdot OEt_2$," J. Polym. Sci., Part A1, 7, 1517-1536 (1969).
82. H. Yuki, K. Hatada, K. Ohta, and T. Sazaki, "Stereospecific polymerization of allyl vinyl ether by $BF_3 \cdot OEt_2$," Bull. Soc. Chem. Japan, 43, 890-897 (1970).
83. D. J. Sikkema and H. Angad-Gaur, "Isotactic poly(allyl vinyl ether), 1. A study of the polymerization and a proposal of a mechanism of stereocontrol," Makromol. Chem., 181 2259-2266 (1980).

84. H. Angad-Gaur and D. J. Sikkema, "Isotactic poly(allyl vinyl ether), 2. An NMR study of stereoregularity and relaxation times," Makromol. Chem., 181, 2385-2393 (1980).
85. F. Heublein, "The role of monomer solvation and counterion complexation in the cationic polymerization of vinyl monomers," J. Macromol. Sci., Chem., A16, 563-577 (1981).
86. Yu. E. Eizner, S. S. Skorokhodov (Skorochodov), and T. P. Zubova, "Electron-density distribution in unsaturated ethers and esters and their reactivity in cationic polymerization," Eur. Polym. J., 7, 869-878 (1971).
87. Yu. E. Eizner and B. L. Erusalimskii, The Electronic Aspect of Polymerization Reactions [in Russian], Leningrad (1976), p. 81.
88. Y. Imanishi, Transfer Reactions in Cationic Polymerization, Kyoto University, Kyoto (1964), pp. 234-240.
89. B. A. Dolgoplosk, Diene Polymerization I [in Russian], Moscow (1972), pp. 697-714.
90. A. Priola, N. Passerini, M. Bruzzone, and S. Cesca, "Cationic cyclization of cis-1,4-polybutadiene. II. Physicochemical characteriziation of the polymer," Angew. Makromol. Chem., 88, 21-35 (1980).
91. C. Corno, A. Priola, and S. Cesca, "Cationic copolymerization of isobutylene, 6. NMR investigation of the structure and sequences of isobutylene-2,3-dimethylbutadiene copolymers," Macromolecules, 15, 840-844 (1982).
92. C. Corno, A. Priola, and S. Cesca, "Cationic copolymers of isobutylene. 1. Nuclear magnetic resonance investigation of the structure and monomer distribution on isobutylene—butadiene copolymers," Macromolecules, 12, 411-418 (1979).
93. C. Corno, A. Proni, A. Priola, and S. Cesca, "Cationic copolymerization of isobutylene, 2. Nuclear magnetic resonance investigation of the structure of the isobutylene—isoprene copolymers," Macromolecules, 13, 1092-1099 (1980).
94. W. Gebert, J. Hinz, and H. Sinn, "Umlagerungen bei der durch Lithiumbutyl initiierten Polyreaktion der Diene Isopren und Butadiene," Makromol. Chem., 144, 97-115 (1971).
95. A. Kh. Bagdasar'yan, B. A. Dolgoplosk, and V. M. Frolov, "The stereoregulating mechanism in the polymerization of dienes in anionic systems," Vysokomol. Soedin., A11, 2191-2196 (1969).
96. D. J. Worsfold and S. Bywater, "Lithium alkyl-initiated polymerization of isoprene. Effect of cis—trans isomerization of organolithium compounds on the polymer microstructure," Macromolecules, 11, 582-586 (1978).
97. M. Schlosser and J. Hartmann, "2-Alkenyl anions and their surprising endo preference. Facile and extreme stereocontrol over carbon—carbon linking reactions with allyl-type organometallics," J. Am. Chem. Soc., 98, 4674-4676 (1976).
98. J. Sledz, F. Shue, B. Kaempf, and S. Libs, "Étude de la Microstructure des oligomères desactivés du butadiène-1,3, Eur. Polym. J., 10, 1207-1215 (1974).

99. S. B. Texeira-Barriera, R. Mechin, and C. Tanielian, "Étude de la structure des oligomères de l'isoprène desactivés par les halogénures d'alcoyle," Eur. Polym. J., 15, 677-683 (1979).
100. A. Garton and S. Bywater, "Anionic polymerization of butadiene in tetraahydrofuran. I. Isomerization of polybutadienyl salts," Macromolecules, 8, 694-697 (1975).
101. A. Garton and S. Bywater, "Anionic polymerization of butadiene in tetrahydrofuran. II. Ion-pair propagation rate," Macromolecules, 8, 697-700 (1975).
102. S. Bywater, D. J. Worsfold, and G. Hollingworth, "Structure of oligomeric polybutadienyllithium and polybutadiene," Macromolecules, 5, 389-393 (1972).
103. V. N. Zgonnik, N. I. Nikolaev, E. Yu. Shadrina, and L. V. Nikovna, "Copolymerization of butadiene with styrene on butyllithium complexes with tetramethylethylenediamine and 2,3-dimethoxybutane," Vysokomol. Soedin., B15, 684-686 (1973).
104. R. Ohlinger and F. Bandermann, "Kinetics of the propagation reaction of butadiene—styrene copolymerization with organolithium compounds," Makromol. Chem., 181, 1935-1947 (1980).
105. E. Yu. Melenevskaya, V. N. Zgonnik, E. R. Dolinskaya, and B. L. Erusalimskii (Erussalimsky), "Structurelle Effekte in Systemen M_nLi—Butadien—Styrol," Makromol. Chem., 179, 2759-2764 (1978).
106. A. Essel, R. Salle, and Q. T. Pham, "Polymérisation anionique des diènes. IV. Contribution aux études des méchanismes de propagation stereospecifique des isoprene et butadiene par les organo-alcalins," J. Polym. Sci., Polym. Chem. Ed., 13, 1869-1877 (1975).
107. R. Salle and Q. T. Pham, "Polymérisation des diènes. VI. Microstructure des polybutadiene et polyisoprène par resonance magnétique protonique à 250 MHz et méchanismes de propagation," J. Polym. Sci., Polym. Chem. Ed., 15, 1799-1810 (1977).
108. R. Salle, "Contribution à l'étude des méchanismes de propagation anionique des diènes par les pairs d'ions en contact et les ions libres," Thesis, Lyon (1976).
109. C. J. Dyball, D. J. Worsfold, and S. Bywater, "Anionic polymerization of isoprene in diethyl ether," Macromolecules, 12, 819-822 (1979).
110. L. E. Forman, Elastomers from catalysts of alkali metals, in: High Polym., J. P. Kennedy and E. G. M. Törnquist (editors), Part 2, 23 (1969), pp. 491-596.
111. V. N. Zgonnik, N. I. Nikolaev, L. V. Vinogradova, N. S. Dimitrieva, K. K. Kalnin'sh, A. P. Koroleva, N. V. Smirnova, E. Yu. Shadrina, and V. M. Borodulina, "Effect of the nature of complexes of organolithium compounds with electron donors on the kinetics of butadiene polymerization and the polymer structure," 5th Int. Conf. Organomet. Chem., 1, 549-550, Moscow (1971).
112. B. L. Erusalimskii, "Unresolved problem in ionic polymerization," in: Advances in Ionic Polymerization [in Russian], Z. Jedlinski (editor), Warsaw (1975), pp. 9-23.

113. A. F. Halasa, D. F. Lohr, and E. Hall, "Anionic polymerization to high vinyl polybutadiene," J. Polym. Sci., Polym. Chem. Ed., 19, 1357-1360 (1981).
114. D. J. Worsfold, S. Bywater, F. Shué, J. Sledz, and V. Marti-Collet, "1,2-Dipiperidinoethane as a complexing agent in the anionic polymerization of butadiene and isoprene," Makromol. Chem. Rapid Commun., 3, 239-242 (1982).
115. L. V. Vinogradova, N. I. Nikolaev, and V. N. Zgonnik, "The nature and reactivity of the active centers in the system butadiene—n-butyllithium—tetramethylethylenediamine—hydrocarbon medium," Vysokomol. Soedin., A18, 1756-1761 (1976).
116. T. Suzuki, Y. Tsuji, and Y. Takegami, "Microstructure of poly(1-phenylbutadiene) prepared by anionic initiators," Macromolecules, 11, 639-644 (1978).
117. T. Suzuki, Y. Tsuji, Y. Takegami, and H. J. Harwood, "Microstructure of poly(2-phenylbutadiene) prepared by anionic initiatiors," Macromolecules, 12, 234-239 (1979).
118. T. Suzuki, Y. Tsuji, Y. Watanabe, and Y. Takegami, "Characterization of the living anion chain end of oligomeric 1-phenyl-1,3-butadienyllithium," Polym. J., 11, 651-660- (1979).
119. T. Suzuki, T. Tsuji, Y. Watanabe, and T. Takegami, "Characterization of the living anion chain end of oligomeric 2-phenyl-1,3-butadienyllithium," Polym. J., 11, 937-945 (1979).
120. Y. Tsuji, T. Suzuki, Y. Watanabe, and Y. Takegami, "Active species in anionic polymerization of phenylbutadienes. Reactivity of model anions," Polym. J., 13, 1099-1110 (1981).
121. K. F. Elgert and W. Ritter, "Struckurbestimmung eines 1,4-Poly(1,3-pentadien)s durch ^{13}C-NMR-Spektroskopie," Makromol. Chem., 177, 2021-2030 (1976).
122. P. Aubert, J. Sledz, F. Shué, and J. Prud'homme, Etude structurale du poly(pentadiène-1,3) par RMN du proton à 100 et 220 MHz," Eur. Polym. J., 16, 361-369 (1980).
123. D. H. O'Brien, A. J. Hart, and C. R. Russel, "Carbon-13 magnetic resonance of allyl, pentadienyl and arylmethyl carbanions. Empirical calculations of π-electron densities," J. Am. Chem. Soc., 97, 4410-4412 (1975).
124. Y. Ysuji, T. Suzuki, Y. Watanabe, and Y. Takegami, "Anionic polymerization of 1-(2-methoxyphenyl)-1,3-butadiene and 1-(4-methoxyphenyl)-1,3-butadiene. Microstructure of polymers and characterization of living anion chain ends," Polym. J., 13, 651-656 (1981).
125. J. E. Bartmess, W. J. Hehre, R. T. McIver, and L. E. Overman, "Gas phase acidities of 2-butenes. Regarding the use of organometallics as models of free anions," J. Am. Chem. Soc., 99, 1976-1977 (1977).
126. M. J. S. Dewar and D. J. Nelson, "Ground states of molecules. 60. A MINDO study of conformations of crotyl anion and a diaza analogue, of their BeH derivatives and of the interconversion of cyclopropyl anion and allyl anion," J. Org. Chem., 47, 2614-2618 (1982).

127. J. C. Brosse, Z. A. Biu Maidung, and J. C. Soutif, "Contribution à l'étude de la métallation de structure polyisoprène," Makromol. Chem., 183, 123-129 (1982).
128. R. P. Burford, "Polymerization of butadiene using Ziegler–Natta catalysts – recent developments," J. Macromol. Sci., Chem., A17, 123-139 (1982).
129. B. A. Dolgoplosk, "Organometallic catalysis in stereospecific polymerization and the nature of the active centers," Usp. Khim., 46, 2027-2065 (1977).
130. B. A. Dolgoplosk and E. I. Tinyakova, "The nature of the active centers and the mechanism of coordination polymerization. I. The nature of the active centers in the stereospecific polymerization of dienes and the mechanism of stereoregulation," Vysokomol. Soedin., A19, 2441-2463 (1977).
131. M. I. Lobach and V. A. Kormer, "The introduction of diene hydrocarbons on transition-metal–ligand bonds," Usp. Khim., 48, 1416-1447 (1979).
132. H. G. Marina, Yu. B. Monakov, S. R. Rafikov, and B. I. Ponomarenko, "The connection between the components of Ziegler systems which contain titanium and the mechanism of stereoregulation in the polymerization of dienes," Usp. Khim., 52, 733-753 (1983).

Chapter 6

Conclusion

The format used here, which differs from that usually found in books on ionic polymerization, is the result of our intention of selecting problems of a general nature and preferring to examine the reactivity and the stereospecificity of active sites separately rather than jointly. It was also decided that a general summary should be given rather than detailed conclusions for each chapter. In such an undertaking we considered it necessary, together with the points already mentioned, to consider certain facts which have, for one reason or another, been omitted from the basic material.

It is natural that those important questions regarding the mechanisms by which macromolecules are formed in ionic systems and which remain open to question should arouse the greatest interest. An evaluation of the situation from this point of view was carried out at the beginning of the sixties by Morton and Mayo [1], to some extent by the present author about ten years later [2], and to a considerable extent at the Japanese—American seminar in the mid-seventies [3]. It is easy to be convinced that exactly the same problems remain which are connected in one way or another with the fact that certain other problems are not completely solved or that other possible solutions may be valid. Of course in-depth studies eliminate some of the former vagueness but this is accompanied by the appearance of new facts which are difficult to fit into the framework of the existing concepts or which even contradict them. As examples which illustrate this situation we choose the results of recent studies of the initiation mechanism in some systems which, it would seem, have not given rise to questions which are especially puzzling. If we do not concern ourselves with the more complex reagents of the type Al—N or with Cr—N derivatives (see Section 4.1.4), the mechanism of various anionic initiators considered earlier seemed to be quite clear. The cases to which we now address ourselves indicate the reverse.

The first of these is the polymerization of acrylonitrile (AN) in dimethylformamide (DMF) under the action of lithium alkoxides.

Lehmann, Riedel, et al. [4-6], using labelled atoms, established that the noncorrespondence between the values of the efficiency of initiation calculated from radiochemical data (F*) and that obtained in the usual way (F), i.e., from the average values of MM, referred to the initiator concentration. Only the value of F*, and not F, proved to be particularly sensitive to variation in the substituent R in ROLi initiators. The difference between F and F* was found to be especially marked for lithium t-butoxide: there are practically no fragments of labelled initiator to be found in the macromolecules. In studying the nature of this phenomenon various model reactions were carried out, labelled terminating reagents were used and, what is particularly important, complexes of lithium t-butoxide with DMF which were previously unknown were isolated and described.

An analysis of the results obtained led the authors to a hypothesis regarding the particular mechanism of initiation, according to which $(t\text{-BuLi})_3 \cdot 3DMF$ complexes function as the active reagents. The existence of such stoichiometry is demonstrated, and their specific features are most easily made to accord with the alternative dispositions of the initial components in the complex. Omitting intermediate stages, we present the scheme which reproduces the initial and final states proposed by the authors:

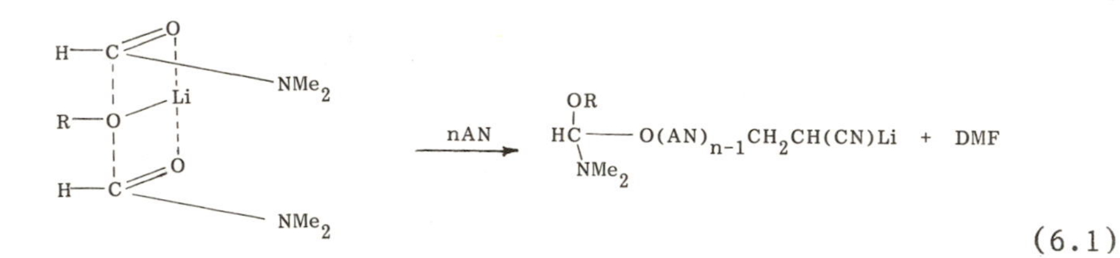

(6.1)

Of the arguments which favor scheme (6.1) which are described in detail in [4], apart from the absence of t-butoxy groups in the polymer, we note the presence of hydroxyl groups in the deactivated macromolecules. Their presence was demonstrated by treating these polymers with labelled benzoylchloride.

The features of lithium t-butoxide which distinguish it from the related initiators may be coordinated with a series of features (6.2) which coincide with the course of change of the corresponding values of F*:

Groups R in alkoxide	t-Bu	i-Pr	n-Pr	$PhCh_2$
F*	0	5.6	25.6	34.4

(6.2)

The latter coincides with the value of F obtained from the MM characteristics. From this the authors conclude that the mechanisms are fundamentally different for the extreme members of series (6.2) and that the processes initiated by the intermediate members proceed by a mixture of these mechanisms.

The experimental effects described are so far without parallel but the possibility of their being reproduced in the polymerization of other polar monomers in analogous conditions is not excluded. In order to clear this up it is necessary to develop further the studies which have been initiated in the Dresden school.

The second example concerns systems which include anion radicals (AR), the study of which both generally and in connection with initiation has considerable history. Research directed to the determination of the mechanism of their action has been recently carried out by Podolsky et al. [7]. Their research was aimed at discovering anion radicals of the initial unsaturated monomer whose existence has so far not been established experimentally. However, ideas about their formation at the initiation stage of polymerization under the action of AR are generally accepted. Interaction products of stilbene (S) and sodium naphthalenide (SN) were chosen as materials for study. The anion radical component of the latter will be designated as AR_s. The research was aimed at obtaining an anion radical of stilbene (AR_{st}) and describing it more fully.

In the study of the reaction between the initial reagents in THF at room temperature using UV and ESR spectroscopy the following unusual phenomena were discovered:

Molar ratio S/SN	UV Characteristics	ESR Characteristics
0.5	Disappearance of the absorption band AR_s (370 nm); the appearance of a band at 510 nm for the stilbene dianion (DS)	Disappearance of the AR_s signals with the absence of new signals
1.0	Weakening of the intensity of the appearance of bands at 485 and 700 nm	Appearance of a paramagnetic signal
1.5	Disappearance of the DS band; the twofold increase in intensity of the 485 and 700 nm bands	Increase in the intensity of the signal (almost twofold as compared with the original SN signal)

The disappearance of the ESR signal with a ratio S/SN = 0.5 is taken as evidence for the absence of the formation of a stable stilbene anion radical (AR_{st}). In its turn, the ESR signal, which grows gradually as the relative fraction of S increases, is taken as characteristic of the adduct of the stilbene disodium anion with stilbene, but not for the signal of AR_{st}. The events described above are in agreement with the following scheme, which is presented in [7] (counterion omitted):

$$AR_s + S \longrightarrow AR_{st} + N \tag{6.3}$$

$$AR_{st} \xrightarrow{AR_s} DS + N \tag{6.4}$$

$$DS + S \longrightarrow (DS)\cdot S \tag{6.5}$$

in which N is naphthalene and DS is the stilbene dianion.

According to this scheme AR_{st}, which is formed at stage (6.3), is subjected to instantaneous transformation (6.4) which excludes the possibility of determining its existence in the ESR spectrum. This conclusion runs counter to those of other workers who have studied the ESR spectra of systems which include stilbene. For bibliography see [7].

The point of view of Podolsky et al. regarding the nature of the product of the interaction of DS and S is supported by the quantitative features of the ESR spectrum at stage (6.5) with a ratio S/SN = 1.5. The considerable rise in its intensity as compared with the original signal of SN may be explained only by ascribing this effect to the participation of both phenyl nuclei of the stilbene which has been incorporated into the (DS)·S complex in accepting dianion electrons.

We note yet another feature of the (DC)·S complex, which became clear in the study of its interaction with polymerizing monomers. It was established using the labelled atom technique that the initiating component of the paramagnetic adduct which is formed by the reaction (6.5) is only DS and not S. This follows from the comparison of the results of radiochemical analysis of polyisoprene and poly-α-methylstyrene obtained under the action of adducts (DS)*·S and (DS)·S* (the asterisk symbolizes the radioactive label). The fact that the adducts are not equivalent indicates the absence of disproportionation between these components which might have been expected on the basis of the known views on anion-radical systems.

The discovery of fundamentally new phenomena in systems which did not promise anything surprising shows the danger of premature generalizations. Essentially the search for common points considering related systems sometimes deprives the phenomena of their individual features. In this it is possible to see one of the important factors hindering the transition from hypothesis to well-founded and complete conclusions. In this sense the search for the limits of applicability of firm conclusions may prove to be more fruitful.

Naturally, the increase in the volume of information can cause considerable variations in the interpretation of experimental results. For example the evolution of concepts on cocatalysis in cationic polymerization may be recalled. Here effects which appeared to be exclusive (in the forties) were soon regarded as universal (in the sixties). Later, however, the necessity of a differentiated approach to the corresponding cationic agents became obvious.

Of the concepts of recent years which illustrate the rejection of the previous generalizations, we note again the carbenium–oxonium problem. The development of studies of processes of cationic polymerization of oxygen-containing heterocyclic compounds provided a basis for the various solutions of this problem depending upon whether the monomer was an oxide or an acetal (see Section 4.1.4). It may be assumed that the differentiated approach would also bring benefit in the precise definition of a mechanism of stereoregulation in the polymerization of dienes, in particular the limits of applicability of scheme (5.30). We have preferred to manage without this by taking the above scheme as being of universal applicability (see Section 5.2), although possible individual exceptions must be considered. In particular they are possible in conditions of the complete absence of 1,4 units in polydiene macromolecules. Such a high selectivity of the direction of the growth reactions prevents the complete rejection of the concept of the existence of active sites of the 1,2 form. For one such case (the polymerization in the butadiene–dipiperidinoethanelithium–hydrocarbon medium system, see Section 5.2.2) ^{13}C-NMR spectra of living chains were obtained (data from Halasa et al. [8]) which these authors ascribe to just this end-unit structure. Holding to such a conclusion is as difficult as refuting it but even if this is correct the effect discovered remains specific to a given real system.

We emphasize that the use of certain facts for one conclusion or another often exceeds the "interpretation level" which is really permitted by the existing data. This is due for example to the assumption that certain model compounds are adequate for the active sites which actually function in real processes. Sometimes the assumptions which are made are connected with comparatively arbitrary preferences for some characteristics which come into the set of contradictory results. In discussing such cases (see Sections 4.1.1.3 and 5.2) we have emphasized the difficulties of uniform interpretation of the results of experimental studies of the active reagents of the type RMt and the correlation of these data with the calculated characteristics; it is also sometimes possible to find such a situation in the case of stable compounds which can exist in different forms. The attempt to find the correlation between experimental and theoretical parameters of stilbene isomers is just such a case. Agreement between these and other features (which favor the greater stability of the trans configuration) has been obtained only comparatively recently [9].

The particular cases, real processes and definite compounds, which we have used here in our discussion are in fact examples which illustrate the situation over the whole field which is discussed in detail in Chapters 4 and 5. The question of the possibility of connecting the reactivity and stereospecificity of the active sites which are often discussed in isolation is still neglected. We have had cause to touch upon this question earlier noting that in a certain sense these properties are "bridged." The function of a bridge is fulfilled by the structure characteristics of the growing chains [10]. It must be borne in mind that the most reactive active sites (free ions and separated ion pairs) are usually distinguished by less stereochemical directionality than are contact ion pairs, the reduced kinetic activity of which is often combined with high stereospecificity. The stereospecific effects in the polymerization of dienes may be evaluated from a similar point of view. Thus the spatial accessibility and the charges on the C_α and C_γ atoms of the end groups of the growing chains determine the rate constants of the corresponding growth reactions and hence the fraction of 1,4 and 1,2 units in the macromolecules.

The correctness of attempting to find a correlation between the kinetic and stereochemical effects was recently examined by Johnson [11] for different organic reactions. Some of the concepts given in this work are quite general and provide a basis for an analogous approach to the processes of stereospecific polymerization which have not been examined in detail here from this point of view.

Completing this work the author would like to point out that the range of the various processes encompassed in a comparatively small monograph has led to the selection of material which may disappoint the reader who is interested in a detailed examination of some particular field. Such detailed examination is usually the subject of works of a specialized nature. The limitations in their range of content is combined with considerable detail in discussion. Among the most modern books of this type are the recent monographs [12-14] which cover the processes of cationic, anionic, and "transition-metal" polymerization. The difference in scale of examination of general and individual features of a certain group of ionic processes, active reagents, or model systems is not so fundamental as the differences in the views of the nature or significance of these or other facts. This phenomenon which is natural for the comparatively young branch of the science of macromolecules is partly connected with the differences in meaning invested by various workers in the concepts and terminology. One, but by no means the only example of such a type, is the concept of a complex which is sometimes applied to cases in which the question is concerned more with the transition state. Such a lack of precision and other terminological deficiencies in understanding lead to fruitless discussions. The necessity of greater clarity in this respect which is found in the most important fundamental concepts in this field is beyond argument. This is indicated by the modern approach to "chemical binding" which takes into consideration the specificity of the interaction of the reagents at the stage of formation of the various bonds from covalent to metallic. A brief but very informative description of the state of this question was given by Del Re [15].

It seemed expedient to note this circumstance since it is almost completely overshadowed by points of greater significance. The most important of these are the structure of the active sites which function in real systems, their relative contributions (monocentered systems are a rare exception), and the detailed paths of the corresponding growth reactions.

REFERENCES

1. F. R. Mayo and M. Morton, "Ionic polymerization," Khim. Tekhnol. Polym., No. 7, 12-15 (1965).
2. B. L. Erusalimskii, The Polarization of Polar Monomers [in Russian], Nauka, Leningrad (1970), pp. 280-286.
3. Japan—U.S. Seminar on Unsolved Problems in Ionic Polymerization, J. Macromol. Sci., Chem., A9, Nos. 5, 6 (1975), pp. 641-1083.
4. D. Lehmann, "Radiochemische Untersuchungen zum Mechanismus der Anionschen Polymerisation im System Acrylnitril-N,N-Dimethylformamid-Lithium-tert-Butanolat," Dissertation, Dresden (1982).
5. S. Riedel, G. Wunderlich, D.Lehmann, R. Dreyer, W. Berger, and H.-J. Adler, "Radiochemische Untersuchungens über den Initiatoreinbau bei der Anionischen Polymerisation von Acrylnitril in DMF, V. Int. Mikrosymp. Fortschr. in Ionenpolymerisation, Prague (1982), prepr. 45.
6. W. Berger, S. Riedel, H.-J. Adler, D. Lehmann, G. Wunderlich, and O. Vogel, "Radiochemical investigations on the mechanism of the initiation of the acrylonitrile polymerization by lithium alkoxides in dimethylformamide," J. Macromol. Sci., Chem., A20, 299-307 (1983).
7. A. F. Podol'skii (Podolsky), A. G. Boldyrev, I. Yu. Sapurina, V. V. Shamanin, I. L. Ushakova, and N. G. Orlova, "The anomalous behavior of a paramagnetic sodium adduct of stilbene in disproportionation and initiation of polymerization in THF," Acta Polym., 33, 181-184 (1982).

8. A. F. Halasa, D. N. Schulz, D. P. Tate, and V. W. Mochel, "Organolithium catalysts and diene polymerization," Adv. Organomet. Chem., 18, 55-95 (1980).
9. A. Wolf and H. H. Schmidtke, "All-electron SCF LCAO MO calculations on various conformations of cis- and trans-stilbene," Theor. Chim. Acta (Berl.), 48, 37-45 (1978).
10. B. L. Erusalimskii (Erussalimsky), "Structure and reactivity of anionic active centers," J. Polym. Sci., Polym. Symp., 62, 29-50 (1978).
11. C. D. Johnson, "The reactivity—selectivity principle: Fact or fiction?," Tetrahedron, 36, 3461-3480 (1980).
12. J. P. Kennedy and E. Marechal, Carbocationic Polymerization, Wiley, New York (1982).
13. M. Szwarc, "Living polymers and mechanisms of anionic polymerization," Adv. Polym. Sci., 49, Springer Verlag, Berlin-Heidelberg (1983).
14. B. A. Dolgoplosk and E. I. Tinyakova, Organometallic Catalysis in Polymerization [in Russian], Nauka, Moscow (1982).
15. G. Del Re, "Binding: A unifying notion or a pseudoconcept?," Int. J. Quantum Chem., 19, 979-984 (1981).